*Treatise on Materials Science
and Technology*

VOLUME 12

*Glass I: Interaction with
Electromagnetic Radiation*

Fleck

TREATISE ON MATERIALS SCIENCE AND TECHNOLOGY

VOLUME 12

GLASS I: INTERACTION WITH ELECTROMAGNETIC RADIATION

EDITED BY

MINORU TOMOZAWA

Materials Engineering Department
Rensselaer Polytechnic Institute
Troy, New York

ROBERT H. DOREMUS

Materials Engineering Department
Rensselaer Polytechnic Institute
Troy, New York

 1977

ACADEMIC PRESS   New York   San Francisco   London

A Subsidiary of Harcourt Brace Jovanovich, Publishers

ACADEMIC PRESS, INC.
111 Fifth Avenue, New York, New York 10003

United Kingdom Edition published by
ACADEMIC PRESS, INC. (LONDON) LTD.
24/28 Oval Road, London NW1

LIBRARY OF CONGRESS CATALOG CARD NUMBER: 77-182672
ISBN 0—12—341812—7

PRINTED IN THE UNITED STATES OF AMERICA

Contents

Introduction

Robert H. Doremus

Optical Absorption of Glasses

George H. Sigel, Jr.

Photochromic Glass

Roger J. Araujo

Anomalous Birefringence in Oxide Glasses

Takeshi Takamori and Minoru Tomozawa

Light Scattering of Glass

John Schroeder

Resonance Effects in Glasses

P. Craig Taylor

Dielectric Characteristics of Glass

Minoru Tomozawa

List of Contributors

Numbers in parentheses indicate the pages on which the authors' contributions begin.

ROBERT J. ARAUJO (91), Research and Development Laboratories, Corning Glass Works, Corning, New York

ROBERT H. DOREMUS (1), Materials Engineering Department, Rensselaer Polytechnic Institute, Troy, New York

JOHN SCHROEDER (157), Department of Chemistry, School of Chemical Sciences, and Materials Research Laboratory, University of Illinois, Urbana, Illinois

GEORGE H. SIGEL, JR. (5), United States Naval Research Laboratory, Washington, D.C.

TAKESHI TAKAMORI (123), IBM Thomas J. Watson Research Center, Yorktown Heights, New York

P. CRAIG TAYLOR (223), United States Naval Research Laboratory, Washington, D.C.

MINORU TOMOZAWA (123, 283), Materials Engineering Department, Rensselaer Polytechnic Institute, Troy, New York

Preface

Scientific understanding of glass has lagged behind that of other engineering materials such as polymers, metals, and crystalline ceramics. Shortly after World War II and into the fifties there was a surge of interest in fundamental studies of these materials, and understanding of the relationships between their lattice and microstructure and their properties increased enormously. Glass did not share in this golden age, perhaps because of its greater complexity and poorer characterization. However, within the last fifteen years, basic studies on glass have increased substantially.

These differences are strikingly characterized by the continuing usefulness in both classroom and laboratory of books on metals and crystalline ceramics written in the fifties and early sixties. On the other hand, only a small fraction of a recent book on glass ("Glass Science," R. H. Doremus, Wiley, 1973) could have been written fifteen or twenty years ago.

It is to document this rapidly expanding new understanding of glass that we have initiated a series of review volumes on the science of glass. The present book focuses on the interaction of electromagnetic radiation with glass. Goaded by applications in fiber optics, lasers, electronics, photosensitivity, and many other areas, much research work is being carried out to understand better interactions of light and electricity with glass.

Special thanks are due to Professor Herbert Herman, Editor of "Treatise on Materials Science and Technology," for stimulating us to organize this volume, and to Academic Press for their interest and help. The authors of the individual chapters are the ones who should be credited with the usefulness we hope this book will fulfill.

M. TOMOZAWA
R. H. DOREMUS

Contents of Previous Volumes

Introduction

ROBERT H. DOREMUS

Materials Engineering Department
Rensselaer Polytechnic Institute
Troy, New York

The clarity and color of glass were for many centuries its most valued properties. In fact if it were not for our tendency to value things for their cost, glass objects would still be the most desirable, because of their ease of formation into countless shapes and designs, and their beauty and variability of color, refraction, and translucency.

The ancient glassmaker knew how to vary the color and transparency of his products by the addition of certain compounds and by crude control of melting temperatures. Modern understanding of the color of glass and also of its absorption of ultraviolet and infrared radiation is summarized in the chapter by Dr. George Sigel, Concepts of electronic excitation and vibrational spectroscopy developed for crystalline materials are successfully applied to glass, emphasizing that there is short-range order in the atomic structure of glass much like that in crystals, even though long-range atomic regularity is absent in glass. Dr. Sigel was trained in physics and has spent the last several years at the Naval Research Laboratory in Washington, D.C., where his studies of ultraviolet absorption, defect centers, and optical effects of radiation damage in glass have been of special distinction.

Many imaginative new materials and applications based on glass have come from the Corning Glass Works. Among these innovations are glass ceramics and various kinds of photosensitive glasses. Glasses that change their absorption of light when exposed to optical radiation, called photochromic glasses, are described in the chapter by Dr. Roger Araujo of Corning. Dr. Araujo has been active in the development of photochromic glasses at Corning, particularly those incorporating fine precipitated particles of silver halides. In his chapter he describes the formation of these and other photochromic glasses, and mechanisms for the induction and fading of absorption by irradiation with light.

Optical birefringence results from anisotropic microstructural elements in a solid, or from elastic strains introduced in cooling, possibly with applied stresses. Recent new experiments, and understanding of birefringence in glass resulting from these causes, are summarized in the chapter by Dr. Takamori and Professor Tomozawa. Dr. Takamori is located at the IBM Research Center, where his interests have been in thermomechanical processes of glasses.

The scattering of light in a solid results from variations in refractive index over distances within about two orders of magnitude of the wavelength of the light. Thus light scattering is well suited to studies of phase separation and density differences in glasses. Phase separation can influence chemical, electrical, and optical properties of glass and is the first step in the manufacture of Vycor high silica glass. Density differences in glass at low temperatures reflect high-temperature composition fluctuations frozen in as the mobility of glass constituents decreases on cooling. The study of these properties by light scattering is summarized in Chapter V by Dr. John Schroeder, presently at the Department of Chemistry of the University of Illinois. Dr. Schroeder studied for his Ph.D. degree in the Vitreous State Laboratory at Catholic University in Washington, D.C., where much work on fluctuations, relaxations, and light scattering in glass has been carried out.

Electromagnetic resonance techniques have been used to examine many aspects of solid structure. Nuclear magnetic resonance has revealed coordination numbers of ions in lead and borate glasses, and Mössbauer spectroscopy shows the coordination number and aggregation state of various ions in glass. Electron spin resonance is valuable for studies of defect centers in glass. These resonance techniques are discussed by Dr. Taylor of the Naval Research Laboratories, who has used resonance techniques to study glass at NRL and in his graduate work at Brown University.

Electrical properties of glass are important in applications as insulators, glass electrodes, and dielectrics. Dielectric loss in glass has been extensively studied, and the predominant effect is a loss peak whose activation energy is the same as the activation energy for electrical conduction. There is also a similar mechanical loss. A number of theories for this dielectric loss have been proposed and are summarized by Professor Tomozawa. In addition, he outlines a new theory for the loss designed to overcome the deficiencies of previously proposed ones.

The common theme of these chapters is the interaction of electromagnetic radiation with glass. As the above summary suggests, this interaction can be used as a probe to study glass structure and to satisfy one's scientific curiosity about glass, but several kinds of interaction are also of direct practical importance. Both of these aspects are brought out in subsequent chapters.

Oxide glasses are emphasized in this volume, because they are by far the most important commercially, and have been most studied. Application of the work described to other glasses, especially nonoxides, is uncertain because of their quite different chemical nature, for example, chalcogenides, nitrates, simple and polymeric organic compounds, and amorphous metallic alloys.

A selective description of general books on glass may help to guide one interested in learning more about other aspects of our knowledge of glass. An introduction to atomic structure and certain properties of glass is provided by "The Physical Properties of Glass," by Holloway (1973). Useful summaries of glass properties are given in "Properties of Glass" by Morey (1954) and "Glas" (in German) by Scholze (1965). Discussions of the science of glass are given in "Glass Science" by Doremus (1973) and "Introduction to Glass Science" edited by Pye *et al.* (1972).

"A History of Glassmaking" by Douglas and Frank (1972) combines the art and technology of glass in an elegant way. In Volume 7 of "Silicate Science" by Eitel (1975), there is an extensive summary of literature on glass from 1962 through 1972. The three volumes of "Modern Aspects of the Vitreous State," edited by J. D. Mackenzie (1960–1964), include review articles on many aspects of glass science up to 1964. More specialized references are given in the following chapters.

References

Doremus, R. H., (1973). "Glass Science." Wiley, New York.
Douglas, R. W., and Frank, S. (1972). "A History of Glassmaking." Foulis, Henley-on-Thames, England.
Eitel, W. (1975). "Silicate Science." Vol. 7. Academic Press, New York.
Holloway, D. G., (1973). "The Physical Properties of Glass," Wykeham Publ., London.
Mackenzie, J. D. (ed.) (1960–1964). "Modern Aspects of the Vitreous State," Vols. I to III. Butterworths, London.
Morey, G. W., (1954). "The Properties of Glass," 2nd ed. Van Nostrand–Reinhold, Princeton, New Jersey.
Pye, L. D., Stevens, H. J., and LaCourse, W. C. (1972). "Introduction to Glass Science." Plenum Press, New York.
Scholze, H. (1965). "Glass." Vieweg, Braunschweig, Germany.

Optical Absorption of Glasses

GEORGE H. SIGEL, JR.

United States Naval Research Laboratory
Washington, D.C.

I. Introduction

From ancient times to the present day the transparency and color of glasses have remained perhaps their most attractive physical properties. The successful utilization of glasses for optical applications such as windows, lenses, containers, filters, lasers, and waveguides often depends on the precise control of material transparency in a selected wavelength interval, and hence a thorough knowledge of optical properties and a good understanding of absorption processes. This chapter will provide a review of the optical absorption measured in a variety of inorganic glasses (primarily the oxide glasses) over the ultraviolet (uv), visible, and infrared (ir) spectral regions. Material properties have been emphasized, although applications are mentioned whenever possible. Since the optical properties of glasses are closely tied to the glass structure and composition, stress has been placed on relating the chemical bonding, symmetry, and coordination within the glass network to the observed optical properties. Discussion is provided for both the absorption mechanisms intrinsic to the glasses as well as those which result, for example, by the selected doping of transition metals to achieve a desired color or from the presence of trace impurities. Detailed theoretical treatments have been avoided in favor of a discussion of the absorption mechanisms. Recent reviews on the optical properties of glasses which may serve as useful references include those by Wong and Angell (1971), and Kreidl (1972) and Krause (1974). In addition, there is a review by Simon (1960) on the infrared absorption of glasses, the early work of Weyl (1951) on colored glasses, the reviews of Bates (1962) and Bamford (1962b) on the ligand field theory and spectra of transition metal ions in glasses, and the book of Patek (1970) and the article by Snitzer (1966) on glass lasers, which review rare earth ion spectral absorption in various host glasses.

The existence of short-range order in glasses has been well-established by the x-ray diffraction studies of Warren and coworkers (1936, 1938) and the numerous other studies which have followed. Urnes (1960) has provided an excellent review of x-ray diffraction studies of glass. It will become clear as this chapter progresses, that the basic building blocks of the glass networks such as the SiO_4 tetrahedral and BO_3 trigonal units in silicate and borate glasses, and the manner in which they are linked, determine to a large extent the intrinsic optical absorption measured in the glasses. The absence of long-range periodicity simply insures the isotropy of the optical properties. Thus the knowledge of structure is essential to the intelligent interpretation of absorption in glass.

A. Measurement Methods and Spectroscopic Units

Optical absorption measurements are most commonly performed using double-beam grating or prism-type scanning monochromators which compare absorption in any given sample to that of a reference. For measurements of materials with extremely low optical loss (such as fiber-optic grade bulk glasses), laser calorimetric techniques have recently been developed (Rich and Pinnow, 1972; White and Midwinter, 1973; Skolnik, 1975; Hass et al., 1975). The latter approach measures the temperature rise of a thermally isolated sample through which a light beam is passing, and relates this to the optical absorption at the laser wavelength. Regardless of the method employed, optical absorption in a homogeneous passive material can be understood in simple terms. The attenuation of light passing through a material of thickness x (cm) is given by the expression

$$I = I_0 \exp(-\alpha x) \tag{1}$$

where I is the transmitted intensity, I_0 the incident intensity, and α the (linear) absorption coefficient expressed in units of inverse centimeters (cm^{-1}). The absorption coefficient is typically both wavelength and temperature dependent, and uniquely defines the strength of the optical absorption. Values of α in high-purity oxide glasses range from about 10^5 to 10^{-5} cm^{-1}.

Spectrometers may record absorption as percent transmission $[(I/I_0) \times 100]$ or as optical density (OD) which is defined as

$$OD = \log_{10}(I_0/I) \tag{2}$$

For example 10% transmission corresponds to 1 OD absorbance and 1% transmission to 2 OD absorbance.

The absorption coefficient α is related to the optical density unit by the expression

$$\alpha = \ln(I_0/I)/x = 2.303(OD)/x \tag{3}$$

Finally, the use of the decibel (dB) as a unit of optical loss is more common today because of the influence of optical waveguide technology. It is also simply related to OD as

$$10 \text{ dB} = 1 \text{ OD} \tag{4}$$

Fiber-optic losses are usually quoted in units of dB/km. Note that both the OD and dB units of attenuation are only meaningful if the material thickness is specified.

In practice, reflective losses at the surfaces are also encountered. In a double-beam system these are easily eliminated by the use of two specimens of different thicknesses in the sample and reference beam. In a single-beam system, samples of multiple thicknesses can also be used to separate the constant surface reflective loss contribution from the thickness-dependent bulk absorption. In regions of extremely intense optical absorption, the reflection itself is used to determine α. Strong reflectance bands correspond to regions of strong absorption. This can be concluded from the fact that the inspection of the expressions for normal incidence reflectivity and optical absorption in a dielectric medium of complex index $\bar{n} = n - ik$, where n is the real component (the index of refraction) and k is the imaginary component. The attenuation of light beam in such a medium is related to Eq. (1) by the expression

$$I = I_0 \exp(-4\pi k/\lambda)x = I_0 \exp(-\alpha x) \tag{5}$$

The reflectivity R, at normal incidence at the first surface, is given by the expression

$$R = \frac{(n-1)^2 + k^2}{(n+1)^2 + k^2} \tag{6}$$

In regions of strong absorption, i.e., for large k, R becomes an appreciable fraction of unity, so high reflectance is expected when the absorption is large.

A Kramers–Kronig analysis (Kramers, 1929; Kronig, 1926; Powell and Spicer, 1970) is used to derive the absorption coefficient and the index of refraction from the reflectivity data. This is accomplished by the integration of reflectivity over a broad range of energies. This approach is normally used in the far ir and uv spectral regions beyond the absorption edges of the glasses.

The strength of the optical absorption produced by a given concentration of dissolved impurities in a glass is generally expressed in terms of a (molar) extinction coefficient ε which is related to the optical density by

$$\varepsilon = \mathrm{OD}\,[M]x \qquad \text{(liters/mole-cm)} \tag{7}$$

where $[M]$ is the concentration of the absorbing ion in gm-moles/liter and x is the thickness. Units of liters/mole-cm have been dropped in the text when quoting ε values.

Optical absorption in glasses is measured as a function wavelength or energy of the incident photons. The visible wavelengths correspond to the region of sensitivity of the human eye, i.e., approximately $0.4 < \lambda < 0.7\ \mu m$. Unfortunately, the units chosen by various authors differ. Wavelength is measured in units as microns (μm), millimicrons (mμm), angstroms (Å), and nanometers (nm), where 1 mμm = 1 nm = 10 Å = $10^{-3}\ \mu m$. The conversion of energy E to wavelength is given by E (eV) = $1239.8/\lambda$ (mμm). Energy and frequency units are preferred in data analysis, e.g., in the assignment of transitions between levels, but spectrometers typically provide a wavelength measurement. Since energy (or frequency) is inversely related to wavelength, an alternative unit chiefly used in the infrared range is that of wavenumber v (cm^{-1}) = $1/\lambda$ (cm). For example, the location of the sodium D line can be expressed as 5893 Å, 589.3 mμm, 589.3 nm, 0.5893 μm, 2.102 eV, or 16,967 cm^{-1}. The units of micron and electron volt have been used in the text for the most part but figures are reproduced from the original references and therefore contain a variety of the units mentioned

B. General Spectral Features of Glasses

The common oxide glasses, such as silicates, phosphates, and borates, typically exhibit an absorption minimum in the visible or near ir. Electronic processes such as valence-to-conduction band transitions are responsible for the high energy absorption cut-off observed in the uv. In the infrared region, the low energy cut-off is produced by the strong molecular and atomic vibrations of the glass. Optical absorption in the visible is typically a composite of the tails of the electronic and vibrational edges of the glass plus the contributions from impurities such as transition metals and OH ions. Figure 1 shows a typical absorption spectrum for a pair of commercial lead silicate glasses, with and without cerium (Evans and Sigel, 1975). The dashed lines indicate extrapolation of the absorption edges. Visible absorption in this glass results principally from the presence of impurities. Cerium produces an effective shift of the uv edge of the glass, imparting a faint yellow color to this glass. Recent work on high-purity silicate glasses for fiber optic applications has demonstrated that absorption coefficients of less than 10^{-5} cm^{-1} are achievable in the 0.8–1.1 μ range, making oxide glasses among the most transparent of all optical materials.

Because of length restrictions, it has naturally been necessary to limit the scope of this chapter. Emphasis has been placed on reviewing the data on silicate, borate, and phosphate glasses. Optical absorption associated with radiation effects and photochromism have been omitted. Rare earth ion absorption spectra in glasses have been treated only briefly, but references

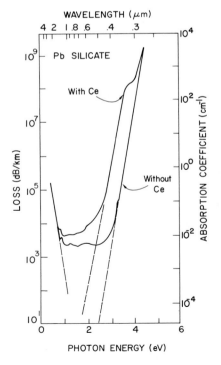

Fig. 1. Typical spectral transmission curves for oxide glasses exhibiting high energy electronic edge and low-energy vibrational edge with impurity absorption in visible and near ir regions. These data are for a commercial lead silicate with and without CeO$_2$. (After Evans and Sigel, 1975.)

have been provided for those seeking more detail on the subject. Finally, the indices of refraction and dispersion of glasses have not been addressed although these are related to the absorptive properties of glasses.

II. Ultraviolet Absorption

The availability of higher purity glasses during the past few years has stimulated work on the ultraviolet absorption of glasses. Strong impurity cation absorption, especially that associated with transition metals such as iron, chromium, and copper, tends to dominate the near-uv absorption of most commercial glasses, obscuring the intrinsic tail of the fundamental absorption edge. Thus, high-purity materials are absolutely necessary for the uv spectroscopy of glasses.

Stevels (1947) was the first to suggest that the intrinsic absorption edge of an oxide glass corresponded to the transition of a valence electron of an oxygen ion in the glass network to an excited state. It was assumed that electrons which participate in strong chemical bonds would require the higher energy (or shorter wavelengths) for excitation. For the simple glass-forming oxides, Stevels predicted an edge shift to shorter wavelengths in the order B_2O_3, SiO_2, and P_2O_5, which corresponds to the experimental values of 0.17, 0.16, and 0.145 μm measured in these materials, respectively. If, however, the glass-forming network is disrupted by the introduction of network modifying alkali or alkaline earth ions which produce singly bonded (nonbridging) oxygen (NBO) ions, the average chemical bond strengths of the network will be reduced and electron excitation will require less energy. The general weakening of the bonding should manifest itself as a shift of the ultraviolet edge to longer wavelengths.

Ultraviolet absorption measurements on binary alkali and alkaline earth oxide glasses in recent years tend to support many of Stevels' early ideas. Kordes and Worster (1959), however, observed that borate glasses transmitted to shorter wavelengths than silicates, contrary to predictions. It is now known from the NMR work of Bray and coworkers (1960, 1963) that boron will four-coordinate with oxygen upon the addition of alkali oxide, thereby eliminating the formation of NBO ions in the more transparent borate glasses. Scholze (1959), Hensler and Lell (1969), and Sigel (1971) observed that in high-purity silicate glasses and alkali-doped silicas the ultraviolet edge shifts as expected. Phosphate glasses (Kordes and Nieder, 1968), on the other hand, show an oscillation of the ultraviolet edge back and forth with increasing alkali or alkaline earth content, due to changes in bonding which take place in the glass. In general it can be concluded that prediction of the location of the ultraviolet absorption edge of oxide glasses is substantially more complex than envisioned by Stevels since structural rearrangements with composition must be considered.

A. Silicate Glasses

The ultraviolet absorption of silicate glasses arises from three principal sources:

(1) absorption intrinsic to electronic excitations of the Si–O network,
(2) absorption arising from the introduction of network modifying and/or network forming cations, and
(3) absorption resulting from the presence of impurities, particularly charge transfer spectra of transition metal ions (Sigel, 1973, 1974).

Each of these sources will be treated in turn.

1. SiO_2

Fused silica and crystal α-quartz both consist of Si atoms tetrahedrally bonded to four O atoms, each O atom being common to (bridging) two such SiO_4 tetrahedra. The Si–O bond distance is roughly 1.6 Å in both materials. The Si–O–Si bond angle in crystal α-quartz is 144°. Bell and Dean (1966) found that a distribution of bond angles with a mean value of 150° was obtained when a fused silica model was constructed, allowing for the proper constraints such as density, entropy, stoichiometry, and bonding. Recent x-ray diffraction data on vitreous silica by Mozzi and Warren (1969) indicate that the Si–O–Si bond angle varies from 120° to 180° with a maximum at 144°, while the data of Konnert and Karle (1972) indicate the possibility of ordering beyond 20 Å. This structural information can be used to understand the electronic absorption of SiO_2.

The ultraviolet cutoff wavelength λ_0, beyond which SiO_2 glass no longer transmits, lies near 0.16 μm (8.0 eV) in the vacuum ultraviolet region of the spectrum. This absorption edge results from the excitation of valence band electrons in the Si–O network to unoccupied higher energy states such as exciton or conduction band levels. The intensity of the absorption is determined by the density of occupied states in the valence band and of the unoccupied states in the conduction band which are within energy E_g of each other, as well as by the transition probability of the initial to final state transition. In most insulators, the resulting absorption spectrum from this type of excitation is a continuum of intense absorption at short wavelengths bounded by a steep absorption edge beyond which the material is relatively transparent.

The exciton (Dexter and Knox, 1965; Knox, 1965) level is associated with bound electron-hole pairs which are created by photons of energy greater than the gap E_g. Since an electron and hole have an attractive Coulomb interaction for each other, it is possible for stable bound states of the two particles to be formed. The photon energy required to create such a pair state will be less than the energy gap E_g. It is difficult to produce excitons in sufficient concentration to observe directly transitions among exciton levels but it is quite common to observe transitions from the valence band edge to exciton levels. Whereas interband transitions normally manifest themselves as broad peaks or absorption edges of various types, exciton transitions generally produce sharp peaks and at energies less than the band gap so that the lowest energy electron transitions are frequently to exciton levels.

As discussed in Section I,A, reflectance spectroscopy is employed to measure absorption beyond the edge where the sample is opaque. Philipp (1966) was the first to measure the room temperature reflectance of crystalline quartz and fused silica. These data are shown in Fig. 2. Platzoder (1968) studied the temperature dependence of the reflectance. Similar measurements on fused silica alone were reported by Loh (1964) and Sasaki et al. (1965). Absorption peaks obtained from a Kramers–Kronig analysis of the reflectance data lie at 10.2, 11.7, 14.3, and 17.2 eV. The similarity of the reflectance results on crystalline and glassy SiO_2 indicate that the absorption mechanisms are not strongly dependent on the long-range periodicity of the network. This suggests that the absorption arises from electronic transitions characteristic of the short-range order, perhaps the SiO_4 tetrahedron or clusters of neighboring tetrahedra. The only apparent difference observed between the two spectra is a broadening of the fused silica spectra which results in a slight shift in its absorption edge to longer wavelengths compared to α-quartz (Sigel, 1973, 1974).

Numerous theoretical approaches have been employed to interpret the uv spectra of SiO_2. Reilly (1970) utilized a molecular orbital approach based on a Si–O–Si molecule and predicted that the lowest energy transitions which determine the edge position arise from oxygen lone pair valence orbitals. Reilly proposed that the 10.2-eV peak arose from a transition to an exciton state with the hole orbital corresponding to the oxygen $2p_x$ (nonbonding orbital orthogonal to the Si–O bonding orbital) and the electron orbital resembling the oxygen 3s.

Ruffa (1968) used the localized valence bond approach to interpret the spectra in terms of bond-breaking processes. Ruffa identified the sharp (0.120 μm) 10.2-eV peak with a Wannier (1937) exciton formed by the breaking of a single Si–O bond. The next broad peak at 11.7 eV was attributed to band-to-band transitions. He proposed that the silicon 3p electron

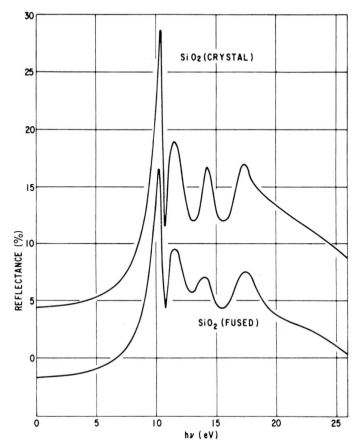

Fig. 2. The reflectance spectra of α-quartz and fused silica. The values of fused silicia have been lowered by 5%. (After Philipp, 1966.)

freed by the breaking of the Si–O bond would go into a relatively large radius orbit about the silicon atom. The exciton binding energy was calculated to be 1.27 eV, provided that the electron–hole reduced mass μ was assumed to be 0.5 and a dielectric constant of K of 2.31 was used. The experimentally observed separation of the two lowest energy reflection peaks is 1.5 eV.

Other more recent and detailed theoretical treatments of the electronic structure of SiO_2 include those by Bennett and Roth (1971), Johnson and Smith (1971, 1972), Slater and Johnson (1972), Tossell *et al.* (1973), and Yip and Fowler (1974). The last of these contains a review of recent experimental and theoretical investigations. These authors treat more rigorously the energy levels and orbital compositions of molecular clusters such as a Si–O–Si unit and have verified many of Reilly's (1970) intuitive ideas such as the existence of the nonbonding O (2p) highest energy valence level. Transition eigenvalues of 10.2, 11.7, 14.3, and 17.2 eV were obtained by Johnson and Smith (1971, 1972) using a SiO_4^{4-} cluster, in good agreement with the optical reflectance data. These theoretical advances are significant in that it has been possible to generate the fundamental electronic absorption spectra of SiO_2 by working with only the small basic building blocks of the network.

Other experimental techniques have also provided information which has helped to clarify the electronic structure and uv reflectance spectra of amorphous and crystalline SiO_2. DiStefano and Eastman (1971) have determined a band gap for SiO_2 of 8.9 eV from photoconductivity measurements on amorphous SiO_2 films, and the position of the valence band relative to the photoemission threshold. The structure and width of the valence band have been probed by x-ray and ultraviolet photoelectron spectroscopy (DiStefano and Eastman, 1971; Rowe and Ibach, 1973; Ibach and Rowe, 1974a,b). X-ray absorption and emission spectra by Ershov et al. (1966, 1967) and Wiech (1967) were found to be identical in glassy and crystalline SiO_2. Nagel (1970) provides an excellent review on interpretation of x-ray spectra, using SiO_2 as an illustration. Energy level models for both SiO_2 and GeO_2 (which is structurally similar to SiO_2) have been obtained by Rowe (1974) by the combination of uv photoemission spectroscopy (UPS) and electron energy loss spectroscopy (ELS). The energy levels obtained were shown to be consistent with photoconductivity, photoelectric threshold, optical reflectance, and x-ray photoemission data. These are shown in Fig. 3. The filled valence states correspond to localized bonding or nonbonding O (2p) molecular orbitals. The higher empty states correspond to excitons formed from either localized antibonding Ge–O or Si–O levels or a "conduction band" state involving both M–M as well as M–O interactions, where M is Si or Ge. This band is broadened more in GeO_2 than SiO_2 and hence the band edge is approximately 2.4 eV lower in GeO_2 (Cohen and Smith, 1958). The results of Rowe support the localized exciton picture for interband transitions previously put forth by Reilly (1970), Philipp (1971), and Phillips (1974) for SiO_2 and GeO_2. One interesting conclusion to the work of Rowe (1974) is that the antibonding final states should be sensitive to short-range order changes such as doping with impurities.

In summary, although some minor disagreements still persist, (Koma and Ludeke, 1975) the optical absorption and reflectance data of SiO_2 and GeO_2 in the ultraviolet is reasonably well understood in terms of the electronic transitions taking place within the fundamental SiO_4 tetrahedra. This localization accounts for the similarlity of the glassy and crystalline spectra of SiO_2 and GeO_2.

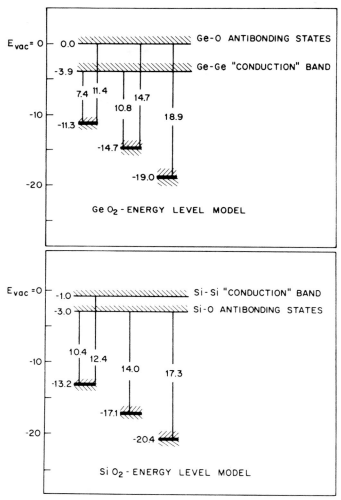

Fig. 3. Energy level models for SiO_2 and GeO_2 consistent with experimental data, including the optical reflectance data of Fig. 2. Transitions from various valence state levels to antibonding and conduction band states are shown. (After Rowe, 1974.)

2 EFFECTS OF NETWORK MODIFIERS ON UV ABSORPTION OF SILICATES

The introduction of network-modifying cations such as alkali and alkaline earths into SiO_2 produces a breakup of the continuous Si–O network. The x-ray diffraction results of Warren and Biscoe (1938) on a series of sodium silicate glasses indicated the presence of a random network of SiO_4 tetrahedra in which some oxygens were bonded to two silicon ions (bridging oxygens) and some were bonded to one silicon (nonbridging oxygens, NBO). The distribution of alkali and alkaline earth ions in many glasses does not appear uniform and evidence of clustering has been reported (Urnes, 1969; Milberg and Peters, 1969). The breakup of the Si–O network by the network modifiers is reflected in the ultraviolet absorption of the glass.

Sigel (1971) has shown that the introduction of small amounts (0.05–0.5 mole %) Li, Na, and K produced a significant shifting of the uv edge to longer wavelengths. Optical absorption in the 0.15–0.20-μm wavelength range was observed to increase linearly with alkali content while the absolute intensity increased in the order Li, Na, and K for a given alkali concentration consistent with the ideas of Stevels (1947) discussed earlier. If aluminum was introduced simultaneously with the alkali into the silica, the ultraviolet edge was observed to move back to higher energies as shown in Fig. 4.

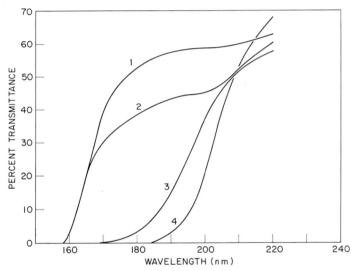

Fig. 4. The effect of aluminum on the uv absorption of alkali-doped silica. Edge shifts to shorter wavelengths upon introduction of aluminum. Sample thickness is 1 mm. Fused silica (1) 0.5 mole % Na + 0.5 mole % Al; (2) 0.2 mole % K + 0.2 mole% Al; (3) 0.5 mole % Na; (4) 0.2 mole % K. (After Sigel, 1971.)

Lell (1962) and Kats and Stevels (1956) had previously seen evidence of alkali–aluminum pairing in glasses. Aluminum can substitute for silicon, requiring a nearby alkali for charge compensation and producing a reduction in the NBO concentration as shown schematically in Fig. 5. Substitutional aluminum alone in SiO_2 did not produce any observable uv absorption in the silica in amounts up to 0.5%.

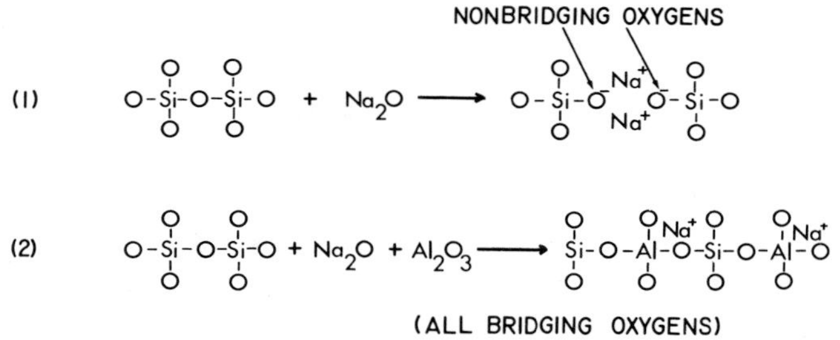

NONBRIDGING OXYGENS

(1)   O-Si-O-Si-O + Na₂O ⟶ O-Si-O Na⁺ O-Si-O

(ALL BRIDGING OXYGENS)

(2)   O-Si-O-Si-O + Na₂O + Al₂O₃ ⟶ Si-O-Al-O-Si-O-Al-O

Fig. 5. Schematic representation in two dimensions of (1) the creation of NBO ions by the introduction of Na₂O into the SiO₂ network and (2) the ability of Al to restore the bridging in spite of the presence of alkali. (After Sigel, 1971.)

Hensler and Lell (1969) correlated the ultraviolet transmission of silicate glasses in the 4–6-eV range with the number of NBO ions and to the bond strength between the NBO ions and the glass modifier ions. The field strength of the modifier ions and therefore the bond energy between the cation and the oxygen was found to have a stronger influence than the cation concentration in this energy range. It was postulated by Dietzel (1949) that cations with high field strength, and therefore high polarizing power, form strong bonds to the oxygen ion which is manifested by a short wavelength absorption cutoff. Lack of complete correlation with this mechanism may be partially resolved by symmetry considerations such as the number of oxygens clustered about a given cation. Hensler and Lell (1969) measured the absorption edge for Li_2O–SiO_2 at 6.6 eV, for Na_2O–SiO_2 at 5.97 eV, and for K_2O–SiO_2 at 5.8 eV. A corresponding effect was also observed with alkaline earth ions. The edge increased to longer wavelengths by the addition of MgO, CaO, SrO, and BaO as shown in Fig. 6. B_2O_3 addition was observed to shift the edge to shorter wavelengths, apparently because of the reduction of the NBO concentration in the glasses by the formation of four-coordinated boron. There is evidence of iron absorption in the glasses used in this study but it is sufficiently small to permit interpretation of the data.

Smith and Cohen (1963) utilized glasses of composition $Na_2O \cdot 3SiO_2$, $Li_2O \cdot 3SiO_2$, and $K_2O \cdot 3SiO_2$ as host glasses to study the ultraviolet absorption of cation impurities. The edge positions of the respective glasses are difficult to measure accurately because of the presence of iron impurities, but all lie near 5.8 eV (0.22 μm). Sigel and Ginther (1968) prepared $Na_2O \cdot 3SiO_2$ glass with sufficiently low iron content that the uv transmission edge remained unchanged at 5.9 eV (0.21 μm) when prepared under either oxidizing or reducing conditions. Similar behavior was observed in K_2O–$3SiO_2$ and soda lime glasses.

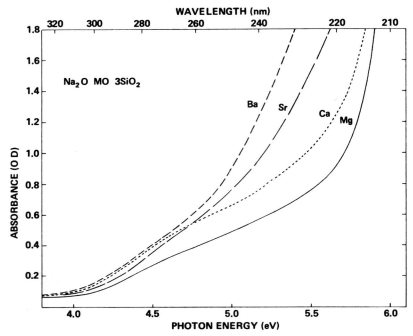

Fig. 6. Effect of alkaline earth ions on the uv absorption edge of a sodium silicate glass. Sample thickness is 2 mm. (After Hensler and Lell, 1969.)

Sigel (1971) measured the uv reflectance of silicate glasses of higher alkali content including $Li_2O \cdot 2SiO_2$, $Na_2O \cdot 2SiO_2$, $Na_2O \cdot 3SiO_2$, $Na_2O \cdot 6SiO_2$, $Na_2O \cdot CaO \cdot 3SiO_2$, and $Na_2O \cdot Al_2O_3 \cdot 3SiO_2$ glasses. The spectra of the glasses in the 6–12-eV range were characterized by reflection peaks near 8.5, 9.3, and 11.5 eV. The data indicate that a new band or continuum of states arises near 8.5 eV which is responsible for the shift of the uv edge of the glasses to longer wavelengths upon the introduction of network modifiers. Sigel showed evidence that the 10.2-eV exciton peak of SiO_2 was perturbed but that the 11.7-eV interband peak appeared largely unchanged in the alkali silicate glasses. As mentioned in Section II,A,1, Rowe (1974) predicts the upset of final antibonding level states by the introduction of impurities. The exact nature of the lower energy transitions in alkali and alkaline earth silicates, however, has yet to be established.

Bagley *et al.* (1976) have recently reported uv reflectance measurements on a high-purity soda–lime–silicate (21.3 wt % Na_2O, 5.2 wt % CaO, 73.5 wt % SiO_2) out to energies of 22 eV. Strong reflection peaks were observed at 16.2, 13.3, and 11.2 eV, similar to the corresponding bands in vitreous SiO_2.

Additional bands were observed at 9.8 and 9.3 eV and were tentatively ascribed to the presence of network modifiers. The reflectance spectrum of this glass is shown in Fig. 7. Lower energy bands on the edge shoulder at 7 and 7.5 eV are attributed to oxygen atoms in a modified state, but have not been identified conclusively.

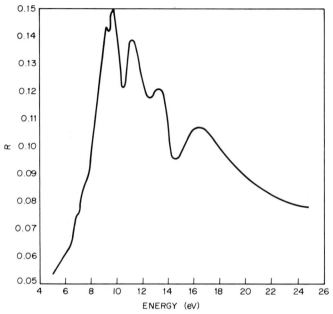

Fig. 7. Reflectance spectra of a high-purity soda–lime silicate glass. (After Bagley *et al.*, 1976.)

In high-purity silicate glasses the tails of the fundamental electronic absorption bands limit the optical absorption in the near uv, throughout the visible, and into the ir regions. Figure 8 shows the data of Pinnow *et al.* (1973) comparing the tailing observed in a soda–lime–silicate glass (20% Na_2O, 10% CaO, 70% SiO_2) with that of a synthetic fused silica (Suprasil W1). Similar results for a soda–lime glass were reported by Takahashi *et al.* (1974). Work by Miyashita *et al.* (1974) indicates that the uv edge in Suprasil W1 is substantially different from that in other high-purity silicas and may be influenced by impurities, particularly iron which they measured at 1.3 ppm in Suprasil W1. Thus even in high-purity fiber-optic grade materials, it can be difficult to establish the slope of the intrinsic absorption tail of the fundamental edge. Tauc (1975) has recently reviewed the subject of band tailing in glasses.

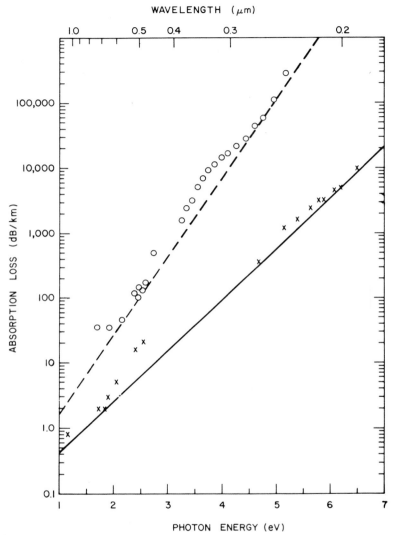

Fig. 8. Effects of electronic absorption band-tailing on the optical absorption of high-purity fused silica and soda–lime glass. Lines are lower bounds to experimental data. O = soda lime–silicate; X = fused silica. (After Pinnow *et al.*, 1972.)

3. Effects of Network Formers on UV Absorption of Silicates

The addition of network forming oxides to silica such as PbO, B_2O_3, TiO_2, GeO_2, and Al_2O_3 to silica may also produce a shifting of the ultraviolet edge since bond strengths and coordination numbers may change.

Rather than attempting a comprehensive review of this subject, a few examples of recent work on high-purity glasses have been selected as illustrations.

a.   PbO.   Glasses containing lead absorb very strongly in the ultraviolet. Stroud (1971) compared the transmission of $Na_2O \cdot 3SiO_2$ glass with that of $Na_2O \cdot 3SiO_2 \cdot 2PbO$ glass prepared under oxidizing and reducing conditions. Both lead glasses were characterized by a structureless edge with $\alpha = 1$ cm^{-1} at $\lambda = 0.35$ μm (3.55 eV). "Reduced lead" glass was obtained by adding 1 wt % of Sb_2O_3 to the batch of the oxidized glass in order to reduce traces of polyvalent cations. Paul (1970a) has shown that the absorption edge of lead oxide glasses is determined by transitions associated with the Pb^{2+} ions and their ligands. Smith and Cohen (1963) noted a substantial edge shift upon the introduction of 0.08% Pb into $Na_2O-3SiO_2$. Smirnova (1965) measured the reflection spectra of various lead silicates containing 50–70 mole % PbO. A triplet reflection peak attributed to Pb^{2+} was observed between 0.26 and 0.29 μm (4.7–4.3 eV). Cohen *et al.* (1973) studied a range of $PbO-SiO_2$ glasses containing 24.6–65 mole % PbO and observed a cutoff shift with increasing PbO concentration from 0.34 to 0.40 μm, the high lead glasses exhibiting the well-known yellow color from the tail of the Pb^{2+} band.

b. B_2O_3.   The introduction of B_2O_3 into silicate glass tends to improve the ultraviolet transmission. Hensler and Lell (1969) measured the ultraviolet transmission of glasses of composition $Na_2O \cdot N(B_2O_3) \cdot 2SiO_2$ for values of N ranging from 0.0 to 4.0. These exhibited absorption edges ranging from 5.8 to greater than 6.5 eV are shown in Fig. 9, with the edge continuously shifting to shorter wavelengths with increasing boron content. Ginther and Sigel (1975) have studied fiber-optic grade binary borosilicate glasses of composition $B_2O_3 \cdot 3SiO_2$ which exhibited an edge near 0.17 μm, very close to that reported by McSwain *et al.* (1963) for B_2O_3. Binary borosilicates have been studied by Van Uitert *et al.* (1973) and Wemple *et al.* (1973) for use as low-index claddings for fused silica-core fiber-optic waveguides. Binary borosilicates are also of interest from a structural point of view since they are composed of tetrahedral SiO_4 units and planar BO_3 units which do not posses any NBO. This apparently accounts for the excellent uv transmission and high melting point (the Ginther glass was melted at 2200°C in iridium crucibles).

Wemple *et al.* (1973) have used a single Sellmeir oscillator model, as discussed by Wemple (1973), to analyze the bandgap of borosilicates as compared to SiO_2 and soda–lime glasses. In glassy systems the precise energy gap is not easily defined because the absorption generally increases exponentially with energy. It has been observed that the single Sellmeir oscillator energy E_0 scales linearly with the bandgap for a variety of semiconductors and insulators so that it can be used to compare bandgaps of different materials. Wemple

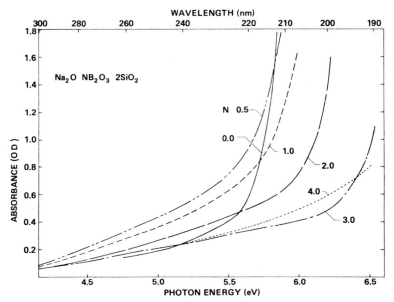

Fig. 9. Absorbance of sodium borosilicate glasses with different concentrations of boron oxide. Sample thicknesses are 2 mm. (After Hensler and Lell, 1969.)

et al. (1973) found that $E_0(SiO_2) = 13.38$ eV; $E_0(6SiO_2 \cdot B_2O_3) = 13.1$ eV; $E_0(B_2O_3) = 12.1$ eV, and that $E_0(SiO_2 \cdot 0.2Na_2O \cdot 0.1CaO) = 11.7$ eV, which indicates the excellent uv transmission of the binary borosilicates. Commercial borosilicates such as Pyrex rarely show high uv transparency because of the presence of trace impurities.

    c. TiO₂. Ti^{4+} enters substitutionally for Si^{4+} when introduced into SiO_2. Carson and Maurer (1973) have studied the uv transmission of SiO_2–TiO_2 glass (Corning Code 7971). The absorption edge of the glass near 4 eV (0.31 μm) is determined by Ti^{4+} ions if care is taken to eliminate Ti^{3+} by annealing in oxygen. Reduced titanium was shown to produce a broad absorption band in the glass centered near 3 eV (0.41 μm) which imparts a yellow tint to many titania containing glasses. Schroeder (1965) utilized the uv absorbing character of Ti to produce protective coatings on glasses. Sigel (1968) doped $Na_2O \cdot 3SiO_2$ glass with small concentrations (25–50 ppm) of Ti^{4+} and observed a peak near 6 eV (0.2 μm) with a tail extending well into the near uv. Smith and Cohen (1963) added 2.2% TiO_2 into $Na_2O \cdot 3SiO_2$ glass prepared under both oxidizing and reducing conditions and observed the edge to be near 3 eV in both cases but with substantial visible and near uv absorption in the reduced glass. The absorption arising from both Ti^{3+} and Ti^{4+} is discussed in more detail in Section III,A,1.

d. GeO₂. Vitreous GeO_2 is built up of randomly oriented GeO_2 tetrahedra with a Ge–O interatomic distance of 1.65 Å (Zarzycki, 1956). The addition of GeO_2 into SiO_2 results in a slight shifting of the uv edge to longer wavelengths. Many present day fiber-optic waveguides employ germanium-doped silica core glasses. Pajasova (1969) measured the uv reflectance of GeO_2 in the region 0.5–25 eV, and reported the first reflective maxima at 6.7 eV as shown in Fig. 10 for both crystalline and glassy GeO_2. A low-energy component below 6 eV (0.245 μm) is apparently associated with the presence of Ge^{2+} impurity as observed by Garino-Canina (1956) and Cohen and Smith (1958). Wemple (1973) has compared the interband optical transition strengths in SiO_2 and GeO_2 and indicates a 2.3-eV shift of the GeO_2 bandgap relative to SiO_2. Much of the earlier optical work on GeO_2 is of little value because of the presence of impurities and the use of nonstoichiometric samples containing large concentrations of Ge^{2+}. Sigel (1975) has measured the uv absorption of fiber-optic grade SiO_2 doped with several mole % GeO_2 (courtesy of R. Maurer of Corning Glass Works). This glass exhibited a very weak band near 5.14 eV (0.24 μm), presumably due to reduced ger-

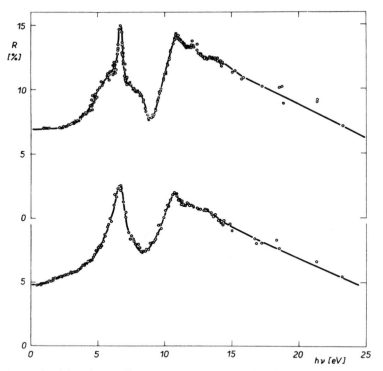

Fig. 10. Reflectivity of crystalline and glassy GeO_2 as a function of photon energy. (After Pajasova, 1969.)

manium, and an edge near 6 eV. Germanate glasses typically exhibit absorption edges in the 5-eV region but are known to undergo a coordination change from 4- to 6-coordinated Ge with addition of alkali (Riebling, 1963). Cohen and Smith (1958) observed a strong uv absorption in good quality fused GeO_2 at 0.245 μm arising from Ge^{2+} and an absorption edge at 5.6 eV (0.222 μm) as shown in Fig. 11. The experimental value of 5.6 eV for λ_0 in GeO_2 vs 8.0 eV in SiO_2, a 2.4-eV difference, is in good agreement with prediction of Wemple (1973).

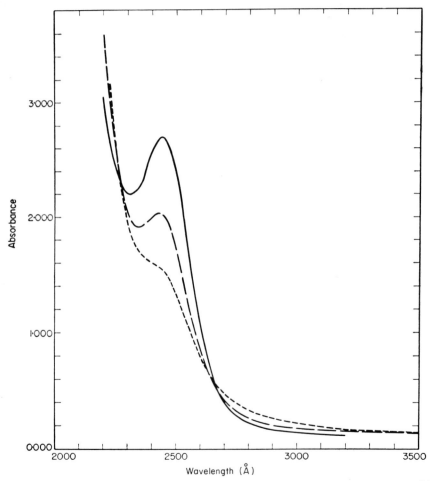

Fig. 11. Absorption band observed at 0.245 μm in glassy GeO_2 associated with the Ge^{2+} ion(——). Ultraviolet light will bleach band at room temperature; (— —) 0.5 hour bleaching, (– – –) 16 hours bleaching. (After Cohen and Smith, 1958.)

e. Al_2O_3. Aluminum can serve as a network modifier or former in silicate glasses. Loh (1964) measured the first reflectance peak in Al_2O_3 at 8.6 eV. Arakawa and Williams (1968) set the main absorption edge at 8.55 eV for single-crystal Al_2O_3 and 6.8 eV for anodized Al_2O_3 films and also provide a good review of previous uv edge measurements on Al_2O_3. Sigel (1971) measured the reflectance of a $Na_2O·Al_2O_3·3SiO_2$ glass and found the lowest energy reflection peak to be at 8.5 eV, as well as the characteristic SiO_2 peaks at 10.2 and 11.5 eV. No movement of the uv edge was observed upon addition of Al_2O_3 into the soda–silica glass (Sigel, 1973/1974). Hensler and Lell (1969) also studied glasses in the system $Na_2O·N(Al_2O_3)·3SiO_2$ where N varied from 0 to 1. These glasses all exhibited a transmission cutoff near 5.8 eV (0.21 μm) similar to the $Na_2O·3SiO_2$ glass, although the spectra show evidence of Fe^{3+} contamination. Thus aluminum does not affect the uv edge of silicates at energies below 6 eV.

4. Effect of Impurities on UV Edge of Silicate Glasses

Most commercial silicate glasses such as borosilicates and soda–limes possess poor ultraviolet transmission because of the presence of impurities. In particular, the 3d electrons of the transition metal ions such as iron, chromium, and copper and the 4f electrons of the rare earths such as cerium can produce charge transfer spectra sufficiently intense that a few ppm can easily be detected in a 1-mm thickness of glass. The transition metal spectra in glasses are discussed in more detail in Section III.

a. Iron, Fe^{2+}, and Fe^{3+}. The most serious problem in many air-melted glasses is associated with the Fe^{3+} ion which absorbs strongly near 0.23 μm (extinction coefficient $\varepsilon = 7000$ liters per mole-cm). Swarts and Cook (1965), Steele and Douglas (1965), and Sigel and Ginther (1968) have studied this band in various silicate glasses. The effects of 1000 ppm Fe^{3+} on the uv absorption of a thin (0.05 mm) section of $Na_2O·3SiO_2$ glass are shown in Fig. 12. In conventional thicknesses (1 $\sim$ 10 mm), the effect of the iron contamination is to produce a substantial edge shift to longer wavelengths. For example, a 1-mm thick soda–silica glass containing 1000 ppm Fe^{3+} cuts off near 0.32 μm. Fe^{2+} also produces a strong absorption near 0.20 μm which is usually masked by the ferric band, and which has an extinction coefficient of 3000 liters/mole-cm.

b. Copper, Cu^+, Cu^{2+}. Cuprous and cupric ion uv absorption is also intense in the ultraviolet, but copper trace impurity levels do not typically approach those of iron. Ginther and Kirk (1971) studied effects of copper doping on the uv edge position of high-purity $Na_2O·3SiO_2$ glass melted under both oxidizing and reducing conditions. Figure 13 demonstrates the edge shift produced by Cu^{2+} ions which also exhibit a charge transfer band

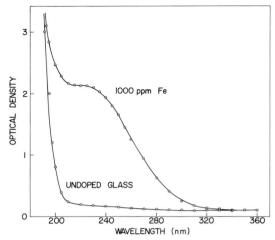

Fig. 12. Effects of 1000 ppm Fe^{3+} contamination on the uv edge of a soda–silica glass. Sample thicknesses are 25 μm. (After Sigel and Ginther, 1968.)

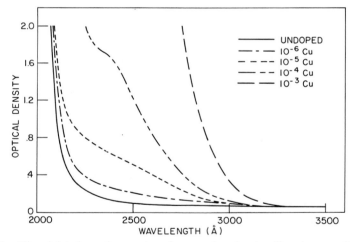

Fig. 13. Ultraviolet absorption spectra of copper-doped soda–silica glasses melted in air. (After Ginther and Kirk, 1971).

near 0.23 μm with an extinction coefficient $> 10^3$ liters/mole-cm, the exact value depending on the host glass. Cu^+ also shows an absorption in the same region (Parke and Webb, 1972).

    c. Chromium, Cr^{6+}. The Cr^{6+} ion has no d electrons (d^0) to produce ligand field-type intershell transitions but does possess strong charge transfer absorption in the uv. Nath *et al.* (1965) have identified two strong uv absorption bands in sodium silicate glasses at 0.37 μm with an extinction coefficient of 4200 liters/mole-cm and at 0.27 μm. Paul and Douglas (1968b) linked these bands to $[HCrO_4]^-$ and $[CrO_4]^{2-}$, respectively. The uv spectra of a chromium-doped glass is shown in Fig. 14.

Other transition metal ions also can contribute to uv absorption in oxide glasses as discussed in Section III.

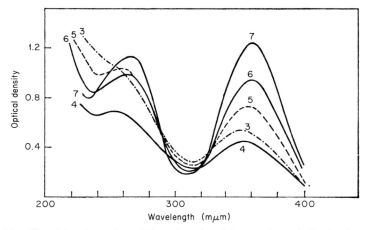

Fig. 14. Ultraviolet absoprtion of chromium-doped borate glasses indicating bands associated with $[HCrO_4]^-$ at 0.37 μm and $[CrO_4]^{2-}$ at 0.27 μm. (After Paul and Douglas, 1968b.)

Glass	K$_2$O (mole %)	BaO	B$_2$O$_3$
3	9.6	10.1	80.2
4	9.3	13.0	77.7
5	9.0	15.8	75.1
6	8.7	18.4	72.9
7	8.5	20.8	70.7

B. *Borate Glasses*

Borate glasses do not as a group have widespread commercial applications as do the silicate glasses. However, they exhibit many interesting structural variations which in turn affect optical properties.

1. B$_2$O$_3$

B$_2$O$_3$ is the basic network-forming oxide of borate glasses. The structure of vitreous B$_2$O$_3$ has been investigated using large-angle x-ray diffraction by numerous workers including Warren *et al.* (1936), Richter *et al.* (1954),

Zarzycki (1956), Herre and Richter (1957), Despujols (1958), Meller and Milberg (1960), and Mozzi and Warren (1970). All x-ray results indicate that boron is surrounded by three equidistant oxygens (average B–O distance 1.37–1.39 Å) and the centers of which form a regular triangle. Each oxygen is tied to two borons. Nuclear magnetic resonance (NMR) data of Silver and Bray (1958) confirmed the 3-coordinated structure of boron in B_2O_3 glass. The BO_3 triangles, like the SiO_4 tetrahedra are interlinked to form a continuous three-dimensional network but the structure of B_2O_3 is more complex than that found in SiO_2. Mozzi and Warren (1970) employed a high sensitivity fluorescence excitation method (Warren and Mavel, 1965) in the most recent study and concluded that planar B_3O_6 boroxol groups were present in vitreous B_2O_3. The spectroscopic data of Fajans and Barber (1952), Goubeau and Keller (1953), and Krogh-Moe (1965, 1969) had earlier indicated this result. Thus rather than a simple random network structure, a linked ring structure is most probable. Neighboring boroxol groups are not always directly connected by B–O or O–B bonds. A typical boroxol ring structure as envisioned by Mozzi and Warren (1970) is shown in Fig. 15. Krogh-Moe (1969) has provided an excellent review of the structure of boron–oxide glass. Bray and Silver (1960) have reviewed the NMR data in B_2O_3.

The fundamental optical properties of B_2O_3 in the uv will be dependent on the strength of the B–O bond. The sp^2 planar bond is covalent in character. The highest valence band should be the nonbonding oxygen 2p orbital as in the case of SiO_2. However, there does not presently exist any theoretical band structure calculations or uv reflectance data on B_2O_3 such as in the case of SiO_2. The ultraviolet transmission cutoff value λ_0 of B_2O_3 has been reported by Kordes and Worster (1959) at 0.186 μm, by Poch (1964) at 0.200 μm, and by McSwain et al. (1963) at 0.172 μm with the latter probably

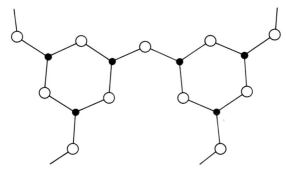

Fig. 15.   Schematic model of boroxol groups linked by a shared oxygen in B_2O_3; ● = boron, ○ = oxygen. (After Mozzi and Warren, 1970.)

representing the most accurate value. Higher purity materials available today may still yield an even lower value for the uv cutoff of B_2O_3.

2. EFFECTS OF NETWORK MODIFIERS ON UV ABSORPTION OF BORATES

The structural rearrangements which take place upon the addition of network modifiers to B_2O_3 are much different than the breakup of the glass network which occurs in silicates. The initial effect upon the introduction of alkali into B_2O_3 is the conversion of sp^2 planar BO_3 units into more stable sp^3 tetrahedral BO_4 units, thereby preserving the B–O bonding without the creation of NBO (nonbridging oxygen) ions. Biscoe and Warren (1938) had first seen evidence of this change of boron coordination in a study of the x-ray diffraction of a series of alkali borate glasses. Silver and Bray (1958) and Bray and O'Keefe (1963) employed NMR to establish that the conversion of 3- to 4-coordinated borons peaks at levels of about 30 mole % alkali oxide, and then reverses. This would imply that NBO ions should be created in large numbers at alkali concentrations above the 30 mole % level. It was also observed that BO_4 units are not linked directly to each other in alkali borate glasses but are screened by BO_3 triangles. Recent results, however, indicate that larger scale changes in structure also take place. Konijnendijk (1975) concluded that in the composition range 0–20 mole % alkali oxide, the boroxol groups are to a major extent gradually replaced by 6-membered borate rings with one BO_4 tetrahedron. The six-membered borate rings are to a large extent ordered to form tetraborate groups in sodium and potassium-borate glasses. In the 20–35 mole % alkali oxide, the 6-membered borate rings with one BO_4 tetrahedron are replaced by rings with two tetrahedra. In those glasses the rings are connected to form diborate groups. Figure 16 shows schematically the structure of some typical borate groups.

The NMR work of Rhee (1971) indicate that the borate groups in glasses are quite similar to those observed in crystals. Above 30 mole % alkali oxide, reasonable amounts of NBO ions are produced, again in large ortho-borate, pyroborate, and metaborate ring structures. In the more practical borosilicate glasses, similar types of borate structures are observed.

The ultraviolet absorption of binary borate glasses have been investigated by Kordes and Worster (1959) and McSwain et al. (1963). The position of the absorption edge of sodium borate glasses as a function of Na_2O molar concentration is shown in Fig. 17. A sharp increase in the edge shift was observed in the 15–20 mole % Na_2O composition range. These results were interpreted by McSwain et al. (1963) to mean that there is initially a con-version of 3- to 4-coordinated boron up to 15–20 molar % alkali oxide and that beyond this region, there is a sharp increase in the NBO concentration, followed by a more gradual increase at still higher Na_2O levels. However,

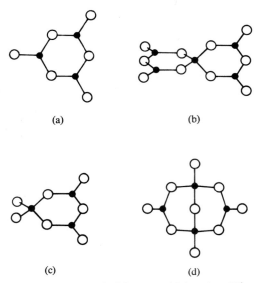

Fig. 16. The borate groups found in alkali borates with less than $33\frac{1}{3}$ mole % alkali; (a) the boroxol group, (b) the pentaborate group, (c) the triborate group, and (d) the diborate group. (After Krogh-Moe, 1965.)

this intepretation is not supported by NMR, Raman, and ir data which indicate that NBO formation does not take place until about 30 mole % Na_2O. One possible explanation is that the formation of an alkali-rich phase in the glasses did contain NBO ions which produced the edge shift. Another possible problem could have resulted from iron contamination of the glasses. Large-scale structural changes may also play a role. In the same paper, a $Li_2O\cdot3B_2O_3$ glass was observed to have a λ_0 below 0.180 μm compared to that for B_2O_3 of 0.172. This seems more consistent with the absence of NBO. The absorption edge of Li_2O, Na_2O, and K_2O borate glasses moves to longer wavelengths with increasing size of the cation or decreasing field strength (Z/r_i^2; Z, atomic number; r_i, ionic radius) as in the silicate case.

Paul and Douglas (1967a) reported a uv cutoff for a Na_2O–B_2O_3 host glass at 0.190 μm somewhat below that of McSwain *et al.* (1963) for the same composition. Hensler and Lell (1969) reported a λ_0 value for $Na_2O\cdot2SiO_2$ glass of less than 0.190 μm. Bishop (1961) reported 77% transmission in a $K_2O\cdot Al_2O_3\cdot4.5B_2O_3$ glass at 0.200 μm. These data seem to indicate that the edge of most borates does not shift to longer wavelengths nearly so much as the silicates at alkali levels below 25 mole %. The introduction of B_2O_3 into alkali silicates is also observed to produce a movement of the uv edge to shorter wavelengths (Hensler and Lell, 1969). Therefore most pure borate

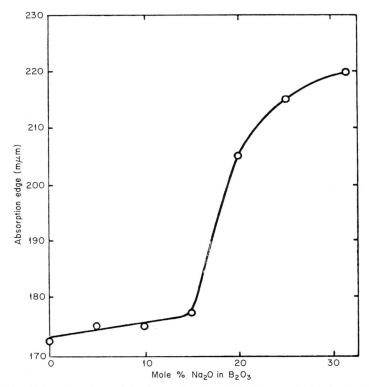

Fig. 17. Absorption edge position λ_0 as a function of Na_2O concentration in $Na_2O-B_2O_3$ glasses. Break in curve at 15 mole % Na_2O does not correspond to any known structural change in the glass. (After McSwain *et al.*, 1963.)

glasses are expected to exhibit substantially higher uv transmission than their silicate counterparts. The electronic absorption does not appear to be sensitive to the large scale changes in ring structures taking place in borate glasses but most likely depends on the excitation of electrons from B–O and M–O bonds.

C. Phosphate Glasses

Phosphate glasses also tend to exhibit excellent uv transmission, and find applications in this regard. The book of Van Wazer (1958) and the review of Westman (1960) are the classic references on phosphate glasses.

1. P_2O_5

The network-forming oxide in phosphate glasses is P_2O_5. The basic structural unit is the PO_4 tetrahedron. Virtually all phosphates, both amorphous and crystalline are composed of chains, rings, or branched polymers of interconnected PO_4 tetrahedra. In P_2O_5, three of the oxygens in a given tetrahedron are shared with a neighboring phosphorus ($\frac{3}{2}$ O atoms) while one oxygen is unshared (1 O atom) to yield the $2\frac{1}{2}$ O atoms per P atom required for stoichiometry. The P–O shared bond is 1.62 Å, while the unshared bond is 1.39 Å, and the P–O–P bond angle is 123.5°. Crystalline P_2O_5 exists in several structures including a hexagonal P_4O_{10} molecular form and two infinite sheet polymer structures. Vitreous P_2O_5 is composed of a three-dimensional network with varying ring and chainlike links.

The basic units of a P_2O_5 structure are described by Van Wazer (1958) as branching points, middle groups, and end groups and are shown schematically in Fig. 18.

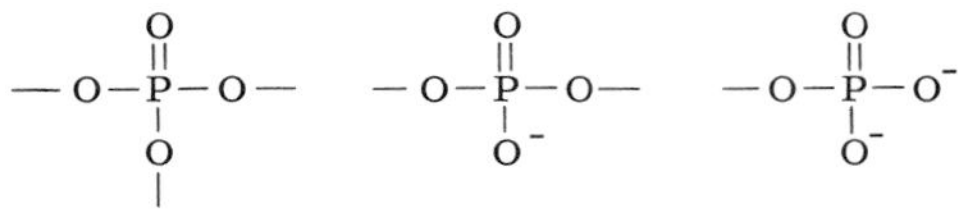

$$
\begin{array}{ccc}
\overset{\displaystyle O}{\underset{\displaystyle \underset{O}{|}}{\overset{\|}{-O-P-O-}}} & \overset{\displaystyle O}{\underset{\displaystyle \underset{O^-}{|}}{\overset{\|}{-O-P-O-}}} & \overset{\displaystyle O}{\underset{\displaystyle \underset{O^-}{|}}{\overset{\|}{-O-P-O^-}}} \\[4pt]
\text{Branching point} & \text{Middle group} & \text{End group} \\
\text{(a)} & \text{(b)} & \text{(c)}
\end{array}
$$

Fig. 18. Schematic representation of the three basic structural units in P_2O_5; (a) branching point, (b) middle group, and (c) end group. (After Van Wazer, 1958.)

Each unit shows the top oxygen in an unshared π bond configuration corresponding to the short bond length. This bond is stronger than the other P–O bonds so that it is not expected to produce a shift of the uv edge to longer wavelengths.

The only experimental data found in the literature on the ultraviolet transmission of glassy P_2O_5 is that of Kordes and Nieder (1968). A λ_0 cutoff value of 0.145 μm was measured. Previously Kordes and Worster (1959) had measured the cutoff of HPO_3 at 0.273 μm and this value has mistakenly been quoted by several reviewers as the P_2O_5 value.

2. EFFECTS OF NETWORK MODIFIERS ON UV ABSORPTION OF PHOSPHATES

The structure of alkali and alkaline earth phosphates is quite complicated, and experimental x-ray diffraction data exists for only a few compositions. Biscoe *et al.* (1941) have confirmed the presence of PO_4 tetrahedra in a calcium phosphate glass with a P–O distance of 1.57 Å. This distance indicates that in the glass the amount of π bonding between the phosphorus and oxygen atoms is, on the average, nearly $\frac{1}{4}\pi$-bond per σ-bond. The coordination measured from the area under the first peak in the radial distribution function is 4.2 oxygens per phosphorus in glasses with CaO/P_2O_5 mole ratios of 0.731 and 0.964. The average P–O–P bond angle was about 140°.

The introduction of M_2O or MO modifying oxides into P_2O_5 will introduce M–O–P links in place of P–O–P links, attaching themselves to the O^- ions of the middle and end groups shown in Fig. 18. The doubly bonded oxygen bond, being stronger, remains intact in the alkali phosphate glasses.

The introduction of alkali or alkaline earths should therefore produce a breakup of the basic P–O–P glass forming network much in the same manner as observed in silicate glasses. This weakening of the network by the introduction of M–O bonds should produce an edge shift to longer wavelengths similar to the silicates since NBO ions will be formed immediately upon addition of modifiers.

The ultraviolet absorption of alkali and alkaline earth phosphates has been studied in a series of papers by Kordes and Worster (1959), Kordes (1965), and Kordes and Nieder (1968). A wide range of network-forming oxides including H_2O, Li_2O, Na_2O, K_2O, Rb_2O, and Cs_2O were introduced into P_2O_5 in the range 0%–50% mole M_2O. In addition, MgO, CaO, SrO, BaO, ZnO, CdO, PbO, HgO, and AgO were introduced. A few examples are shown in Fig. 19. In all cases the initial movement of the edge was a movement to longer wavelengths. For up to 20 mole % M_2O, the cutoff moved to $\lambda_0 = 0.20$–0.25 μm, similar to the silicate glasses. However, as the M_2O and MO content of the modifier is further increased, the uv edge was observed to retreat back to shorter wavelengths, thereby providing the excellent uv transmission for which many phosphate glasses are known. Still further increases in M_2O content result in another movement of the edge to longer wavelengths. These oscillations of the edge are most likely related to the structural rearrangements of the glass and the relative concentrations of the various fundamental units previously discussed.

Van Wazer (1958) has shown that the increase of Na_2O in an alkali phosphate glass will result in a linear drop in the number of branching units and a simultaneous increase in the middle unit concentration. This conversion is apparently responsible for the initial movement of the edge. The higher λ_0 of the edge at very large modifier concentrations is probably explained by the increase in end and orthophosphate units (3 NBO) where more M–O replacement of P–O bonds takes place. The intermediate high-uv transmission zone of phosphate glasses is more difficult to explain, and may be related to formation of stable ring structures, but requires further investigation.

Another interesting structure in phosphate glasses is the so-called four-way branching structure which can be achieved by the introduction of aluminum or other trivalent species into a phosphate glass. In theory, the Al^{3+} and the P^{5+} should charge compensate for one another to produce a glass network analogous to SiO_2. Binary $FePO_4$, $AlPO_4$, BPO_4, etc., do not exist as glasses but aluminum can be introduced into alkali and alkaline earth

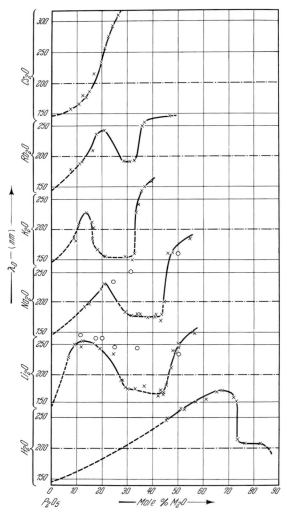

Fig. 19. Effects of the introduction of network modifiers on the uv cutoff of phosphate glasses. (After Kordes and Nieder, 1968.)

phosphate glasses to achieve good commercial-quality uv transmitting glasses. For example, Hensler (1962) measured a λ_0 value of 0.216 in a reduced $Al(PO_3)_3$–MgO–Li_2O glass.

Duffy (1972) employed a uv high-transparency host glass (Na_2O–P_2O_5) for study of the ultraviolet absorption of transition metals. The λ_0 value for this glass was near 50,000 cm^{-1} (0.20 μm). Edwards et al. (1972) employed glasses of the composition $MO–P_2O_5$, where $M = M_g$, Ca, Sr, and Ba as base glasses for a study of iron absorption. These glasses also showed high transmission down to 0.20 μm. Van Wazer (1958) has provided a list of several phosphate glass compositions which have been utilized as uv transmitting materials.

III. Visible Absorption

As discussed in Section II, the bond strengths of oxide glasses are sufficiently large that the fundamental band gaps lie in the far uv with only weak tails extending into the visible. A discussion of the visible absorption in these materials therefore reduces largely to a review of impurity absorption.

A. Absorption of Impurities in Oxide Glasses

Impurities may enter into the glass structure in a variety of positions. For example in a silicate glass, the possibilities generally fall into three major categories: (1) a network-modifying cation position surrounded by six or more oxygens, (2) a network-forming cation position with four-fold oxygen coordination, replacing Si^{4+}, and (3) a network-forming anion position, replacing an oxygen. Some impurities can enter the glass network in more than one position, in different valence states as well as with different oxygen coordination numbers.

1. TRANSITION METAL IONS

The most important classes of impurities found in oxide glasses are the transition metal elements (Ti, V, Cr, Mn, Fe, Co, Ni, and Cu) which introduce optical absorption bands associated with excitation of 3d electrons in partially filled inner shells. When these ions are introduced into the glass network, energy levels of the free ions are split and shifted by the electrostatic fields of the nearest-neighbor anions, typically oxygens. Since the field of a metal ion falls off extremely rapidly with distance ($\propto 1/r^5$), only interactions with adjacent atoms, molecules, and ions (termed ligands) need to be considered. Ligand field theory or crystal field theory (Bethe, 1929; Van Vleck, 1932; McClure, 1959; Orgel, 1960) was developed to calculate the splittings and transitions which will arise for a given electronic configuration in a specified symmetry. In glasses, ligand field theory has been used with good success to identify most of the absorption bands associated with transition metal ions (Bates, 1962; Bamford, 1962b) in various valence states and coordinations. The mathematical details of the ligand field theory of glasses are available in the above references and will not be derived here, but a few elementary concepts are useful for discussing the absorption of colored glasses.

The absorption spectra of transition metal complexes arise from internal transitions between the d electron levels of the central ion as modified by the ligand field. These bands usually lie in the ir, visible, and near uv and exhibit weak molar extinction coefficients, $\varepsilon \approx 0.01-200$. The electronic configuration is determined by the number of electrons in the d shell of the ion and is written as d^n. Typical transition metal ions that might be encountered in an oxide glass and their specific electronic configurations are shown in Table I. Normally the symmetry in oxide glasses will be octahedral (six-fold coordination) or tetrahedral (four-fold coordination), often with varying degrees of distortion.

The energy levels (terms) arising for a given d^n configuration depend on the orbital angular momentum L and the spin multiplicity $2S + 1$. Angular momentum values of a given configuration are discrete numbers 0, 1, 2, 3, 4, and 5 and are designated as S, P, D, F, G, and H in conventional quantum mechanics. The free ion levels are denoted then as ^{2S+1}L, e.g., ^3P, ^1D, etc. In a crystal or glass in octahedral or tetrahedral coordination, the energy levels of the free ion are split, the number and degeneracy of the new levels depending on L and S. The new levels are denoted by $^{2S+1}\Gamma_n(L)$, where S and L refer to the values of the free ion, and Γ_n indicates the orbital degeneracy of the level. The splitting of the new levels will vary as a function of ligand field strength Δ which typically takes on values between 10,000 and 30,000 cm^{-1}. Optical absorption arises from electronic absorptions between these new levels subject to the selection rules $\delta L = \pm 1$, and $\delta S = 0$, with

TABLE I

Transition Metal Ions in Glasses

Configuration	Ion	Color	Configuration	Ion	Color
d^0	Ti^{4+}	Colorless	d^4	Cr^{2+}	Faint blue
	V^{5+}	Faint yellow to colorless		Mn^{3+}	Purple
			d^5	Mn^{2+}	Light yellow
	Cr^{6+}	Faint yellow to colorless		Fe^{3+}	Faint yellow
			d^6	Fe^{2+}	Blue-green
d^1	Ti^{3+}	Violet-purple		Co^{3+}	Faint yellow
	V^{4+}	Blue	d^7	Co^{2+}	Blue-pink
	Mn^{6+}	Colorless	d^8	Ni^{2+}	Brown-purple
d^2	V^{3+}	Yellow-green	d^9	Cu^{2+}	Blue-green
d^3	Cr^{3+}	Green	d^{10}	Cu^+	Colorless

all transitions initiated from the ground state. The selection rules break down in glasses to some extent resulting in weak forbidden bands. In a classic paper by Tanabe and Sugano (1954), the energy levels of the d^n configurations were calculated for both octahedral and tetrahedral symmetries as a function of the ligand field strength. This permits a fitting of the experimental optical absorption data of a given ion to a specific Δ value. After one absorption band is chosen to calculate Δ, the other bands serve as a check on the correctness of the value selected. The fitting of the data to the ligand field theory provides information on the symmetry, coordination and respective valence states of the ion producing the absorption, and to the extent that selection rules are violated, some information on the degree of distortion present at the ion site.

While the transitions between the d electron levels of the transition metal ions produce relatively weak absorption bands, there is another class of transitions which arise from the transfer of charge from the central ion to ligands or to other ions of the same element in different valence states. These so-called "charge transfer" bands usually exhibit molar extinction coefficients above 10^3 and occur chiefly in the uv region (as discussed in Section II,A,4) but the tails of such bands can extend well into the visible producing strong coloration. Charge transfer bands also occur in d^0 (empty shell), and d^{10} (full shell) systems where there are no ligand field bands. Because of the similarity of transition metal ion spectra in all oxide glasses, results on silicate, borate and phosphate glasses have been combined in the discussions which follow on the absorption of specific ions.

a. Titanium. Ti^{4+} is an example of an empty shell d^0 system which in theory is not expected to exhibit ligand field bands. This is confirmed experimentally by the absence of any visible coloration in most titania containing glasses. Bamford (1962b) introduced 1 wt % TiO_2 into a series of binary sodium silicate and sodium borate glasses melted under oxidizing conditions and failed to observe any coloration. Similar results on a variety of other glass systems have been obtained by Jahn (1966), Stroud (1971), and Paul (1975). However, Ti^{4+} does introduce strong ultraviolet absorption near 44,200 cm^{-1} (0.23 μm) as measured by Duffy (1972) in $Na_2O \cdot P_2O_5$ glass. Smith and Cohen (1963) observed a shift of the uv edge of a $Na_2O \cdot 3SiO_2$ glass from 0.22 to 0.30 μm upon the introduction of 2.2 wt % TiO_2 and noted that Ti^{4+} produces a 5.4 eV (0.23 μm) absorption in an HCl solution. Carson and Maurer (1973) measured the edge of a silica glass containing 7.4 wt % TiO_2 to also lie near 0.30 μm. All of the foregoing results are consistent with the presence of a strong charge transfer band in a d^0 system.

TiO_2 has been introduced into silica to achieve a higher index core material for low-loss fiber-optic waveguides. This has only been possible because of the absence of any Ti^{4+} absorption bands in the visible and near ir regions. Studies by Schultz and Smyth (1972) and Evans (1970) in TiO_2–SiO_2 glass have concluded that the Ti^{4+} enters substitutionally for Si^{4+} in a fourfold coordination.

Ti^{3+} possesses a d^1 electronic configuration and in octahedral symmetry is expected to exhibit a single ligand field transition $^2\Gamma_5 \rightarrow {}^2\Gamma_3$. Reduced titanium is not easily formed in most silicate glasses. Ti^{3+} in solution as the $[Ti(H_2O)_6]^{3+}$ ion is known to absorb at 20,300 cm^{-1} (0.492 μm) with a shoulder near 17,400 cm^{-1} (0.574 μm) (Bates, 1962). The shoulder arises from a tetragonal distortion of the octahedral symmetry of the Ti^{3+} ion' as shown schematically in Fig. 20. The optical absorption of Ti^{3+} has been

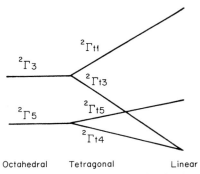

Fig. 20. Effect of tetragonal distortion on the octahedral levels of a d^1 system. Splitting of levels of the Ti^{3+} ion in glasses results in the presence of a shoulder on the low-energy side of the main band. Transitions always arise from the ground state only. (After Bates, 1962.)

measured in a number of glasses. Smith and Cohen (1963) observed absorption bands near 0.55 and 0.80 μm (weak) in reduced Na_2O–$3SiO_2$ glasses containing 2.2 mole % Ti. Carson and Maurer (1973) observed a broad band in reduced SiO_2–TiO_2 (1971) glass near 0.41 μm which was attributed to Ti^{3+}. This assignment was confirmed by Friebele et al. (1974) and Kurkjian and Peterson (1974) by use of electron spin resonance (ESR). The latter authors also measured a broad optical absorption in SiO_2–TiO_2 glass (Corning Code 7971) peaking near 22,000 cm^{-1} (0.455 μm) with a shoulder near 12,000 cm^{-1} (0.833 μm). Bates (1962) observed Ti^{3+} absorption in a borosilicate glass at 17,500 cm^{-1} (0.572 μm) with a shoulder at 14,300 cm^{-1} (0.70 μm), as shown in Fig. 21. Paul (1975) has recently reported Ti^{3+} absorption in reduced Na_2O–B_2O_3 and Na_2O–P_2O_5 near 20,000 cm^{-1} (0.50 μm) and 14,000 cm^{-1} (0.71 μm). All of the above studies on Ti^{3+} indicated octahedral symmetry of the oxygen ions clustered about the central ion, but with some tetragonal distortion apparent.

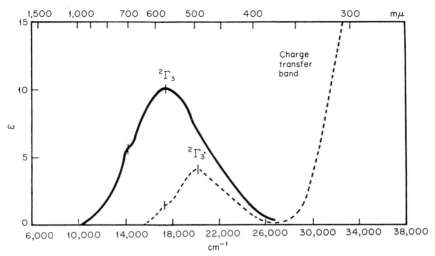

Fig. 21. Absorption spectrum of a Ti-doped borosilicate glass (solid line) with band at 0.57 μm (17,500 cm^{-1}) and a shoulder at 0.70 μm (14,300 cm^{-1}). The absorption is characteristic of the Ti^{3+} ion in distorted octahedral symmetry. The dashed line is associated with a $^2\Gamma_5 \rightarrow {}^2\Gamma_3$ transition of $[\text{Ti}(\text{H}_2\text{O})_6]^{3+}$ in a crystal. (After Bates, 1962.)

b. Vanadium. Glasses containing vanadium can exhibit a range of colors from green to yellow to almost colorless as the alkali oxide content is increased. Spectra can be complex since vanadium can exist in one of four valence states, $V^{5+}(d^0)$, $V^{4+}(d^1)$, $V^{3+}(d^2)$, or $V^{2+}(d^3)$ depending on glass composition and melting conditions. Ideally, a d^0 configuration will contribute only a charge transfer band, a d^1 configuration a single strong band, and a d^2 and d^3 configuration several possible bands.

Kumar (1959) observed that vanadium-doped alkali silicate glasses exhibit a very strong uv absorption near 0.35 μm (28,500 cm^{-1}), two strong visible bands near 0.425 μm (23,500 cm^{-1}) and 0.625 μm (15,500 cm^{-1}), and a strong band in the near ir at 1.12 μm (8900 cm^{-1}), all of which were assigned to V^{3+}. However, only the two visible bands appear to be associated with octahedrally bonded V^{3+}, the transitions being denoted by $^3\Gamma_4 \rightarrow {}^3\Gamma_5$ (15,500 cm^{-1}) and $^3\Gamma_4 \rightarrow {}^3\Gamma_4(P)$ (23,500 cm^{-1}). This assignment is consistent with the work of Weyl (1951), Bates (1962), Bamford (1962b), Johnston (1965), and Smith and Cohen (1963) on silicate glasses containing vanadium. The spectra of V^{3+} in glass, a solution of $V(\text{H}_2\text{O}_6)^{3+}$, and a crystal are shown in Fig. 22. The shift of the bands to lower frequencies is typical of the spectra of transition metal ions in glasses. These same two bands were also observed by Kakabadse and Vassiliou (1965) in a series of careful experiments on alkaliborophosphate glasses and were assigned to the same

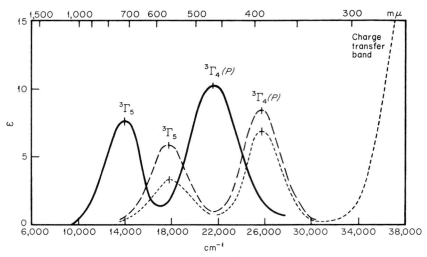

Fig. 22. Absorption spectrum associated with V^{3+} in octahedral symmetry in a glass (———). Ligand field bands in glasses shift to lower frequencies than that found in $[V(H_2O)_6]^{3+}$ in solutions (— —) and crystals (-–-). (After Bates, 1962.)

two transitions, but substantial shifting of band positions with composition was noted.

The band in the near ir of some glasses containing vanadium appears to be associated with V^{4+} as first claimed by Weyl (1951). Bates (1962) suggested that the 8900-cm^{-1} band in silicate glasses corresponds to a $^2\Gamma_{t4} \to {}^2\Gamma_{t5}$ transition of Fig. 20. with the V^{4+} ion in distorted octahedral symmetry having a tetragonal distortion slightly smaller than for VO^{2+} in aqueous solution. However, Orgel (1960) has identified the transition in VCl_4 near 9000 cm^{-1} as being due to V^{4+} in tetrahedral coordination, $^2\Gamma_3 \to {}^2\Gamma_5$. Hecht and Johnson (1967), Toyuki and Akagi (1972) and Paul and Assabghy (1975) have employed both optical and ESR spectroscopy to study the $V^{3+} \to V^{4+}$ valence changes with composition and melting conditions in a number of glass hosts, lending further support that the ir band is associated with V^{4+}. Studies of the magnetic susceptibility and optical spectra of V^{4+} in borate glasses by Kumar (1964) and the ESR and optical data on alkali-rich borates and germanates by Toyuki and Akagi (1972) indicate the presence of the vanadyl ion (VO^{2+}) with absorptions near 1.6 and 1.0 μm. However, these bands shift substantially with glass composition producing colors from pale blue, through green to pale yellow-ish brown as the alkali oxide concentration of the host glass is increased. These changes in the ligand field strength parameter Δ have been related to the increase in the coordination of the germanium and boron network-forming ions. Bates (1962) has noted that if V^{2+} occurs in octahedral symmetry in glass, bands are predicted at 10,500, 16,000, and 24,600 cm^{-1}, assuming Δ to be about 12% smaller than the Δ for H_2O ligands. These bands are low in intensity in solutions and overlap with the stronger V^{3+} and V^{4+} bands.

The strong uv charge transfer band at 41,700 cm^{-1} in some vanadium-doped glasses was not observed in glasses containing only V^{3+} such as the alumino–borophosphate glass shown in Fig. 22. It is most likely due to V^{4+} ion although V^{5+} might also contribute in this region since it is a d^0 configuration.

    c. Chromium. The name chromium is derived from the Greek word "chromos," color, suggesting immediately that chromium compounds typically possess strong visible absorption. The chromium ions commonly encountered in glasses are $Cr^{3+}(d^3)$ and $Cr^{6+}(d^0)$, with the latter valence state expected to exhibit only uv charge transfer spectra such as the Ti^{4+} and V^{5+} ions discussed earlier. Chromium-doped glasses for the most part exhibit a characteristic green color because of the presence of strong absorption bands in both the blue and red regions of the spectrum.

In a series of chromium-doped soda–silica glasses, Bamford (1962b) observed a relatively composition insensitive triple-peaked band centered near 0.65 μm (0.635, 0.650, and 0.675), and an intense uv band near 0.365 μm which increased sharply with higher alkali content. Bates and Douglas (1959) observed a similar triplet band in soda–lime glass as well as a broad band at 0.45 μm. Bamford (1962b) observed virtually identical spectra in reduced soda–silica glass, as shown in Fig. 23. The visible bands are associated with

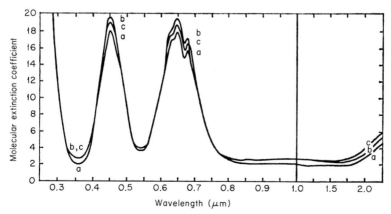

Fig. 23. Absorption spectrum of chromium in reduced soda–silica glass. The strong Cr^{3+} triplet band in the red (0.65 μm) combined with the blue absorption band near 0.45 μm impart the characteristic green color to chromium-doped glasses. (After Bamford, 1962b.)

Cr^{3+} in octahedral coordination (Marsh, 1965) while the uv absorption apparently arises from Cr^{6+}.

Nath and Douglas (1965) investigated the Cr^{3+}/Cr^{6+} equilibrium in binary alkali silicate glasses. The Cr^{6+}/Cr^{3+} ratio was observed to increase with alkali content and in the order Li, Na, and K for a given molar alkali content. The 0.65 μm Cr^{3+} band ($\varepsilon = 18.6$) in sodium silicates was found at 0.63 μm ($\varepsilon = 16.3$) in lithium silicates and at 0.654 μm ($\varepsilon = 19.07$) in potassium silicates. Nath *et al.* (1965) have determined that Cr^{6+} produces two ultraviolet absorption bands in sodium silicate glasses at 0.37 μm ($\varepsilon = 4200$) and at 0.27 μm as shown earlier in Fig. 14. Similar bands in aqueous solutions involving $(CrO_4)^{2-}$ ions suggest that the Cr^{6+} ion in the glass possesses tetrahedral symmetry.

Stroud (1971) compared chromium coloration in oxidized and reduced lead silicate glasses with that found in soda–silica glass. In oxidized lead glass the uv absorption band due to Cr^{6+} ions moved to 0.384 μm, producing a yellow glass due to the long tail of this band. The reduced lead glass exhibited the composite band Cr^{3+} near 0.65 μm and a well-defined band at 0.43 μm. In soda–silica glass addition of 1% Sb_2O_3 reduced all chromium to the 3+ state and eliminated the 0.37 μm band, leaving only the Cr^{3+} bands at 0.45 and 0.64 μm. The lead glasses were shown by Stroud (1971) to contain more intense coloration per unit weight of chromium than the soda–lime glasses.

While the green color of soda–silica glasses containing chromium remains relatively constant as the alkali content is increased, borate glasses become less colored with increasing Na_2O (Bamford, 1962b). Alkali borates doped with chromium possess bands at 0.35, 0.43, and 0.60 μm. The intense uv band also increases with alkali content in the borates, again apparently due to an increase in the fraction of chromium ions in the Cr^{6+} valence in the high-alkali glass. ESR measurements (Loveridge and Parke, 1971) also indicate the change in symmetry and valence of the Cr^{3+} ion with the increase of alkali content. Paul and Douglas (1968b) examined the ultraviolet absorption of Cr^{6+} in alkali and alkaline earth borates and concluded that chromium was present as a borochromate group in acid glasses similar to $[HCrO_4]^-$ and as a chromate group $[CrO_4]^{2-}$ in a basic glass, with the former being responsible for the 0.36-μm band and the latter the 0.26-μm band, quite similar to the silicate glasses.

Duffy (1972) observed bands at 0.65 μm (a triplet band similar to that seen in silicates) and at 0.45 μm in chromium-doped $Na_2O\cdot P_2O_5$ glass. A uv band was observed near 0.25 μm, possibly due to the chromate ion, as well as an absorption edge at 0.20 μm. Landry *et al.* (1967) found octahedral symmetry in a Cr-doped aluminum zinc phosphate glass. Tischer (1968) employed pressures up to 50 kbars on a variety of Cr-doped glasses to show that only small distortions from cubic symmetry occur.

Paul (1974) has recently reported the measurement of the optical absorption of Cr^{2+} in several silicate glasses melted under reducing conditions. Four absorption bands were observed near 0.50, 0.59, 0.73, and 1.0 μm. It was concluded that Cr^{2+} was present in a high-spin distorted octahedral site.

d. Managanese. Manganese-doped glasses can range in color from pale or deep purple to various shades of brown. The use of manganese purple glasses date back at least to the early Egyptians. Manganese has also been used to decolor green-tinted iron-containing glasses because of its complementary absorption in the visible. Manganese can be present in oxide glasses as $Mn^{3+}(d^4)$ or $Mn^{2+}(d^5)$. The energy diagram for d^4 systems in octahedral symmetry predicts only a single allowed transition together with several weak spin-forbidden lines and bands (Bates, 1962). Fuwa (1923) measured the absorption in soda–lime silicate and potash–lime–silica glasses containing Mn^{3+} and observed a single band in each at 0.47 and 0.52 μm, respectively. Turner and Weyl (1935) added arsenic and antimony (reducing agents) to Na_2O–CaO–SiO_2 glass to obtain Mn^{2+} alone in the glass, which exhibited several bands with very weak extinction coefficients including a narrow band near 0.43 μm ($\varepsilon = 1$) and evidence of weak broad bands near 0.63 and 1.0 μm ($\varepsilon < 0.1$).

Bamford (1962b) measured the absorption of about 1 wt % Mn_3O_4 introduced into oxidized silicate glasses containing between 17.2 and 40.8 wt % Na_2O. A single band was observed near 0.49 μm which was associated with Mn^{3+} in octahedral coordination. A large increase in the intensity of the band with increasing alkali content indicated an increasing proportion of the Mn^{3+} ions as the Na_2O content increased. Reduced silicate glasses behaved similarly, and in addition showed evidence at low alkali content of the Mn^{2+} band near 0.42 μm as shown in Fig. 24. Schultz (1974) measured the Mn^{3+} absorption in SiO_2 at 0.46 μm while Stroud (1971) noted that the Mn^{3+} band is shifted to 0.505 μm in oxidized lead alkali glass. Bamford (1962b) also studied manganese in a series of sodium borate glasses. These also exhibit a single visible band near 0.45 μm, which increases with alkali content and shifts to 0.48 μm at 26.6 wt % Na_2O. Thus a broad Mn^{3+} band absorbs in the green, and passes some blue and some red to give the characteristic purple color of many manganese-doped glasses.

Bingham and Parke (1965) measured the absorption spectra of $3K_2O \cdot 7P_2O_5$, $3Na_2O \cdot 7P_2O_5$, $CaO \cdot P_2O_5$, and $Na_2O \cdot 4B_2O_3$ glasses containing Mn^{2+}. The Mn^{2+} absorption was extremely weak with $\varepsilon \approx 0.2$ in all cases. Bands were measured at 0.372, 0.342, 0.309, and 0.228 μm. The Mn^{2+} ion was shown to be in octahedral symmetry in the phosphate glasses and predominantly tetrahedral symmetry in silicates. The extinction coefficient for Mn^{2+} in $3Na_2O \cdot 7P_2O_5$ is shown in Fig. 25. Mn^{2+} also is responsible for strong fluorescence in oxide glasses which shows considerable variation

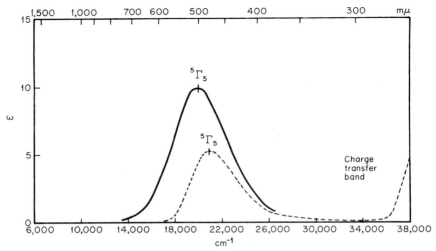

Fig. 24. Absorption spectrum of Mn^{3+} in soda lime–silica glass (solid line) and of

$$[Mn(H_2O)_6]^{3+}$$

in crystal (dashed line). Shift in frequency in glass is consistent with Mn^{3+} in a distorted octahedral symmetry. This band in the green is responsible for the characteristic purple color of many manganese-doped glasses. (After Bates, 1962.)

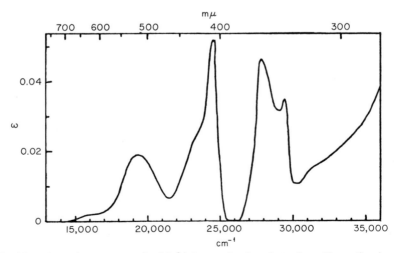

Fig. 25. Absorption spectrum for Mn^{2+} in a soda–phosphate glass. Absorption is much weaker than for the Mn^{3+} ion. (After Bingham and Parke, 1965.)

with host glass, and Mn^{2+} produces an easily recognizable ESR signal, both of which provide complementary data for interpretation of the bonding of the ion in glasses (Lunter *et al.*, 1968; Bingham and Parke, 1965; Parke *et al.*, 1970).

   e. Iron. Iron is universally present in almost all oxide glasses as trace impurity, and is responsible for the undesireable greenish tint of many commercial glasses. Iron is the most important and most widely investigated of the transition metals in glasses, being present in oxide glass as either $Fe^{2+}(d^6)$ or $Fe^{3+}(d^5)$. Redox conditions and the glass composition are important in the melting of most optical glasses so that the Fe^{2+}/Fe^{3+} ratio may be controlled. Fe^{2+} exhibits a broad, strong infrared absorption ($^5\Gamma_5 \rightarrow {}^5\Gamma_3$) near 1.0 μm, which extends into the red, producing the characteristic blue-green of some iron-containing glasses. Fe^{3+} produces strong uv absorption. The addition of iron into reduced glasses in the presence of sulfur results in the formation of the well-known amber color used commercially in bottles. Weyl (1951) is a good source for much of the early work on iron-containing glasses. A more recent review of iron in glass was given by Bamford and Hudson (1965).

   Bamford (1960a,b, 1961, 1962a), in a series of four papers, studied the absorption of iron in silicate, borate and phosphate glasses. In the silicate glasses, weak bands at 0.38, 0.425, and 0.44 μm, plus intense absorption at shorter wavelengths were associated with Fe^{3+} in octahedral symmetry. Colloidal ferric iron was observed in borates due to limited solubility, giving rise to a yellow, brown, or reddish brown color. In phosphates 0.505- and 0.71-μm bands were attributed again to Fe^{3+} ions, as well as the intense uv band. Steele and Douglas (1965) suggested that tetrahedral coordination of Fe^{3+} is most consistent with the optical data on silicates. Kurkjian and Sigety (1968) employed a combination of Mössbauer, optical, and paramagnetic resonance techniques to determine that Fe^{3+} is predominantly tetrahedrally coordinated in silicate glasses and octahedrally coordinated in phosphate glasses.

   The strong ultraviolet absorption of both Fe^{3+} and Fe^{2+} has been studied by Steele and Douglas (1965), Swarts and Cook (1965), and Sigel and Ginther (1968) in various host glasses. In silicate and borate glasses, Fe^{3+} exhibits a charge transfer band near 0.23 μm with an ε of 7000. Fe^{2+} also exhibits a charge transfer band near 0.20 μm or less with an ε of about 3000 as shown in Fig. 26. Both bands have long tails which limit the uv absorption in most common glasses. A band near 2.0 μm is also observed in the ir, and has been linked to Fe^{2+} in tetrahedral coordination. Swarts and Cook (1965) observed that uv irradiation (excitation at 0.254 and 0.366 μm) of reduced iron silicates induced peaks at 0.45, 0.34, and 0.26 μm. Fe^{2+} absorption was observed at

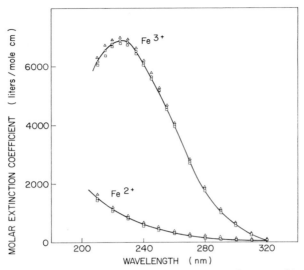

Fig. 26. Ultraviolet absorption spectra of the overlapping Fe^{2+} and Fe^{3+} charge transfer bands measured in a soda–silica host glass. (After Sigel and Ginther, 1968.)

1.06 and 2.0 μm. Paul and Douglas (1969b) studied the spectra of Fe^{3+} in a series of binary $Na_2O–B_2O_3$ glasses with and without NaCl added. As the Na_2O concentration increased from 11 to 33 mole %, the position of the uv charge transfer bands shifts from 0.225 to 0.210 μm. The possibility of two bands related to 4-and 6-coordinated ferric iron was suggested, tetrahedral coordination being most likely in borates with high alkali contents.

Bishay and Kinawi (1964) and Edwards *et al.* (1972) measured the effects of iron in a wide range of phosphate glasses. Fe^{2+} bands were observed in the regions of 0.95–1.1 μm and 2.05–2.2 μm depending on composition; the former being attributed to octahedral symmetry ($^5\Gamma_5 \rightarrow {}^5\Gamma_3$) and the latter to tetrahedral ($^5\Gamma_3 \rightarrow {}^5\Gamma_5$), as shown in Fig. 27. Fe^{3+} bands were observed at 0.375–0.365 μm, 0.425–0.410 μm, 0.540–0.510 μm, and 0.700–0.780 μm, arising from iron in both octahedral and tetrahedral coordination. Charge transfer bands were observed at 0.195 and 0.240 μm and attributed to Fe^{3+}. Hirayama *et al.* (1968) studied iron-doped alkaline earth phosphates of composition $MO–2P_2O_5$, where M is Mg^{2+}, Ca^{2+}, or Ba^{2+}. An ir band of Fe^{2+} was observed at 1.17, 1.12, and 1.05 μm, respectively, for the Mg^{2+}, Cu^{2+}, and Ba^{2+} with ε of 42, 22, and 14 liters/mole-cm.

Loveridge and Parke (1971), Bishay and Markar (1969), and Tucker (1962) have studied the ESR of Fe^{3+} in various glasses, but it is difficult to unambiguously assign the $g = 4.3$ or $g = 2$ resonance to a particular iron coordination, as discussed by Kurkjian and Sigety (1968).

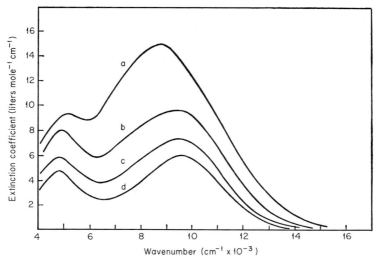

Fig. 27. Absorption spectra associated with Fe^{2+} in tetrahedral (2.1-μm band) and octahedral (1.05-μm band) symmetry in binary alkaline earth phosphate glasses. (a) $MgO \cdot P_2O_5$, (b) $CaO \cdot P_2O_5$, (c) $SrO \cdot P_2O_5$, and (d) $BaO \cdot P_2O_5$. (After Edwards *et al.*, 1972.)

Iron-doped glasses of substantial commercial interest are the so-called "carbon amber" glasses. A good amber color can be produced by melting alkali silicates containing iron and sulfur under reducing conditions. The mechanism responsible for the coloration was a source of controversy for many years. In recent times, Bamford (1961) measured absorption bands in alkali silicate glasses with sodium sulfate and 0.5 wt % Fe_2O_3 added to exhibit absorption bands in the 0.40–0.44 μm regime as well as at 0.36 and 0.29 μm. Magnetic susceptibility studies ruled out the presence of any colloidal ferric oxide in the amber glasses. Brown and Douglas (1965) suggested that sulfur in the form of sulfide is necessary for the production of amber glass, but early work by Bork (1930) and Weckerle (1933) had shown that sulfur alone would not produce a good amber color. Bacon and Billian (1954) showed that the Fe^{2+} band at 1.05 μm possesses the same extinction coefficient in amber glass as in sulfur-free green iron glasses, suggesting the ferrous complex does not play a major role in the amber coloration process. Brown and Douglas (1965) proposed that a ferric ion surrounded by four sulfurs in the glass network was responsible for the absorption.

In a later paper Douglas and Zaman (1969) demonstrated that only one sulfide ion is linked to each ferric ion. It was shown that iron and sulfur in a $3K_2O \cdot 7SiO_2$ glass are present as Fe^{2+} in octahedral coordination with oxygen, Fe^{3+} in tetrahedral coordination with oxygen, the amber complex containing Fe^{3+} in tetrahedral coordination with one oxygen replaced by a sulfur, sulfur as sulfate and sulfur as sulfide but not associated with an iron ion. The principal absorptions due to the amber complex are centered near 0.295 and 0.425 μm as shown in Fig. 28. The extinction coefficients in this host glass were measured as 1500 (0.295-μm band) and 9000 (0.425 μm). Karlsson (1969) reported a similar charge transfer band in a doped $3Na_2O\cdot7SiO_2$ glass at 0.412 μm (24,300 cm^{-1}) with an $\varepsilon = 3900$. The absorption was also attributed to ferric iron in tetrahedral coordination with oxygen and sulfur. Loveridge and Parke (1971) employed ESR to show that the addition of sulfur to glasses containing fixed amounts of iron produces a decrease in the Fe^{3+} resonance near $g = 4.3$ and an increase in the resonance at $g = 6$. For tetrahedrally coordinated Fe^{3+}, the $g = 6$ ESR resonance results from the substitution of an oxygen by a sulfur, which is consistent with the model suggested by Douglas and Zaman (1969) for the amber complex.

The substitution of sulfur by selenium in $3R_2O\cdot7SiO_2$ glasses (where R can be Li, Na, or K) produces a similar complex, the iron–selenium black glasses as shown by Paul (1973).

f. Cobalt. Cobalt glasses are known to range in color from blue to red, depending on composition. Cobalt is normally found only as $Co^{2+}(d^7)$ in oxide glasses but some evidence of $Co^{3+}(d^6)$ has been reported.

Weyl (1951) noted that the absorption spectra of cobalt glasses were similar to those of aqueous solutions but that as with other ions such as Cr^{3+} and V^{3+}, the absorption bands in the glasses were shifted to longer wavelengths.

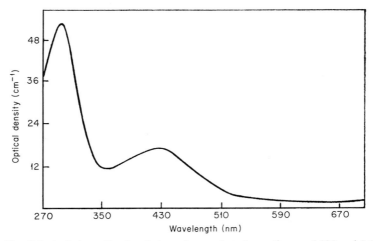

Fig. 28. Principal absorption bands in carbon amber glasses lie near 0.295 and 0.425 μm and have been associated with Fe^{3+} in tetrahedral coordination with one oxygen replaced by a sulfur. ($3K_2O\cdot7SiO_2$ glass, 0.051 wt % iron, 0.125 wt % sulfur). (After Douglas and Zaman, 1969.)

Weyl (1951) attributed the blue color to a CoO_4 complex and the pink color to a CoO_6 complex, but noted that the blue color tends to predominate. Similar conclusions were reached in a comprehensive study by Aglan and Moore (1955) on the colors of cobalt in glass.

Bamford (1962b) found that both oxidized and reduced soda–silica glasses containing Co^{2+} were the same blue color independent of alkali content and showed no change under uv irradiation. The absorption bands consisted of a triply split band at 0.54, 0.59, and 0.64 μm ($\varepsilon = 190$) together with a split ir band at 1.4, 1.6, and 1.8 μm ($\varepsilon = 40$). McClure (1957) observed that tetrahedrally coordinated Co^{2+} ions in ZnO produce two triply split bands near 0.6 μm (16,500 cm^{-1}) and 1.5 μm (7000 cm^{-1}) due to $^4\Gamma_2 \rightarrow {}^4\Gamma_4(P)$ and $^4\Gamma_2 \rightarrow {}^4\Gamma_4(F)$ transitions. The glass absorption is thus consistent with Co^{2+} in a distorted tetrahedral configuration as shown in Fig. 29.

In Co-doped fused silica Schultz (1974) observed a royal blue color arising from a visible triplet band at 0.51, 0.61, and 0.68 μm and a broad ir band near 1.7 μm with a shoulder perhaps near 1.2 μm. This glass appears to show evidence of both octahedral and tetrahedral coordination of Co^{2+}. The extinction coefficient of the tetrahedral 0.68 μm peak ($\varepsilon = 105$) is less than that observed in alkali silicates ($\varepsilon = 190$) by Bamford (1962b) which also suggests the presence of the octahedrally coordinated ion.

Bamford (1962b) found a variation in the color of cobalt-doped alkali borate glasses from pink to blue as the Na_2O content increased from 5 to

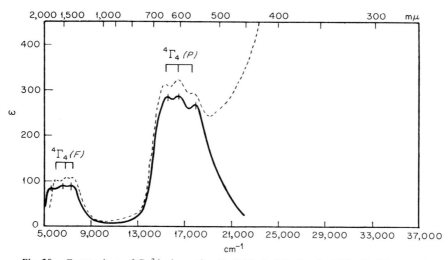

Fig. 29. Comparison of Co^{2+} absorption ZnO (dashed line) and a high alkali borate glass (solid line). Triply split bands near 0.6 μm (16,500 cm^{-1}) and 1.5 μm (7000 cm^{-1}) are associated with Co^{2+} in tetrahedral coordination. (After Bates, 1962.)

28.1 wt %. The low-soda glasses exhibited a 0.55-μm band ($\varepsilon = 15$) with a shoulder at 0.52 μm and ir bands near 1.2 μm (very weak) and 1.5 μm. The high-soda glass exhibited two triply split bands near 0.55 ($\varepsilon = 110$) and 1.5 μm ($\varepsilon = 20$). The presence of the weak 1.2-μm band in glasses containing less than 15 wt % Na_2O is indicative of some octahedral coordination of Co^{2+} ions. The greatly increased Co absorption of the high-soda glasses coupled with the disappearance of the 1.2-μm band indicates a marked increase in the number of tetrahedrally coordinated Co^{2+} ions. Bates (1962) also noted this Co coordination change in going from low to high alkali borates.

Paul and Douglas (1968a) and Juza *et al.* (1966) studied the coordination change of cobalt in more detail in a series of binary $R_2O–B_2O_3$ glasses where R was K, Na, and Li. In the former work, tetrahedrally coordinated Co^{2+} was shown to begin formation at 18 mole % K_2O, 19 mole % Na_2O, and 22 mole % Li_2O. It was concluded that octahedral complexes are formed in acid melts where the bonds are mainly ionic, while more covalency existed in tetrahedral complexes formed in basic glasses.

In phosphate glasses of composition $Na_2O \cdot P_2O_5$ Duffy (1972) reports bands at 0.54 ($\varepsilon = 12$) and 1.43 μm ($\varepsilon = 1.8$) which also suggests octahedral coordination in this glass. Earlier, Veinberg (1962) employing Co^{2+} as a probe in $K_2O–P_2O_5$ and $K_2O–ZnO–P_2O_5$ glasses had attempted to link structural changes in the phosphate glass to coordination of the cobalt ion. Haddon *et al.* (1969) found Co^{2+} to be found entirely in octahedral sites in $Na_2O–P_2O_5$ glass.

Paul and Douglas (1968a) and Juza *et al.* (1966) have studied the effects of the addition of halide ions on the spectra of cobalt-doped alkali borate glasses. Cobalt was observed to move from octahedral coordination to tetrahedral halide complexes in low alkali glasses. Paul (1974) has also studied the octahedral-tetrahedral coordination of Co^{2+} in glass as a function of temperature.

The optical absorption of Co^{3+} has recently been reported by Paul and Tiwari (1974) in $Na_2O–NaBr–B_2O_3$ glass. It was found that in acid glasses containing NaBr, a small fraction of the Co^{2+} is oxidized to a 6-coordinated Co^{3+} complex containing a mixture of bromide and oxide ligands. Co^{3+} optical absorption bands were observed in the blue and near uv region at 0.350, 0.425, and 0.463 μm. Dietzel and Coenan (1961) had earlier reported the existence of Co^{3+} in soda–silica glasses containing upwards of 40 wt % Na_2O.

g. Nickel. Nickel-doped glasses tend to absorb across the entire visible spectrum typically producing unattractive brown-colored glasses. Nickel is normally present in oxide glasses only as $Ni^{2+}(d^8)$. The energy level diagram for d^8 systems in octahedral symmetry predicts that the spectrum of Ni^{2+}

will consist of three spin-allowed transitions $^3\Gamma_2 \rightarrow {}^3\Gamma_5$, $^3\Gamma_2 \rightarrow {}^3\Gamma_4$, and $^3\Gamma_2 \rightarrow {}^3\Gamma_4(P)$ together with several spin-forbidden bands (Bates, 1962). Four-coordinated Ni^{2+} should also exhibit three spin-allowed transitions $^3\Gamma_4 \rightarrow {}^3\Gamma_5$, $^3\Gamma_4 \rightarrow {}^3\Gamma_2$, and $^3\Gamma_4 \rightarrow {}^3\Gamma_4(P)$ with several weak spin-forbidden bands being possible.

The experimental data on nickel-doped oxide glasses indicate that Ni^{2+} is both 4- and 6-coordinated in silicate glasses while exhibiting largely octahedral symmetry in both borate and phosphate glasses. In silicate glasses, Weyl and Thumen (1933) and Moore and Winkelman (1955) concluded that nickel was present as Ni^{2+} in a wide range of compositions in both four- and six-coordination with oxygens. Bamford (1962b) observed five bands in nickel-doped soda–silica glasses as shown in Fig. 30. These were at 0.45 ($\varepsilon = 50$), 0.56, 0.63 ($\varepsilon = 20$), 0.93 ($\varepsilon = 8$), and 1.2 μm. By comparison with known crystalline data, it was concluded that tetrahedrally coordinated Ni^{2+} produces the 0.56 and 0.63-μm bands $[^3\Gamma_4(F) \rightarrow {}^3\Gamma_4(P)]$ and the band at 1.2 μm $[^3\Gamma_4(F) \rightarrow {}^3\Gamma_2(F)]$. Octahedral Ni^{2+} produces the 0.45-μm band, a $^3\Gamma_2(F) \rightarrow {}^3\Gamma_4(P)$ transition, the 0.93-μm band $[^3\Gamma_2(F) \rightarrow {}^3\Gamma_4(F)]$ and absorption in the 1.6–2.0-μm region $[^3\Gamma_2(F) \rightarrow {}^3\Gamma_5(F)]$. The equilibrium between the two coordinations is apparently unaltered by the change in the Na_2O content in the silicate glasses. Fused silica (Schultz, 1974) and lead silicate glasses (Stroud, 1971) doped with nickel show spectra similar to the alkali silicates.

In borate glasses Bates (1962) showed that there is good agreement between the observed spectrum of Ni^{2+} in low alkali borate glasses and the ligand field theory predictions for six-fold coordinated Ni^{2+} with only a slight shift to lower energies in the glass relative to the $[Ni(H_2O)_6]^{2+}$ ion in solution.

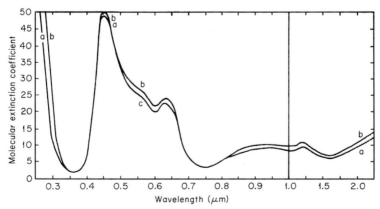

Fig. 30. Absorption spectrum of Ni^{2+} in soda–silica glass. Nickel is present in both octahedral and tetrahedral coordination. (a) 16.6 Na_2O, 83.0 SiO_2, 0.10 NiO (wt %), (b) 39.1 Na_2O, 60.7 SiO_2, 0.24 NiO (wt %). (After Bamford, 1962b.)

Bands were observed at 0.42, 0.65, 0.74, and 1.33 μm. Bamford (1962b) also observed that Na_2O–B_2O_3 glasses containing between 5 and 15 wt % Na_2O showed strong absorption at 0.41 μm, a double-peaked band at 0.64–0.74 μm and ir absorption at 1.4 and 1.55 μm, characteristic of octahedral symmetry. At higher Na_2O concentrations tetrahedrally coordinated Ni^{2+} is apparently responsible for bands at 0.53, 0.62, and 1.55 μm which remain fixed with composition changes, but the relative concentration of 4-coordinated Ni^{2+} ions appears small. The large shift observed in the octahedral spectra was attributed by Bamford (1962b) to d–p orbital mixing. Hecht (1967) used optical and ESR techniques on Ni-doped Na_2O–B_2O_3 glasses to also conclude that most sites possessed six fold coordination. Other recent studies of nickel in borate glasses have included those of Tiedemann (1966), Berkes and White (1966), Paul and Douglas (1967b), Takahashi and Goto (1970), and Hussein and Moustaffa (1972). These authors have evaluated the effects of various alkali and alkaline earth oxides and composition variations on the symmetry and ligand fields of Ni^{2+}.

In phosphate glasses of composition $Na_2O \cdot P_2O_5$, Duffy (1972) reports absorption bands at 0.43, 0.70, 0.83, and 1.37 μm, indicating only the presence of Ni^{2+} in octahedral coordination.

It can be concluded that Ni^{2+} exists primarily in octahedral coordination in borate and phosphate glasses but in both four- and six-fold coordination in most silicate glasses.

h. Copper. The use of copper as an important pigment has a long history. Virtually all early blue and green Egyptian glasses contain copper oxide as a colorant. Copper can be present in glasses as the Cu^{2+} ion (d^9), the Cu^+ ion (d^{10}), and under some circumstances as metallic Cu^0, the latter occurring in the so-called copper–ruby glass as a colloidal suspension. The cuprous ion having a d^{10} configuration will not impart color to the glass but may contribute uv absorption as a charge transfer band. The d^9 configuration of the cupric ion in simple octahedral and tetrahedral coordination consists of only a single transition $^2\Gamma_3(D) \rightarrow {}^2\Gamma_5(D)$, just the inverse of a d^1 system shown in Fig. 20. Copper-containing glasses exhibit a range of colors from blues to greens to browns depending on composition and melting conditions. However Jorgenson (1957) has shown that Cu^{2+} complexes often have a strong tetragonal distortion which can split both the $^2\Gamma_3$ ground state and the $^2\Gamma_5$ upper state each into two states so that these bands are often observed in the visible and near ir regions. If the tetragonal absorption is very large, the three excited states converge and a single band is seen again. Kumar (1959) observed a single band due to Cu^{2+} in a sodium silicate at 12,700 cm^{-1} (0.79 μm) which is similar to the $[Cu(H_2O)_6]^{2+}$ ion in solution. Bates (1962) attributes this to Cu^{2+} in octahedral symmetry with a strong tetragonal distortion. The characteristic blue-green color of many copper-doped glasses arises from the absorption of this broad composite band in the near ir which extends into the red.

Bamford (1962b) observed a similar 0.79-μm band in Na_2O–SiO_2 glasses ($\varepsilon = 8$) as well as a very weak band near 0.45 μm ($\varepsilon < 1$). Reduced soda–silica glasses were colorless but the 0.78-μm band grew quickly for Na_2O above 26.1 wt % due to a change from Cu^+ to Cu^{2+} valence at high alkali content. Photo-oxidation of the reduced glasses was observed in the presence of uv light. Johnston and Chelko (1966) studied Cu(I)–Cu(II) equilibrium in soda–silica glass as a function of oxygen pressure. Stroud (1971) measured the Cu^{2+} band at 0.775 μm in lead glass and at 0.790 μm in soda–silica glass. Edwards *et al.* (1972) measured the absorption of Cu^{2+} in a family of lead silicates of composition 15 $Na_2O\cdot XPbO\cdot(85-X)SiO_2$ as shown in Fig. 31. The intensity of the 12,500-cm^{-1} (0.80 μm) band increased with PbO content, indicating an increase in the covalent bonding in the structure.

Bamford (1962b) observed that sodium borate glasses changed from pale green through turquoise to blue with increasing alkali content, again because of the growth of the Cu^{2+} band near 0.8 μm, which was assigned to a $^2\Gamma_3(D) \rightarrow$ $^2\Gamma_5(D)$ transition of Cu^{2+} in octahedral symmetry.

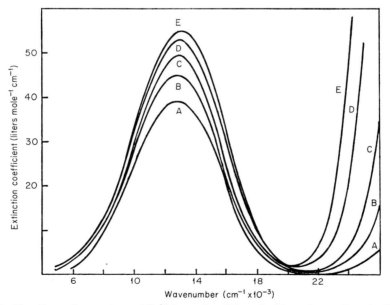

Fig. 31. Absorption spectrum of Cu^{2+} in a series of Pb–soda silicate glasses. Growth of band with PbO content is due to an increase in the covalent bonding in the glass structure (wt % PbO). (a) 37.1, (b) 48.1, (c) 53.5, (d) 67.5, and (e) 70.6. (After Edwards *et al.*, 1972.)

Cu$^+$ absorption is not commonly observed in most oxide glasses. However, Schultz (1974) found that copper entered fused silica prepared by flame hydrolysis in the Cu$^+$ state producing a strong uv absorption and bright yellow-green fluorescence. Parke and Webb (1972) studied the optical absorption and fluorescence of Cu$^+$ in CaO·P$_2$O$_5$, Na$_2$O·3SiO$_2$, and Na$_2$O·4B$_2$O$_3$ glasses. A broad absorption band at 0.237 μm was observed in all three glasses and in the CaO·P$_2$O$_5$ glass was correlated with Cu$^+$ ($\varepsilon = 800$).

As discussed in Section II, A, 4, b Cu^{2+} also exhibits a charge transfer band in the uv. Ginther and Kirk (1971) observed a Cu^{2+} band in Na$_2$O–SiO$_2$ glass at 0.235 μm ($\varepsilon = 8500$). Duffy (1972) also observed a charge transfer band in copper-doped Na$_2$O·P$_2$O$_5$ at 44,700 cm^{-1} (0.224 μm) and a broad band at 0.87 μm, both of which were ascribed to Cu^{2+}.

i. General Conclusion on Transition Metals. Oxide glasses for the most part accept transition metals in the more stable octahedral coordination. Ligand field theory is a useful tool for interpreting the spectra. Although the transition metal ions to a certain extent serve as a probe of the glass structure, it is dangerous to predict structural rearrangements in the glass on the basis of spectral changes unless there is other supporting experimental evidence. Ligand field transitions generally take place at lower frequencies in glasses than in solutions or crystals containing the same ion.

2. RARE EARTH IONS

The rare earth ions (La^{57} through Lu^{71} on the periodic table) are characterized by having partially filled 4f shells in their ground states in contrast with the partially filled 3d shells of the transition metal ions. Optical absorption associated with the rare earth ions consists typically of a spectrum of many very weak, sharp lines ranging from the near ir to near uv. These are associated with transitions between states of the 4f configurations although some strong 4f → 5d charge transfer bands are found in the far uv region. Since the intershell 4f electrons are only weakly perturbed by the ligand fields of neighboring ions, the rare earth spectra of a given ion are similar in all matrices. The rare earth ions found in oxide glasses along with their electronic configurations and characteristic colors are given in Table II.

The study of free ion schemes (Klingenberg 1947; Meggers 1942; Moore 1949; Wybourne, 1965) by analysis of arc and spark spectra permits the identification of most lines in rare earth-doped crystals and glasses. Dieke and Crosswhite (1963) and McClure (1959) have summarized the observed energy levels associated with trivalent and divalent rare earths in crystals. In glasses, interest in rare earth ion absorption and emission spectra has been driven by the development of laser glasses using ions such as Nd^{3+}, Yb^{3+}, Er^{3+}, and Ho^{3+}. Of these, neodymium represents by far the most important

TABLE II

RARE EARTH IONS IN GLASSES

Configuration	Ion	Color	Configuration	Ion	Color
$4f^0$	La^{3+}	None	$4f^7$	Eu^{2+}	Brown
	Ce^{4+}	Weak yellow		Gd^{3+}	None
$4f^1$	Ce^{3+}	Weak yellow	$4f^8$	Tb^{3+}	None
$4f^2$	Pr^{3+}	Green	$4f^9$	Dy^{3+}	None
$4f^3$	Nd^{3+}	Violet-pink	$4f^{10}$	Dy^{2+}	Brown
$4f^4$	Pm^{3+}	None		Ho^{3+}	Yellow
$4f^5$	Sm^{3+}	None	$4f^{11}$	Er^{3+}	Weak pink
$4f^6$	Sm^{2+}	Green	$4f^{12}$	Tm^{3+}	None
	Eu^{3+}	None	$4f^{13}$	Tm^{2+}	None
				Yb^{3+}	None
			$4f^{14}$	Lu^{3+}	None

system with the 1.06-μm emission line of Nd^{3+} being most commonly used. Figure 32 indicates the typical absorption and emission spectra obtained in a Nd-doped glass and gives an indication of the complexity of the energy level scheme. Excellent reviews of laser glasses have been given by Snitzer (1966), Young (1969), and Patek (1970).

While the positions of both the absorption and emission bands of the rare earth ions are similar in all hosts, as for example in Fig. 33, selection rules are usually dependent on perturbations produced by the glass matrix ligand field since many forbidden transitions are observed. Absorption intensities, lifetimes, and fluorescent efficiencies for specific transitions will vary from glass to glass. Lasers containing neodymium have been made in a large number of host glasses including silicates, phosphates, germanates, and borates. Oxide glasses typically offer large sites to the rare earth ions, permitting as much as six to twelve oxygen–ligand coordinations. At high concentrations, the fields of the large rare earth ions begin to interact, perturbing the energy levels and producing the so-called "concentration quenching" of the laser action in the glass. Strong absorption bands in the green and yellow spectral regions are responsible for the characteristic delicate purple color of neodymium-doped glasses.

Another of the rare earth ions which has found commercial importance in glasses is cerium. Depending on the host glass and melting conditions, cerium can be present as either the Ce^{3+} or Ce^{4+} ion. Both ions produce strong uv absorption, Ce^{3+} near 0.31 μm and Ce^{4+} near 0.24 μm as shown in Fig. 34 (Stroud, 1961). CeO_2 addition to oxide glasses provides excellent uv absorption without the introduction of significant visible absorption. Examples of Ce applications include eye protection and laser rod cladding protection to absorb the uv pump light of the flash lamp. CeO_2 is also employed to

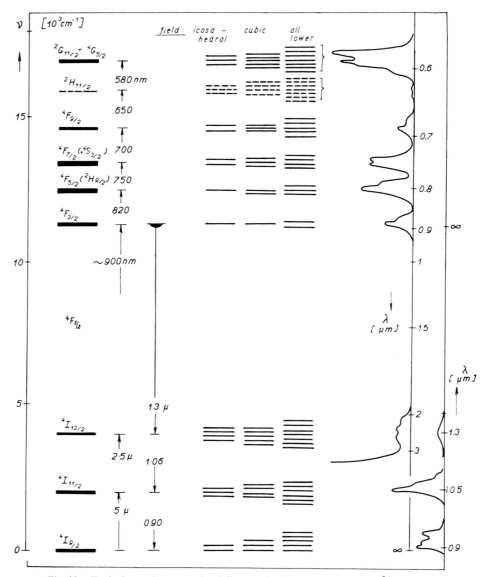

Fig. 32. Typical complex energy level diagram for the rare earth ions, Nd^{3+} in this example, with the optical absorption and emission spectra also indicated. (After Patek, 1970.)

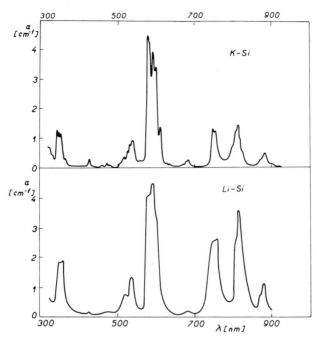

Fig. 33. Comparison of the optical absorption spectra of Nd^{3+} in different host glasses. Ligand field interactions with the 4f electrons are weak so that spectral features of trivalent rare earth ions are similar in all hosts. (After Patek, 1970.)

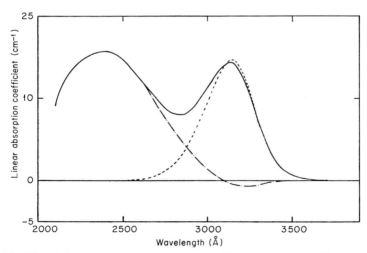

Fig. 34. Ultraviolet absorption associated with Ce^{3+} (0.31 μm) and Ce^{4+} (0.24 μm) in a silicate glass. The photosensitivity of the cerium ion in glass is important in numerous applications. (After Stroud, 1961.)

radiation-protect glasses which are exposed to ionizing or nuclear radiation (Stroud, 1961, 1962, 1964; Lell *et al.*, 1966). Evans and Sigel (1975) have measured the tails of the Ce^{3+} and Ce^{4+} bands extending through the visible into the near ir (Fig. 1) and have demonstrated the use of cerium to produce low to moderate loss radiation-protected fiber-optic waveguides. As discussed in Section II,3, CeO_2 is also introduced into photosensitive glasses to permit photonucleation of metallic particles using uv light (Stookey, 1949a,b). CeO_2 can also be introduced into glasses to control the valence equilibrium of other impurities or dopants and thereby control the color of the glass (Weyl, 1951). In some host glasses such as the lead silicates, the tail of the Ce^{4+} absorption band may extend into the blue, imparting a faint yellow color characteristic of many heavily doped Ce glasses.

The spectra of the other rare earth ions in glasses can be found in the references cited in this section.

3. METALLIC PARTICLES

The brillant reds and yellows of certain gold, silver and copper-doped glasses have been well-known since the middle of the seventeenth century. Weyl (1951) provides many interesting details on the history of these glasses. The so-called "gold-ruby" glasses have always held a particular fascination for both the public and glass scientists alike.

a. Gold-Ruby Glasses. Faraday (1857) seems to be the first to establish that metallic gold in a fine state of subdivision can produce red colors, and that colloidal metallic gold particles were responsible for the color of gold-ruby glass. The coloration is not the result of a scattering process but is associated with light absorption within the metal particles themselves which are suspended in the glass. Mie (1908) employed Maxwell's equations to calculate the absorption and scattering of the electromagnetic radiation by a spherical particle and predicted the absorption expected in a gold sol by using the bulk optical properties of gold. Zsigmondy (1909) and Gans (1922) employed ultramicroscope techniques to deduce the approximate spherical shape and size of the metal particles found in the ruby glasses.

Traditionally, gold-ruby glasses have been prepared by dissolving small amounts of gold, typically 0.01–0.02 wt %, in a host glass containing small amounts of a reducing agent such as SnO_2, ZnO, and Sb_2O_3. Although colorless when cooled quickly from the melt to room temperature, the nucleation and growth of gold particles takes place during subsequent annealing near the softening point of the glass, resulting in the development of a striking red color (Zsigmondy, 1909; Badger *et al.*, 1939; Weyl 1945, 1951; Bachman *et al.*, 1946; Dietzel, 1945; Stookey, 1949a,b). A more recent method developed by Stookey (1949a) and studied by Maurer (1958, 1959) and Doremus (1962,

1964, 1965) employs photoreducible agents such as CeO_2 as room-temperature nucleating agents. In the latter case more control of the coloring process is obtained since only growth of the metallic particles without further nucleation takes place during the high-temperature annealing stage.

The color of the gold-ruby glass results from an absorption band near 0.53 μm as shown in Fig. 35, taken from the work of Doremus (1962, 1964, 1973) who undertook a systematic study of the effects of particle size, temperature, and the method of preparation on the color of ruby glass. Doremus (1964) showed that the absorption can be considered as a "plasma resonance" band in which the free electrons of the metal particles are treated as a bounded plasma, and oscillate collectively at a characteristic resonance frequency. By taking in both the bulk optical properties of gold and the size and spherical shape of the metallic particles, the theory and experimental data were shown to be in good agreement.

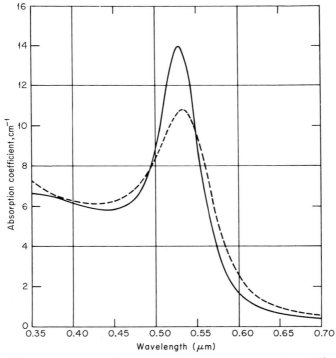

Fig. 35. Absorption spectrum of gold-ruby glass due to metal particles 200 Å in diameter in a photonucleated soda–alumino–silicate glass; (———) 25°C, (– – –) 514°C. (After Doremus, 1964.)

b. Silver-Yellow Glasses.   Similar absorption arises in glasses containing dissolved silver. The so-called "silver-yellow" glasses have been studied by Weyl (1945, 1951), Stookey (1949a,b), Yokota and Shimizu (1957), and Doremus (1965). An example of the metallic silver absorption found by Doremus (1965) in a soda–aluminosilicate glass is shown in Fig. 36. Silver particles were nucleated in a glass of composition by weight (71.5% SiO_2, 23% Na_2O, 4% Al_2O_3, 1% ZnO, 0.02–0.1% CeO_2, and 0.002% Ag) by room-temperature irradiation with uv light. Growth of the metallic particles was achieved by heating at 580°C. Doremus (1965) utilized the bulk optical properties of silver to calculate that spherical metallic silver particles will produce an absorption near 0.396 μm versus the experimentally observed position of 0.46 μm. This slight shift probably arises from the deviations from spherical symmetry which can take place in the metallic particles (Gans, 1922). Yokota and Shimizu (1957) have observed a similar absorption band in silver-doped glasses. Studies on these types of photosensitive glasses laid much of the groundwork for the later development of silver halide photochromic glasses.

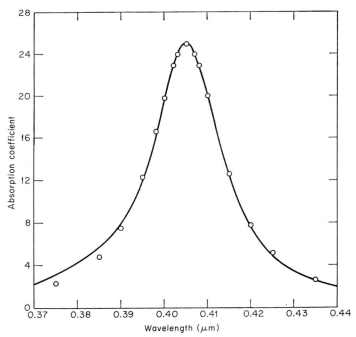

Fig. 36. Absorption spectrum of silver-yellow glass due to photonucleated metallic silver particles [(——) experimental, (○) calculated. (After Doremus, 1965.)

c. Copper-Ruby Glasses. Most investigators have reported that the presence of copper metallic particles also produces a strong green absorption band similar to the gold-ruby absorption band, e.g., Weyl (1945, 1951), Stookey (1949), and Sharma *et al.* (1956). However, Ram and coworkers (1959, 1960, 1962) have shown that the presence of cuprous oxide is also capable of imparting a red color to a glass and may well be the dominant source of color in the glasses. Colloidal suspensions of Cu_2O were found to possess the color and absorption characteristics of copper red glass. X-ray data indicate the presence of both Cu_2O and Cu in copper-ruby glasses and the spectral characteristics of these glasses appear to arise from contributions from both species.

4. OTHER SOURCES OF VISIBLE COLORATION

There are numerous other sources of visible coloration in glasses which are of interest but which will not be treated here because of length restrictions. Weyl (1951) is a good reference for a qualitative description of many of these absorption processes. Included in this group are the blue-sulfur, pink-selenium, and purple-tellurium glasses whose colors are associated with elemental clustering, the colloidal sulphide and selenide glasses including the well known CdS yellow filter glasses, and glasses containing metals such as tungsten, molybdenum, and uranium.

IV.   Infrared Absorption

The optical absorption of oxide glasses in the near and middle ir region is determined by the collective vibrations of molecules, atoms, and ions in the glass network. Because of the complexity of the required calculations, no exact theory presently exists which satisfactorily describes the vibrational modes of a disordered system such as an oxide glass. However, several different approaches have been employed which provide some insight into the interpretation of the ir and Raman spectra of glasses. These include

(1)   the calculation and comparison of free molecular or structural unit vibrational frequencies with the vibrational modes observed in the glasses,

(2)   the comparison of the spectra of crystalline polymorphs with those found in glasses,

(3)   the use of the lattice dynamics approach which considers the coupled long-range vibrations of a solid, and

(4)   the construction of physical models of glass networks in which the measured positions of individual atoms are used as a basis for the calculation of atomic vibrational properties of the glass.

A complementary method to ir absorption which also provides information on the vibrational modes of a glass is Raman spectroscopy. In the Raman effect the incident photons interact with the glass to produce scattered photons which are either increased or decreased in energy by quantized amounts corresponding to the vibrational energies of the glass system. While ir absorption and Raman scattering both arise from transitions involving the same vibrational energy levels, different selection rules apply so that different vibrations can often be monitored by the two techniques. For example, when a molecule has a center of symmetry, all vibrations which are symmetrical with respect to the center are ir inactive while all vibrations which are antisymmetrical with respect to the center are Raman inactive. Since the Raman effect is actually a scattering process rather than an absorption process, Raman data on glasses has only been included here when it provides complementary data which is crucial to structural interpretations such as in the case of borate glasses.

Experimentally, ir absorption measurements in the near and middle ir region ($4000-300$ cm^{-1}) are normally made by transmission through thin blown or drawn films or pressed potassium bromide (KBr) disks which contain small amounts of the sample, or by reflection techniques similar to those discussed earlier for measurement of the electronic absorption of glasses in the uv. The ir region of oxide glasses has been studied extensively. Review articles include those by Simon (1960), Wong and Angell (1971), Neuroth (1972, 1974), and Konijnendijk (1975). The present review will not attempt to provide a comprehensive summary of the theoretical and experimental work on the ir spectra of glasses. Discussion will focus on the absorption of the simple glass-forming oxides along with a few simple binary and ternary glasses of each system. The effects of OH absorption in oxide glasses is included and a brief summary is also given on the ir transmitting chalcogenide glasses.

Oxide glasses typically exhibit their lowest absorption levels in the near ir, making this region the most attractive for fiber optics applications. Figure 37 shows the near ir absorption measured by Maurer (1973) for a low-loss silica fiber. The numerous bands are associated with the presence of OH in the glass. The cutoff wavelengths for most oxide glasses lie in the $4.5-6.5$-μm ($2200-1500$ cm^{-1}) range with a strong OH absorption often evident in the 2.8-μm (3570 cm^{-1}) region.

The vibrations of polyatomic groups in solids may be qualitatively described in terms of stretching and bending vibrations as shown in Fig. 38 for a tetrahedral molecule of the type XY_4. The bending vibrations are characterized by much lower frequencies than the stretching modes which involve larger force constants. The highest frequency vibrational modes of oxide glasses which are associated with cation–oxygen stretching vibrations

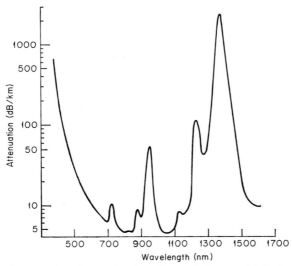

Fig. 37. Total attenuation of a very low-loss silica optical waveguide in the visible and near ir showing the influence of 100 ppm water impurity. Bands are vibrational overtones and combinations linked to the OH⁻ ion. (After Maurer, 1973.)

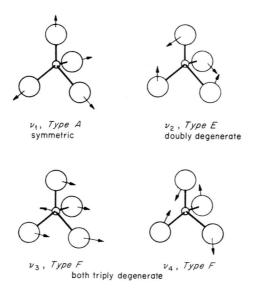

ν_1, *Type A*
symmetric

ν_2, *Type E*
doubly degenerate

ν_3, *Type F*

ν_4, *Type F*
both triply degenerate

Fig. 38. Four normal modes of tetrahedral molecule of the type XY_4. Modes ν_1 and ν_2 are only Raman active, while modes ν_3 and ν_4 are both ir and Raman active. (After Simon, 1960.)

are quite sensitive to coordination and tend to be more localized than the low-frequency modes. These high-frequency stretching modes to a great extent determine the ir cutoff wavelength of the glass. Some typical observed values of stretching frequencies in the glassy oxides are 1100 cm^{-1} for Si—O—Si (Simon and McMahon, 1953a; Jellyman and Procter, 1955; Hanna and Su, 1964), 1250 cm^{-1} for B–O–B (Moore and McMillan, 1956; Parsons and Milberg, 1960; Borelli *et al.*, 1963), 1265 cm^{-1} for P–O–P (Bues and Gehrke, 1956; Williams *et al.*, 1959; Shih and Su, 1965), and 900 cm^{-1} for Ge–O–Ge (Cohen and Smith, 1958; Chen and Su, 1971; Murthy and Kirby, 1964).

A. Silicate Glasses

1. SiO$_2$

The ir transmission of a typical synthetic silica containing about 1200 ppm OH is shown in Fig. 39 (Adams and Douglas, 1959). A very intense OH fundamental is observed at 2.73 μm (3660 cm^{-1}) with a sharp overtone at 1.38 μm (7250 cm^{-1}). These ir data indicate that the water in fused silica is present as unassociated OH groups rather than as water molecules. Bands at 2.22 and 2.6 μm, vary in intensity with the fundamental OH vibration, and have been shown to arise from OH and SiO$_4$ combination frequencies

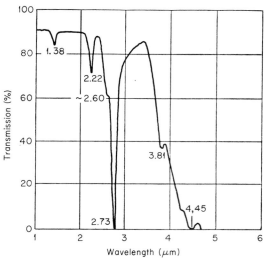

Fig. 39. Transmission curve of a Corning synthetic fused silica containing about 1200 ppm OH. Fundamental O–H stretching band lies near 2.73 μm. Sample thickness is 5.26 mm. (After Adams and Douglas, 1959.)

(Adams and Douglas, 1959; Moore and McMillan 1956; Adams, 1961; Kats and Haven, 1960). Other peaks are due to the glass network itself and have been related by Adams and Douglas (1959) to normal fundamental and combination frequencies at the SiO_4 tetrahedron. As seen in Fig. 37 the second overtone of the OH fundamental lies near 0.945 μm (10,580 cm^{-1}) and is a major concern for low-loss fiber-optic waveguide systems which often employ GaAs light sources.

The strongest fundamental vibrational absorptions of SiO_2 lie at lower frequencies. Simon and McMahon (1953a,b) employed the reflection technique to study the ir spectra of α- and β-quartz, crystobalite, and fused silica in the limited range from 7.14 to 13.4 μm (1400–700 cm^{-1}). A strong peak near 1100 cm^{-1} was observed in all samples and assigned to the Si–O–Si stretching vibration. Lippincott et al. (1958), using the KBr pellet technique, noted that three characteristic group frequencies are common to all polymorphs of SiO_2, lying near 1100, 800, and 480 cm^{-1}. These were, respectively, assigned to stretching modes primarily involving displacement of oxygens, stretching modes primarily involving displacement of silicons, and an Si–O bending mode. It was also observed that fused silica ir absorption most closely resembled that of tridymite. Matossi (1949), Lyon (1962), Zarzycki (1964), and Tarte (1964) have also pointed out the sensitivity of the fundamental Si–O–Si stretching frequency to coordination changes as shown in Fig. 40. More detailed assignments of both ir and Raman vibrational bands in SiO_2 are discussed by Simon (1960), Scott and Porto (1967), and Wong and Angell (1971). However, the presence of long-range collective vibrations in SiO_2 and the absence of crystalline order makes it impossible to interpret many spectral features in terms of simple molecular or structural unit vibrational modes.

Bell and Dean (1966) constructed a physical model of SiO_2 consistent with experimental data on structure and then used the atomic coordinates as a basis for theoretical calculations of the atomic vibrational properties of the glass (Bell et al., 1968; Bell and Dean, 1972). The main bands in the computed vibrational spectra were found at 400–450, 700–800, and 1000–1100 cm^{-1}, in excellent agreement with the experimental results. This approach suggested the following interpretations for the ir spectra of SiO_2. The 1100-cm^{-1} band arose from Si–O–Si vibrations in which the bridging oxygens move in opposite directions to their Si neighbors and parallel to Si–Si lines although these modes were not generally localized on just one group of Si–O–Si atoms. The band near 700–800 cm^{-1} was identified with bond bending modes in which the oxygen atoms vibrate in the plane of the triangle formed with their neighboring silicon atoms, the motion being perpendicular to the Si–Si line again with the modes not in general being highly localized. The 400–450-cm^{-1} band was associated with bond rocking

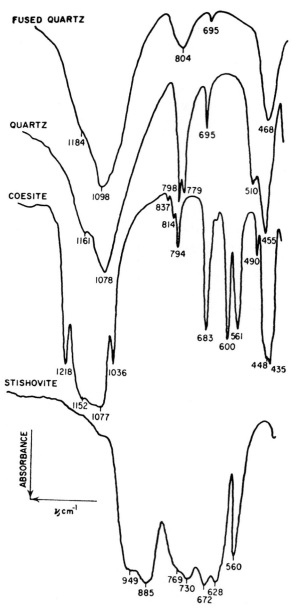

Fig. 40. Infrared absorption spectra of fused silica and three crystalline polymorphs. A change in the coordination of Si in stishovite (6-coordinated) produces a substantial shift in the Si–O stretching frequency, 1078–885 cm^{-1}. (After Lyon, 1962.)

vibrations of oxygen perpendicular to the Si–O–Si planes but these modes extended spatially throughout the whole network.

2. ALKALI SILICATES

The effects of the introduction of alkali on the ir and Raman spectra of SiO_2 have been investigated by numerous authors. More recent studies include those by Simon and McMahon (1953b), Jellyman and Procter (1955), Gross and Kolesova (1958), Su et al. (1962), Hanna and Su (1964), Crozier and Douglas (1965), Gaskell (1967), Sweet and White (1972), Lazarev (1972), and Brawer and White (1975).

Addition of alkali produces distinct changes in the ir spectra of silicate glasses. As the percentage of Na_2O is increased in binary soda–silica glasses, the 1100-cm^{-1} Si–O–Si stretching band decreases in intensity, broadens, and shifts to lower frequencies. This has been interpreted by Simon (1960) and Jellyman and Procter (1955) as an indication of the breakup of the Si–O network and a general weakening of the structural bonding. As the 1100-cm^{-1} peak decreases, a second lower frequency band appears near 940–950 cm^{-1} which has been attributed to the bond stretching vibrations of the Si–O$^-$ or O$^-$–Si–O$^-$ NBO groups. This interpretation is supported by the observation that all of the binary alkali silicate glasses exhibit similar spectra in the 1000-cm^{-1} region, independent of the mass of the alkali cation present (Simon and McMahon, 1953b; Jellyman and Procter, 1955; Gross and Kolesova, 1958). Figures 41 and 42 show the reflection spectra measured by Jellyman and Procter (1955) for both a series of soda–silicates and a group of various alkali silicates with the same mole % alkali. Wilmot (1954), Hass (1970), and Brawer and White (1975) have employed Raman spectroscopy to show that similar changes occur in the low-frequency region. Florinskaya and Pechenkina (1953) have concluded that sodium disilicate crystallites form in glasses beyond 33 mole % range of Na_2O but this has not been substantiated. It has been shown, however, that the vibrational spectra of alkali silicate glasses show substantial similarities to crystalline spectra, supporting the claim that the short range disilicate, metasilicate, and orthosilicate type bonding found in crystals occurs to some extent in most glasses (Brawer and White, 1975).

Bell et al. (1968) observed that the surface (i.e., nonbridging) oxygen (NBO) atoms in their model contributed vibrational modes in the 900-cm^{-1} range. A study of the atomic displacement eigenvalues for modes in the region of the 950-cm^{-1} band shows them to be highly localized Si–O$^-$ stretching vibrations with about 99% of the energy contained in the vibrations of a few close NBO atoms. Further support for the NBO interpretation of the 950-cm^{-1} band was gained by Simon's (1957) observation that neutron irradiation produces a well-defined 920-cm^{-1} band in SiO_2. Since irradiation is also known to produce NBO ions in oxide glasses (Lell et al., 1966).

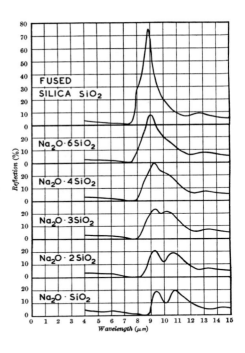

Fig. 41. Infrared spectral reflection curves for a series of soda–silica glasses. Si–O–Si fundamental stretching mode near 9.8 μm (1100 cm^{-1}) is observed to decrease and shift to longer wavelengths (lower frequencies) with increasing Na$_2$O content. (After Jellyman and Procter, 1955.)

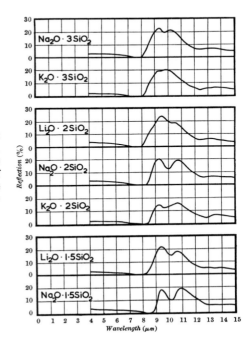

Fig. 42. Infrared spectral reflection curves for various alkali silicate glasses. Band arising near 10.6 μm (850 cm^{-1}) is associated with localized vibrations of NBO ions and is similar regardless of the mass of the alkali ion in the glass. (After Jellyman and Procter, 1955.)

The infrared cutoff of alkali silicates does not differ substantially from that of SiO_2 although absorption bands at lower frequencies appear to differ. Figure 43 shows the ir spectra of a 29.4 mole % Na_2O–70.6 mole % SiO_2 glass as measured by Adams (1961). Scholze (1959) established that the weak bands near 2.75–2.95, 3.35–3.85, and 4.25 μm are due to OH absorption in a study of a large number of binary, ternary, and quaternary silicates. The relationship of the negatively charged NBO ions to the position of the 3.6-μm OH peak of binary silicates has been measured by Scholze (1959) on glasses of composition $20\ Na_2O \cdot XAl_2O_3 \cdot (80—X)SiO_2$, where X varies from 0–20. This band vanishes when the aluminum content equals the alkali content and the NBO concentration goes to zero. Adams (1961) noted that this behavior is typical of hydrogen bonding in that a shift to lower frequencies, broadening, and intensification are expected as the degree of association increases.

Binary silicate glasses with Ca, Sr, and Ba oxides have been studied by Simon and McMahon (1953a). The 940 and 1050 cm^{-1} bands were also observed in these glasses, further supporting the NBO hypothesis. More

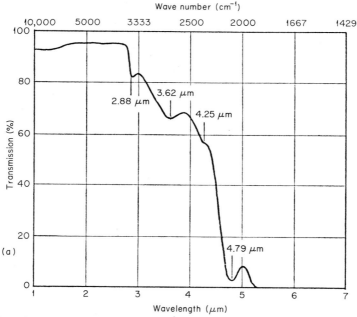

Fig. 43. Infrared transmission of a soda–silica glass with 29.4 mole % Na_2O. The ir cutoff wavelength is not extremely sensitive to the presence of network modifying cations but OH absorption bands are found to differ in alkali silicates as compared to SiO_2. (After Adams, 1961.)

recently Konijnendijk (1975) has measured the absorption of ternary silicate glasses containing CaO, MgO, and Al_2O_3 and observed similar spectra to that of the binary alkali silicates. Data on numerous more complex silicate glasses can be found in the references cited in this section.

B.   Borate Glasses

1. B_2O_3

As discussed in Section II,B,1, glassy B_2O_3 does not possess the simple three-dimensional network structure of SiO_2 but rather a linked-ring structure comprised chiefly of B_3O_6 planar boroxol groups. Evidence for such structures is found in both the ir and Raman spectra which for many years have represented perhaps the strongest physical evidence for the boroxol group structural model of B_2O_3. The ir absorption of B_2O_3 has been reported by numerous workers, including Jellyman and Procter (1955), Moore and McMillan (1956), Dachille and Roy (1959), Parsons and Milberg (1960), Simon (1960), Adams (1961), Borelli et al. (1963), Borrelli and Su (1963), Krogh-Moe (1965, 1969), Tenney and Wong (1972), and Konijnendijk (1975).

The transmission of a thin drawn film of B_2O_3 glass measured by Adams (1961) is shown in Fig. 44. Strong OH bands are evident, just as in the case of SiO_2 discussed earlier. The OH fundamental stretching band occurs at 2.79 μm (3583 cm^{-1}) in B_2O_3 together with a sharp overtone at 1.4 μm (7140 cm^{-1}) which confirms that water is held internally in the glass in an unassociated form such as a $>$B—OH OH—B$<$ configuration. The principal ir reflection peaks of B_2O_3 lie at 710, 1270, and 1420 cm^{-1} (Simon, 1960). Raman lines have been reported at 470, 670, 808, and 1258 cm^{-1} (Borrelli and Su, 1963). Goubeau and Keller (1953) and Krogh-Moe (1960b) first pointed out that the strong Raman line at 808 cm^{-1} indicates the presence of boroxol groups in B_2O_3. Krogh-Moe (1965) suggested that the 1260-cm^{-1} ir band involves the vibration of a B–O–B bond constituting the linkage of the boroxol group to neighboring groups, and was able to qualitatively explain the vibrational spectra of B_2O_3 in terms of a boroxol group structure as shown, e.g., in Fig. 45. This structural model was later confirmed by the x-ray work of Mozzi and Warren (1970). The band near 1420 cm^{-1} has been assigned by Krogh-Moe (1965) to the ring stretching of the boroxol unit $B_3O_3(O^-)_3$. The weaker band at 720 cm^{-1} was attributed to a bond bending motion of the B–O–B centers within the network. Figure 46 shows the ir absorption of a series of binary borosilicate glasses measured by Tenney and Wong (1972) and shows clearly the shift in the fundamental stretching bands as the glass composition goes from pure SiO_2 through borosilicate to pure B_2O_3. Konijnendijk (1975) has recently reported an extensive investigation on the Raman and ir spectra of borosilicate glasses.

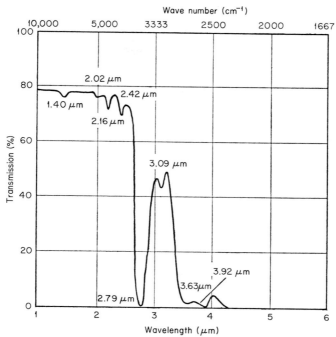

Fig. 44. Infrared transmission of a drawn film of B_2O_3 glass showing the presence of water in the glass with OH fundamental at 2.79 μm. Water contamination is a problem in measuring the ir absorption of borate glasses, especially when using the KBr pellet technique. (After Adams, 1961.)

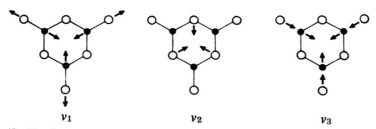

Fig. 45. The three symmetrical vibrations of the boroxol group. For many years the Raman and ir spectrum of glassy B_2O_3 have been cited as strong evidence of boroxol group formation. (After Krogh-Moe, 1965.)

2. ALKALI BORATE GLASSES

The ir and Raman spectra of borate glasses have been investigated by many workers including Jellyman and Procter (1955), Anderson *et al.* (1955), Moore and McMillan (1956), Borrelli *et al.* (1963), Krogh-Moe (1962, 1965), Quan and Adams (1966), and Konijnendijk (1975).

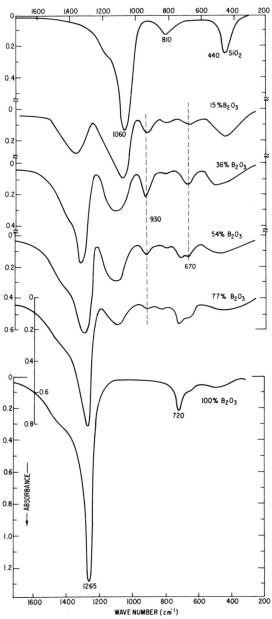

Fig. 46. Infrared absorption spectra of vapor-deposited B_2O_3–SiO_2 glass films from 0 to 100 mole % B_2O_3. The high-frequency Si–O stretching vibration near 1060 cm^{-1} compares with a value of 1265 cm^{-1} observed in B_2O_3. Films were 6000 $\pm$ 100 Å in thickness. (After Tenney and Wong, 1972.)

The reflectivity of a series of soda–borate glasses is shown in Fig. 47 (Simon, 1960). Just as in the case of the silicate glasses, the fundamental B–O–B stretching band near 1265 cm^{-1} is observed to decrease with alkali content of the glass, indicating some modification in the B–O network. Absorption bands arise in the 800–1000 cm^{-1} region in the glasses with greater than 25 mole % Na$_2$O present. Krogh-Moe (1960, 1962) and Bray and O'Keefe (1963) concluded from NMR results that the excess oxygen ions in borate glasses introduced by alkali oxide are used in the formation of BO$_4$ tetrahedra up to about 30 mole %. Borrelli *et al.* (1963) used KBr, thin-film, and reflection techniques on a series of binary borates with 0–33.3 mole % Na$_2$O and obtained similar results except by the KBr pellet technique which is not reliable for hygroscopic borate glasses. As with the silicates the spectral features are not dependent on the type of alkali present in the glass.

As discussed in Section II,B,2 one interpretation of the vibrational data which is also consistent with the NMR results is that boroxol groups are to a major extent gradually replaced by six-membered borate rings containing

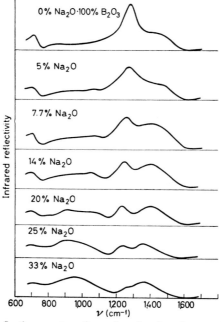

Fig. 47. Infrared reflection spectra of a series of sodium borate glasses. The 1265 cm^{-1} B–O–B stretching frequency decreases and shifts, as does the Si–O mode in silicate glasses, when the alkali oxide concentration is increased. (After Simon, 1960.)

a 4-coordinated boron in the 0–20 mole % alkali oxide range. In glasses with 20–35 mole % alkali oxide, rings with two BO_4 tetrahedra and diborate groups are formed. This interpretation is supported by the observation that the 808-cm^{-1} Raman line characteristic of boroxol groups has disappeared at about 25 mole % alkali oxide (Konijnendijk, 1975) and has been replaced by a strong 770-cm^{-1} peak which has been assigned by Bril (1975) to a symmetric vibration of the six-membered borate ring with a single BO_4 unit. Alkaline earth oxide addition to B_2O_3 produces similar changes in the vibrational spectra of borates.

One interesting system is the alkali–alumino–borates which have been measured using ir and Raman techniques by Konijnendijk (1975). For compositions of 0.2 K_2O–0.8 B_2O_3, the addition of up to 20 mole % Al_2O_3 leads to a decrease of absorption in the 900–1100-cm^{-1} region and an increase in the intensity of the 808-cm^{-1} Raman line, indicating that BO_4 units disappear and boroxol groups reform upon the addition of Al_2O_3. There is also evidence of AlO_4 tetrahedra formation when alkali oxide content exceeds 30 mole %. Further details are found in the above reference.

C. Germanate Glasses

1. GeO$_2$

GeO_2 occurs in two well-defined crystalline forms, the tetragonal rutile structure (six coordinated Ge) and the hexagonal form (four-coordinated Ge) similar to the α-quartz structure. Zarcycki (1957) has shown that GeO_2 glass is built up of randomly oriented GeO_4 tetrahedra with a Ge–O distance of 1.65 Å. The ir reflectance spectra of the two crystalline polymorphs and vitreous GeO_2 is shown in Fig. 48 which clearly indicate the similarity of the glass to the hexagonal crystal form. The vibrational properties of GeO_2 have been studied by many investigators, including Lippincott et al. (1958), Zarzycki (1964), Murthy and Kirby (1964), Bell et al. (1968), Hass (1970), Stolen (1970), and Chen and Su (1971).

The Ge–O–Ge stretching absorption lies near 878 cm^{-1} in both the hexagonal crystal and the glass forms, indicating that the Ge is 4-coordinated. In the tetragonal crystal in which Ge exhibits sixfold coordination, the stretching frequency moves lower to 680 cm^{-1}, reflecting a weakening of the Ge–O bond as discussed by Tarte (1964). The shift of the fundamental stretching band of GeO_2 to 878 cm^{-1} relative to that of SiO_2 at 1100 cm^{-1} gives the former better transparency in the near ir. Lippincott et al. (1958) assigned the band near 880 cm^{-1} to Ge–O stretching, the 550- and 585-cm^{-1} bands to Ge stretching and a low-frequency 332 band to Ge–O bending. More recently Bell et al. (1968, 1972) computed the spectrum of GeO_2

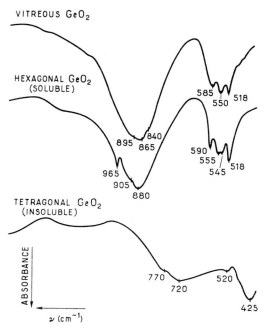

Fig. 48. Infrared absorption spectra of vitreous GeO_2, and the hexagonal (4-coordinated Ge) and tetragonal (6-coordinated Ge). The similarity of the glass to the hexagonal crystalline form is quite apparent. (After Zarcycki, 1964.)

using the physical model approach discussed earlier for SiO_2. Computed vibrational bands were obtained near 150–300, 300–450, 500–650, and 800–1000 cm^{-1}, in good aggreement with the low-frequency Raman data of Obukhov-Denisov et al. (1958) and the ir data of Lippincott et al. (1958). Peaks in the GeO_2 spectrum were shown to be analogous to the SiO_2 computed spectrum but shifted to lower frequencies. Wong and Angell (1971) have reviewed the ir and Raman frequencies observed in GeO_2.

2. ALKALI GERMANATE GLASSES

The addition of alkali oxide to GeO_2 produces an immediate structural reorientation in the glass, the coordination of the Ge going from 4 to 6. This is accomplished by the incorporation of two high-energy Ge d orbitals into the Ge–O bond which goes from a sp^3 tetrahedral to a sp^3d^2 configuration. With further addition of alkali oxide, GeO_4 tetrahedra with NBO ions are formed (Riebling, 1963; Murthy and Ip, 1964; Murthy and Kirby, 1964). This is clearly evident upon the examination of the ir absorption data of alkali germanates

Murthy and Kirby (1964) have studied the ir spectra of binary glasses of the composition $R_2O–GeO_2$ where R was Li, Na, K, Rb, or Cs. Typical results for glasses of the system $K_2O–GeO_2$ are shown in Fig. 49. In all cases, the principal Ge–O–Ge stretching absorption band at 878 cm^{-1} was observed to shift to longer wavelengths with increasing amounts of alkali, and to a much greater extent than the silicates. Second, there is a splitting of the absorption band at about 25–30 mole % alkali oxide concentration which appears to be somewhat dependent on the type of alkali ion added. The large shift of over 100 cm^{-1} of the Ge–O–Ge stretching band arises from a change of the GeO_4 tetrahedra to the GeO_6 groups. The splitting at concentrations above 25 mole % alkali was attributed by Murthy and Kirby (1964) to the formation of NBO groups. Bobovich and Tolub (1962) have reviewed the Raman spectra of alkali germanate glasses.

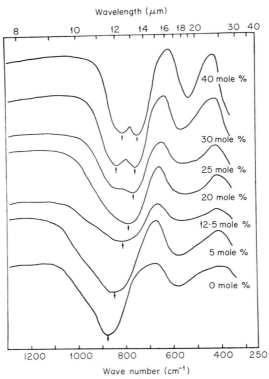

Fig. 49. Infrared absorption spectra of $K_2O–GeO_2$ glasses. The Ge–O–Ge stretching mode near 878 cm^{-1} was observed to shift substantially upon the introduction of alkali. This movement is associated with a change in Ge coordination (4–6) in the alkali germanate glasses. (after Murthy and Kirby, 1964.)

D. Phosphate Glasses

1. P₂O₅

Very little data is available on the vibrational spectra of P_2O_5. Two of the crystalline forms of P_2O_5 which are known to exist include a rhombohedral form composed of groupings of P_4O_{10} molecules, and an orthorhombic form which exhibits a three-dimensional network of ten-membered rings formed by PO_4 tetrahedra. In the latter form, three of the four oxygens of each tetrahedra are bonded to neighboring phosphorus ions. Glassy P_2O_5 appears to resemble the orthorhombic form. Wong and Angell (1971) have used group theoretical analysis to predict the numbers of ir and Raman modes expected in these two polymorphs.

Bobovich (1962) measured the Raman spectra of a number of simple phosphate glasses including P_2O_5. The spectrum was characterized by peaks near 325, 700, 1150, and 1360 cm^{-1}. The strong bands near 700 and 1150 cm^{-1} were attributed to the symmetric and antisymmetric P–O–P vibrations, respectively. Reliable ir absorption data on P_2O_5 do not appear to be available. Kolesova (1957) measured the ir and Raman spectra of P_2O_5 but serious contamination by water makes the data questionable.

2. ALKALI PHOSPHATE GLASSES

The ir and Raman spectra of crystalline and glassy phosphates have been investigated by a number of workers including Corbridge and Lowe (1954), Bues and Gehrke (1956), Kolesova (1957), Williams et al. (1959), Bobovich (1962), Murthy et al. (1963), Murthy and Kirby (1964), and Shih and Su (1965). The latter paper provides a good summary of earlier work on the ir spectra of phosphate glasses.

Bues and Gehrke (1956) used the KBr pellet technique to study the ir spectra of binary alkali phosphate glasses. Absorption bands were assigned to vibrations of PO_2, PO_3, and POP groups. Williams et al. (1959) and Bobovich (1962) both observed that as the alkali oxide content increases, an ir line near 900 cm^{-1} and a Raman line near 1150 cm^{-1}, both moving to higher frequencies. This has been interpreted as a manifestation of the increase in ionicity of the bonding which results in a strengthening of the P–O bonding. In general the vibrational spectra of most alkali and alkaline earth phosphate glasses appear similar except for a slight broadening in the alkaline earth glasses. Strong tendencies toward devitrification and hydration often make measurements difficult.

The reflection spectra of several metaphosphate glasses as measured by Shih and Su (1965) are shown in Fig. 50. In sodium metaphosphate glass bands appear near 7.9 μm (1265 cm^{-1}), 0.2 μm (1090 cm^{-1}), 10.2 μm (982 cm^{-1}), and 11.3 μm (885 cm^{-1}), but all spectra are similar. Transmission

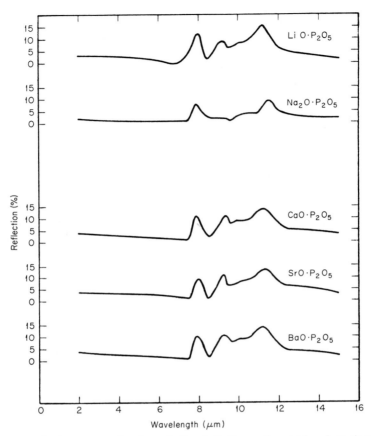

Fig. 50. Reflection spectra of various alkali and alkaline earth metaphosphate glasses. The data are consistent with a zig–zag chain structural model. (After Shih and Su, 1965.)

spectroscopy indicated bands near 1144, 775, and 740 cm^{-1} as well as at lower frequencies. Shih and Su (1965) provide a comprehensive table of observed Raman and ir line frequencies in vitreous sodium metaphosphate glasses. These data have been interpreted structurally in terms of a zig–zag chain of polyphosphate anions with the cations situated between the chains. In view of the close similarity which exists between the ir spectra of all the vitreous metaphosphate glasses, the zig–zag chain model proposed by Shih and Su (1965) for $Na_2O \cdot P_2O_5$ glass, appears to be equally suitable for application to the other phosphate glass systems. Milberg and Daly (1963) found that a similar zig–zag chain model best explained the x-ray scattering pattern of sodium metaphosphate glass fibers. Thus it can be concluded that the introduction of network modifiers into P_2O_5 produces a gradual break-down of the three-dimensional structural network and leads to the formation of ion chain structures in the region near the metaphosphate composition.

E. Other Oxide Glasses

Numerous other glass-forming systems including aluminates, arsenates, titanates, antimonates, tungstates, and vanadates have been studied in the ir region. Wong and Angell (1971) have briefly reviewed some of these systems. However the participation of oxygen in the vibrational modes of all of these glasses limits their application to wavelengths of less than around 5.5 μm (2000 cm^{-1}). For extended ir transmission, nonoxide glasses must be utilized.

F. Chalcogenide Glasses

Sulfur, selenium, and tellurium are other Group VI elements which can be used to replace oxygen as the glass-forming anion in the glass network, thereby achieving transmission at longer wavelengths than is possible with the oxides. This extended ir transmission, however, is accompanied by a sharp decrease in the bandgap of these materials which typically makes them opaque in the visible but often with semiconducting properties. Only a few brief remarks on optical properties will be given here. Recent summaries and papers which are useful include those by Frerich (1950, 1953), Pearson (1964), Savage and Nielson (1964, 1965a,b), Hilton and coworkers (1963, 1964, 1966a,b), Wong and Angell (1971), Hilton (1973), and Marsh and Savage (1974).

The ir transmission of chalcogenide glasses is particularly sensitive to the presence of low atomic mass impurities such as hydrogen and oxygen which must be excluded during melting. Most chalcogenides begin to transmit in the region of about 1–1.5 μm (Hilton and Jones 1966). At shorter wavelengths, the tail of the electronic edge produces strong optical absorption. The low energy of the bonding in these glasses is also reflected in the low hardness and low softening points relative to oxide glasses.

Structural studies by Hilton et al. (1966b) by means of ir, atomic mass, and x-ray spectroscopy indicate the presence of zig–zag Si–Te and Ge–Te chains in Si–As–Te and Ge–As–Te glasses. As is observed to form As–S pyramidal molecules in As_2S_3. The data indicate that Si–Te, Ge–Te, Si–As, and As–Te bonds form in preference to the purely covalent Te–Te, Si–Si, and As–As bonds.

Some of the fundamental vibrational frequencies observed are listed in Table III. These values contrast with values of 1100 cm^{-1} (9.1 μm) for the

TABLE III

FUNDAMENTAL STRETCHING VIBRATIONS
OF CHALCOGENIDE GLASSES[a]

System	Frequency (cm^{-1})	Wavelength (μm)	Bond
Se	356	39	Se–Se
As–S	300–310	32–33	As–S
As–Se	217–226	44–46	As–Se
Ge–S	357–370	27–28	Ge–S
Ge–Se	238	42	Ge–Se
P–S	535	14	P–S
P–Se	363	27.5	P–Se
Si–As–Te	307–323	31–32.5	Si–Te
Si–P–Te	307	32.5	Si–Te
Ge–As–Te	182–5	54–55	Ge–Te
Ge–P–Te	182–212	47–55	Ge–Te
Ge–As–Se	238	42	Ge–Se
Ge–P–Se	356	28	Ge–Se

[a] Data from Hilton and Jones (1966).

Si–O stretching frequency in SiO_2. The typical transmission range of pure sulfide glasses is 0.6–11.5 μm, selenides 1–15 μm and tellurides 2–20 μm (Wong and Angell, 1971). Hilton *et al.* (1966a,b) have been able to develop Se and Ge ternary glasses for use as ir window materials which have relatively high melting points. Two specific chalcogenide glasses which have been employed as practical ir window materials are $Ge_{28}Sb_{12}Se_{60}$ (transmission range 1–14 μm) and $Ge_{33}As_{12}Se_{55}$ (transmission range 0.8–16 μm) (Hilton, 1973). These glasses are often referred to as TI1173 and TI20 glasses and are especially useful for applications in the 8–13 μm atmospheric window (Dimmock, 1972).

Savage and Nielson (1970) have shown that S–S, As–S, Ge–S, and P–S bond vibrations limit the useful transmission of bulk sulfide glasses to 11.5 μm. Glasses in the Ge–As–Se system such as the commercial composition given above, normally transmit beyond 15 μm but are quite sensitive to traces of oxide impurities which limit transmission in the region of As–O and Ge–O vibrations near the 12–13-μm range (Hilton and Jones, 1966). Similar problems exist with the tellurides. In this system the Si–As–Te glasses are attractive due to relatively high mechanical strength and softening temperatures and transmission to at least 20 μm.

The ir transmission of some high-purity chalcogenide glasses has been discussed by Hilton *et al.* (1975). High-purity Ge–Sb–Se glass was found to have an absorption coefficient of 0.01 cm^{-1} at 10.6 μm and Ge–As–Se glass 0.05 cm^{-1}. Oxygen and silicon levels were reduced to less than 5 ppm in these glasses. A typical transmission spectrum for TI1173 with and without Al (used to getter oxygen) is shown in Fig. 51. Similar absorption levels were found by Moynihan *et al.* (1975) in high-purity As$_2$Se$_3$. The two latter papers were concerned with the use of chalcogenide glasses as high-energy laser window materials at the 10.6-μm CO$_2$ laser wavelength. Moynihan *et al.* (1975) concluded that As$_2$Se$_3$ and similar glasses such as those in the Ge–As–Se and Ge–Sb–Se systems appear to be limited in transmission by intrinsic multiphonon processes to α values of the order of 10^{-2} cm^{-1}. This value is several orders of magnitude higher than that measured in pure crystalline materials such as alkali halides (Deutsch, 1974).

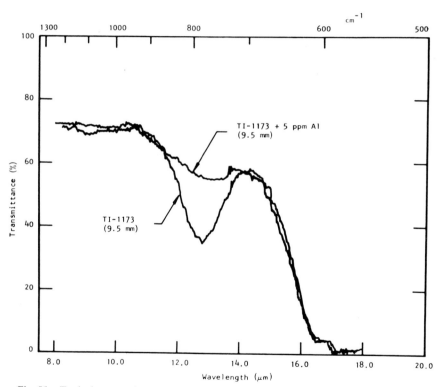

Fig. 51. Typical transmittance curve of a commercial chalcogenide glass, TI1173 (Ge$_{28}$Sb$_{12}$Se$_{60}$). Aluminum is used to getter oxygen which represents one of the principal impurity problems in glasses with extended ir transmission and which is responsible for the absorption near 12.5 μm. (After Hilton *et al.*, 1975.)

V. Summary

This review of optical absorption in glasses has attempted to provide a summary of the optical properties of oxide glasses in the uv, visible and ir regions, and to describe how the electronic, impurity, and vibrational absorptions contribute to the make up of the respective spectral components measured in simple glasses. Every effort has been made to incorporate recent work into the review and to relate the optical absorption to the structure and chemical bonding of the various glasses described. An extensive bibliography has been provided for the reader who desires more detailed information on particular aspects of the material which has been reviewed.

The field of optical properties of glasses is so vast that it has been impossible to include all of the material deserving of mention in this chapter. However, the many other review articles mentioned throughout the text are excellent alternate sources which should help to fill in any gaps. It is also apparent that there still remains much to be learned with regard to the understanding of the fundamental absorption processes in glasses, and it is hoped that this review will serve as a stimulus for further investigations of optical properties.

References

Adams, R. W. (1961). *Phys. Chem. Glasses* **2**, 39.
Adams, R. W., and Douglas R. W. (1959). *J. Soc. Glass Tech.* **43**, 147T.
Aglan, M. A., and Moore, H. (1955). *J. Soc. Glass Tech.* **39**, 351T.
Anderson, S., Bohon, R. L., and Kimpton, D. P. (1955). *J. Am. Ceram. Soc.* **38**, 370.
Arakawa, E. T., and Williams, M. W. (1968). *J. Phys. Chem. Solids* **29**, 735.
Bachman, G. S., Fischer, R. B., and Badger, A. E. (1946). *Glass Ind.* **27**, 399.
Bacon, F. R., and Billian, C. J. (1954). *J. Am. Ceram. Soc.* **37**, 60.
Badger, A. E., Weyl, W., and Rudow, H. (1939). *Glass Ind.* **20**, 407.
Bagley, B. G., Vogel, E. M., French, W. G., Pasteur, G. A., Gan, J. N., and Tauc, J. (1976). *J. Non-Cryst. Solids* **22**, 423 (1976).
Bamford, C. R. (1960a). *Phys. Chem. Glasses* **1**, 159.
Bamford, C. R. (1960b). *Phys. Chem. Glasses* **1**, 165.
Bamford, C. R. (1961). *Phys. Chem. Glasses* **2**, 163.
Bamford, C. R. (1962a). *Phys. Chem. Glasses* **3**, 54.
Bamford, C. R. (1962b). *Phys. Chem. Glasses* **3**, 189.
Bamford, C. R., and Hudson, E. J. (1965). *Proc. Int. Congr. Glass, 7th, Brussels* paper 6.
Bates, T. (1962). *In* "Modern Aspects of the Vitreous State" (J. D. Mackenzie, ed.), Vol. 2, pp. 195–254. Butterworths, London.
Bates, T., and Douglas, R. W. (1959). *J. Soc. Glass Tech.* **43**, 289.
Bell, R. J., and Dean, P. (1966). *Nature (London)* **212**, 1354.
Bell, R. J., and Dean, P. (1972). *In* "Amorphous Materials" (R. W. Douglas and B. Ellis, eds.), pp. 443–452. Wiley (Interscience), New York.
Bell, R. J., Bird, N. E., and Dean, P. (1968). *J. Phys. C.* **1**, 299.

Bennett, A. J., and Roth, L. M. (1971). *J. Phys. Chem. Solids* **32**, 1251; *Phys. Rev. B* **4**, 2686.
Berkes, J. S., and White, W. B. (1966). *Phys. Chem. Glasses* **7**, 191.
Bethe, H. (1929). *Ann. Phys.* **3**, 133.
Bethell, D. E., and Sheppard, N. (1955). *Trans. Faraday Soc.* **51**, 9.
Bingham, K., and Parke, S. (1965). *Phys. Chem. Glasses* **6**, 224.
Biscoe, J., and Warren, B. E. (1938). *J. Am. Ceram. Soc.* **21**, 287.
Biscoe, J., Pincus, A. G., Smith, C. S. Jr., and Warren, B. E. (1941). *J. Am. Ceram. Soc.* **24**, 116.
Bishay, A., and Kinawi, A. (1965). *In* "Physics of Non-Crystalline Solids" (J. A. Prins, ed.). Wiley (Interscience), New York.
Bishay, A., and Markar, L. (1969). *J. Am. Ceram. Soc.* **52**, 605.
Bishop, A. M. (1961). *Phys. Chem. Glasses* **2**, 26.
Bobovich, V. S. (1962). *Opt. Spectrosc.* **13**, 274.
Bobovich, Y. S., and Tolub, T. P. (1962). *Opt. Spectrosc.* **12**, 264.
Bock, J., and Su, G. J. (1970). *J. Am. Ceram. Soc.* **53**, 69.
Bork, A. (1930). *Glastech. Ber.* **8**, 275.
Borrelli, N. F., and Su, G. J. (1963). *Phys. Chem. Glasses* **4**, 206.
Borrelli, N. F., McSwain, B. D., and Su, G. J. (1963). *Phys. Chem. Glasses* **4**, 11.
Brawer, S. A., and White, W. B. (1975). *J. Chem. Phys.* **63**, 2421.
Bray, P. J., and O'Keefe, J. G. (1963). *Phys. Chem. Glasses* **4**, 37.
Bray, P. J., and Silver, A. H. (1960). *In* "Modern Aspects of the Vitreous State" (J. D. Mackenzie, ed.), Vol. I, pp. 92–119. Butterworths, London.
Bril, T. W. (1975). Thesis, Technological Univ., Eindhoven, Netherlands (unpublished).
Brown, D., and Douglas, R. W. (1965). *Glass Tech.* **6**, 190.
Bues, W., and Gehrke, H. W. (1956). *Z. Anorg. Chem.* **288**, 291, 307.
Carson, D. S., and Maurer, R. D. (1973). *J. Non-Cryst. Solids* **11**, 368.
Chen, B. T. and Su, G. J. (1971). *Phys. Chem. Glasses* **12**, 33
Cohen, A. J., and Smith, H. L. (1958). *J. Phys. Chem. Solids* **7**, 301.
Cohen, B. M., Uhlmann, D. R., and Shaw, R. R. (1973). *J. Non-Cryst. Solids* **12**, 177.
Corbridge, D. E. C., and Lowe, E. J. (1954). *J. Chem. Soc.* **493**, 4555.
Crozier, D., and Douglas, R. W. (1965). *Phys. Chem. Glasses* **6**, 240.
Dachille, F., and Roy, R. (1959). *J. Am. Ceram. Soc.* **41**, 78.
Despujols, J. (1958). *J. Phys. Radium* **19**, 612.
Deutsch, J. F. (1974). *J. Phys. Chem. Solids* **34**, 2091.
Dexter, D. L., and Knox, R. S. (1965). "Excitons." Wiley (Interscience), New York.
Dieke, G. H., and Crosswhite, H. M. (1963). *Appl. Opt.* **2**, 675.
Dietzel, A. (1945). *Z. Elektrochem.* **51**, 32.
Dietzel, A. (1949). *Glastech. Ber.* **23**, 1212.
Dietzel, A., and Coenen, M. (1961). *Glastech. Ber.* **34**, 49.
Dimmock, J. O. (1972). *J. Electron. Mater.* **1**, 255.
DiStefano, T. H., and Eastman, D. E. (1971). *Phys. Rev. Lett.* **27**, 1560; *Solid State Commun.* **9**, 2259.
Doremus, R. H. (1962). *In* "Nucleation and Crystallization in Glasses and Melts" (M. K. Reser, ed.), pp. 119–123. Am. Ceram. Soc., Columbus, Ohio.
Doremus, R. H. (1964). *J. Chem. Phys.* **40**, 2389.
Doremus, R. H. (1965). *J. Chem. Phys.* **42**, 414.
Doremus, R. H. (1973). *In* "Glass Science." Wiley, New York.
Douglas, R. W., and Zaman, M. S. (1969). *Phys. Chem. Glasses* **10**, 125.
Duffy, J. A. (1972). *Phys. Chem. Glasses* **13**, 65.
Edwards, R. J., Paul, A., and Douglas, R. W. (1972). *Phys. Chem. Glasses* **13**, 131, 137.
Evans, B. D., and Sigel, G. H. Jr. (1975). *IEEE Trans. Nucl. Sci.* **NS-22**, 2462.

Evans, D. L. (1970). *J. Am. Ceram. Soc.* **53**, 418.

Ershov, O. A., Goganov, D. A., and Lukirskii, A. P. (1966). *Sov. Phys.-Solid State* **7**, 1903.

Ershov, O. A., and Lukirskii, A. P. (1967). *Sov. Phys.-Solid State* **8**, 1699.

Fajans, K., and Barber, S. W. (1952). *J. Am. Chem. Soc.* **74**, 2761.

Faraday, M. (1857). *Phil. Mag.* **14**, 401, 512.

Florinskaya, V. A. (1960). *In* "Structure of Glass" (E. A. Porai-Koshits, ed.), Vol. 2, pp. 154–168. Consultants Bur., New York.

Florinskaya, V. A., and Pechenkina, R. S. (1960). *In* "Structure of Glass," Vol. 2., pp. 135–153. Consultants Bur., New York.

Frerichs, R. (1950). *Phys. Rev.* **78**, 643.

Frerichs, R. (1953). *J. Opt. Soc. Am.* **43**, 1153.

Friebele, F. J., Ginther, R. J., and Sigel, G. H. Jr. (1974). *Appl. Phys. Lett.* **24**, 412.

Fuwa, K. (1923). *J. Japn. Ceram. Assoc.* **366**, 80.

Gans, R. (1922). *Ann. Phys.* **37**, 881.

Garino-Canina, V. (1956). *C. R. Acad. Sci. Paris* **242**, 1982.

Gaskell, P. H. (1967). *Phys. Chem. Glasses* **8**, 69.

Ginther, R. J., and Kirk, R. D. (1971). *J. Non-Cryst. Solids* **6**, 89.

Ginther, R. J., and Sigel, G. H. Jr. (1975). Unpublished.

Goubeau, J., and Keller, H. (1953). *Z. Anorg. Chem.* **272**, 303.

Gross, E. F., and Kolesova, V. A. (1958). *In* "The Structure of Glass" (F. Coleman, ed.), pp. 45–48. Consultants Bur., New York.

Haddon, J. C., Rogers, E. A., and Williams, D. J. (1969). *J. Am. Ceram. Soc.* **52**, 52.

Hanna, R., and Su, G. J. (1964). *J. Am. Ceram. Soc.* **47**, 597.

Hass, M. (1970). *J. Phys. Chem. Solids* **31**, 415.

Hass, M., Davisson, J. R., Rosenstock, H. B., Slinkman, J. A., and Babiskin, J. (1975). *In* "Optical Properties of Highly Transparent Solids" (S. S. Mitra and B. Bendow, eds.), pp. 435–442. Plenum Press, New York.

Hecht, H. G. (1967). *J. Chem. Phys.* **47**, 1840.

Hecht, H. G., and Johnston, T. S. (1967). *J. Chem. Phys.* **216**, 23.

Hensler, J. R. (1962). *In* "Advances in Glass Technology", Part 2 (F. R. Matson and G. E. Rindone, eds.) pp. 25–27, Plenum Press, New York.

Hensler, J. R., and Lell, E. (1969). *Proc. Ann. Meeting Int. Comm. Glass* (S. Bateson and A. G. Sadler, eds.), pp. 51–57. Toronto.

Herre, F., and Richter, H. (1957). *Naturforsch.* **12A**, 545.

Hetherington, G., and Jack, K. H. (1962). *Phys. Chem. Glasses* **3**, 128.

Hilton, A. R. (1973). *J. Electron. Mater.* **2**, 210.

Hilton, A. R., and Brau, M. J. (1963). *Infrared Phys.* **3**, 69.

Hilton, A. R., and Jones, C. E. (1966). *Phys. Chem. Glasses* **7**, 112.

Hilton, A. R., Jones, C. E., and Brau, M. J. (1964). *Infrared Phys.* **4**, 212.

Hilton, A. R., Jones, C. E., and Brau, M. J. (1966a). *Phys. Chem. Glasses* **7**, 105.

Hilton, A. R., Jones, C. E., Dobrott, R. D., Klein, H. M., Bryant, A. M., and George, T. D. (1966b). *Phys. Chem. Glasses* **7**, 116.

Hilton, A. R., Hayes, D. J., and Rechtin, M. D. (1975). *J. Non-Cryst. Solids* **17**, 319.

Hirayama, C., Castle, J. G., and Kuriyama, M. (1968). *Phys. Chem. Glasses* **9**, 109.

Hussein, A. L., and Moustaffa, F. A. (1972). *J. Non-Cryst. Solids* **11**, 64.

Ibach, H., and Rowe, J. E. (1974a). *Phys. Rev. B* **9**, 1951.

Ibach, H., and Rowe, J. E. (1974b). *Phys. Rev. B* **10**, 710.

Jahn, H. A. (1966). *Glastech. Ber.* **39**, 118.

Jellyman, P. E., and Procter, J. P. (1955). *J. Soc. Glass Tech.* **39**, 173T.

Johnson, K. H., and Smith, F. C. Jr. (1971). *Chem. Phys. Lett.* **10**, 219.

Johnson, K. H., and Smith, F. C. Jr. (1972). *Phys. Rev. B* **5**, 831.
Johnston, W. D. (1964). *J. Am. Ceram. Soc.* **47**, 198.
Johnston, W. D. (1965). *J. Am. Ceram. Soc.* **48**, 608.
Johnston, W. D., and Chelko, A. (1966). *J. Am. Ceram. Soc.* **44**, 562.
Jorgensen, C. K. (1957). *Acta. Chem. Scand.* **11**, 73.
Juza, R., Seidel, H., and Tiedemann, J. (1966). *Angew. Chem. Int.* **5**, 85.
Kakabadse, G. J., and Vassiliou, E. (1965). *Phys. Chem. Glasses* **6**, 33.
Karlsson, K. (1969). *Glastek. Tids Kr.* **24**, 13.
Kats, A., and Haven, Y. (1960). *Phys. Chem. Glasses* **1**, 99.
Kats, A., and Stevels, J. M. (1956). *Philips Res. Rep.* **11**, 115.
Klinkenberg, P. F. A. (1947). *Physica* **13**, 1.
Knox, R. S. (1963). *In Solid State Phys. Suppl. 5.* (F. Seitz and D. Turnbull, eds.), Academic Press, New York.
Kolesova, V. A. (1957). *Opt. Spectrosc.* **2**, 165.
Koma, A., and Ludeke, R. (1975). *Phys. Rev. Lett.* **35**, 107.
Konijnendijk, W. L. (1975). *Philips Res. Rep. Suppl. 1* 1–223.
Konnert, J. H., and Karle, J. (1972). *Nature (London) Phys. Sci.* **236**, 92.
Kordes, E. (1965). *Glastech. Ber.* **38**, 242.
Kordes, E., and Nieder, E. (1968). *Glastech. Ber.* **41**, 41.
Kordes, E., and Worster (1959). *Glastech. Ber.* **32**, 267.
Kramers, H. A. (1929). *Phys. Z.* **30**, 521.
Krause, D. (1974). *In* "Nahordnungensfelder in Glasern" (H. Scholze, ed.), pp. 188–218. Deutschen Glastechnischen Gesellschaft E. V., Germany.
Kreidl, N. J. (1972). *In* "Physics of Electronic Ceramics" (L. L. Hench and D. B. Dove, eds.), Part B, pp. 915–961. Dekker, New York.
Kristiansen, L. A., and Krogh-Moe, J. (1968). *Phys. Chem. Glasses* **9**, 96.
Krogh-Moe, J. (1960a). *Phys. Chem. Glasses* **1**, 26.
Krogh-Moe, J. (1960b). *Ark. Kemi* **14**, 451.
Krogh-Moe, J. (1962). *Phys. Chem. Glasses* **3**, 1.
Krogh-Moe, J. (1965). *Phys. Chem. Glasses* **6**, 46.
Krogh-Moe, J. (1969). *J. Non-Cryst. Solids* **1**, 269.
Kronig, R. deL. (1926). *J. Opt. Soc. Am.* **12**, 547.
Kronig, R. deL. (1929). *Phys. Rev.* **30**, 521.
Kumar, S. (1959). *Bull. Cent. Glass Ceram. Res. Inst. Calcutta* **6**, 99.
Kumar, S. (1964). *Phys. Chem. Glasses* **5**, 107.
Kurkjian, C. R., and Peterson, G. E. (1974). *Phys. Chem. Glasses* **15**, 12.
Kurkjian, C. R., and Sigety, E. A. (1968). *Phys. Chem. Glasses* **9**, 73.
Landry, R. J., Fournier, J. T., and Young, C. G. (1967). *J. Chem. Phys.* **46**, 1285.
Lazarev, A. N. (1972). *In* "Vibrational Spectra and Structure of Silicates." Consultants Bur., New York.
Lell, E. (1962). *Phys. Chem. Glasses* **3**, 84.
Lell, E., Kreidl, N. J., and Hensler, J. R. (1966). *Progr. Ceram. Sci.* **4**, 1–93.
Linwood, S. H., and Weyl, W. A. (1942). *J. Opt. Soc. Am.* **32**, 443.
Lippincott, E. A., Valkenburg, A. V., Weir, C. E., and Bunting, E. N. (1958). *J. Res. Nat. Bur. Std.* **61**, 61.
Loh, E. (1964). *Solid State Commun.* **2**, 269.
Loveridge, D., and Parke, S. (1971). *Phys. Chem. Glasses* **12**, 19.
Lunter, S. G., Karapetyan, G. O., Bokin, N. M., and Yudin, D. M. (1968). *Sov. Phys.-Solid State* **9**, 2259.
Lyon, R. J. P. (1962). *Nature (London)* **196**, 266.

Marsh, H. T. (1965). *Proc. Int. Congr. Glass, 7th, Brussels* Vol. 1, paper 46.
Marsh, K. J., and Savage, J. A. (1974). *Infrared Phys.* **14**, 85.
Matossi, F. (1949). *J. Chem. Phys.* **17**, 679.
Maurer, R. D. (1958). *J. Appl. Phys.* **29**, 1.
Maurer, R. D. (1959). *J. Chem. Phys.* **31**, 444.
Maurer, R. D. (1973). *Proc. IEEE* **61**, 452.
McClure, D. S. (1957). *J. Phys. Chem.* **3**, 311.
McClure, D. S. (1959). *Solid State Phys.* **9**, 399–525.
McSwain, B. D., Borrelli, N. F., and Su, G. J. (1963). *Phys. Chem. Glasses* **4**, 1.
Meggers, W. F. (1942). *Rev. Mod. Phys.* **14**, 96.
Meller, F., and Milberg, M. E. (1960). *J. Am. Ceram. Soc.* **43**,
Mie, G. (1908). *Ann. Phys.* **25**, 377.
Milberg, M. E., and Daly, M. C. (1963). *J. Chem. Phys.* **39**, 2966.
Milberg, M. E., and Peters, C. R. (1969). *Phys. Chem. Glasses* **10**, 46.
Miyashita, T., Edahiro, T., Horiguchi, M., and Masuno, K. (1974). *Proc. Int. Congr. Glass, 10th, Kyoto* **6**, 52.
Moore, C. E. (1949). *In* "Atomic Energy Levels," Nat. Bur. Std. Circ. 467, Vol. I; (1952), *ibid.*, Vol. II, (1958), *ibid.*, Vol. III. Washington, D.C.
Moore, H., and McMillan, P. W. (1956). *J. Soc. Glass Tech.* **40**, 97T; *ibid.*, **40**, 66T.
Moore, H., and Winkelmann, H. (1955). *J. Soc. Glass Tech.* **39**, 215T, 250T.
Moynihan, C. T., Macedo, P. B., Maklad, M. S., Mohr, R. K., and Howard, R. E. (1975). *J. Non-Cryst. Solids* **17**, 369.
Mozzi, R. L., and Warren, B. E. (1969). *J. Appl. Cryst.* **2**, 164.
Mozzi, R. L., and Warren, B. E. (1970). *J. Appl. Cryst.* **3**, 287.
Murthy, M. K., and Ip., J. (1964). *Nature (London)* **201**, 285.
Murthy, M. K., and Kirby, E. M. (1964). *Phys. Chem. Glasses* **5**, 144.
Murthy, M. K., Muller, A., and Westman, A. E. R. (1963). *J. Am. Ceram. Soc.* **46**, 530, 558; *ibid.* **47**, 375.
Nagel, D. J. (1970). *Advan. X-Ray Anal.* **13**, 182–236.
Nath, P., and Douglas, R. W. (1965). *Phys. Chem. Glasses* **6**, 197.
Nath, P., Paul, A., and Douglas, R. W. (1965). *Phys. Chem. Glasses* **6**, 203.
Neuroth, N. (1972). *In* "Handbuch der Infrarot Spektroskopie" (H. Volkmann, ed.), Chapter 10. Verlag-Chemie, Weinheim.
Neuroth, N. (1974). *In* "Nahordnungsfelder in Glasern" (H. Scholze, ed.), pp. 140–187. Deutschen Glastechnischen Gesellschaft E. V., Germany.
Obukov-Denisov, V. V., Sobolev, M. N., and Cheremisinov, V. P. (1958). *Bull. Acad. Sci. USSR (Phys. Ser.)* **22**, 1073.
Orgel, L. E. (1960). "Introduction to Transition Metal Chemistry: Ligand Field Theory." Methuen, London.
Pajasova, L. (1969). *Czech. J. Phys.* **19**, 1265.
Parke, S., Bingham, K., and Watson, A. I. (1970). *Phys. Chem. Glasses* **11**, 223.
Parke, S., and Webb, R. S. (1972). *Phys. Chem. Glasses* **13**, 157.
Parsons, J. L., and Milberg, M. E. (1960). *J. Am. Ceram. Soc.* **43**, 326.
Patek, K. (1970). *In* "Glass Lasers" (J. G. Edwards, ed.), Chemical Rubber Co., Cleveland, Ohio.
Paul, A. (1970a). *Phys. Chem. Glasses* **11**, 159.
Paul, A. (1970b). *Phys. Chem. Glasses* **11**, 168.
Paul, A. (1973). *Phys. Chem. Glasses* **14**, 96.
Paul, A. (1974). *J. Non-Cryst. Solids* **15**, 517.
Paul, A. (1975). *J. Mater. Sci.* **10**, 692.

Paul, A., and Assabghy, F. (1975). *J. Mater. Sci.* **10**, 613.
Paul, A., and Douglas, R. W. (1967a). *Phys. Chem. Glasses* **8**, 151.
Paul, A., and Douglas, R. W. (1967b). *Phys. Chem. Glasses* **8**, 233.
Paul, A., and Douglas, R. W. (1968a). *Phys. Chem. Glasses* **9**, 21.
Paul, A., and Douglas, R. W. (1968b). *Phys. Chem. Glasses* **9**, 27.
Paul, A., and Douglas, R. W. (1969a). *Phys. Chem. Glasses* **10**, 133.
Paul, A., and Douglas, R. W. (1969b). *Phys. Chem. Glasses* **10**, 138.
Paul, A., and Gomulka, S. (1975). *Phys. Chem. Glasses* **16**, 57.
Paul, A., and Tiwari, A. N. (1974). *Phys. Chem. Glasses* **15**, 81.
Pearson, A. D. (1964). In "Modern Aspects of the Vitreous State" (J. D. Mackenzie, ed.), Vol. 3, pp. 29–58. Butterworths, London.
Philipp, H. R. (1966). *Solid State Commun.* **4**, 73.
Philipp, H. R. (1971). *J. Phys. Chem. Solids* **32**, 1935.
Phillips, J. C. (1974). *Phys. Rev. B* **9**, 2775.
Pinnow, D. A., Rich, T. C., Ostermayer, F. W. Jr., and DiDomenico, M. Jr. (1973). *Appl. Phys. Lett.* **22**, 527.
Platzoder, K. (1968). *Phys. Status Solidi* **29**, K63.
Poch, W. (1964). *Glastech. Ber.* **37**, 533.
Powell, R. J., and Spicer, W. E. (1970). *Phys. Rev. B* **2**, 2182.
Quan, J. T., and Adams, C. E. (1966). *J. Phys. Chem.* **3**, 1.
Ram, A., Prasad, S. N., and Vaish, V. K. (1960). *Cent. Glass Ceram. Res. Inst. Bull. (India)* **7**, 49.
Ram, A., and Prasad, S. N. (1962). In "Advances in Glass Technology" (Am. Cenam. Soc., ed.) pp. 256–270, Plenum Press, New York.
Reilly, M. H. (1970). *J. Phys. Chem. Solids* **31**, 1041.
Reitzel, J. (1955). *J. Chem. Phys.* **23**, 2407.
Rhee, C. (1971). *J. Korean Phys. Soc.* **4**, 51.
Rhee, C., and Bray, P. J. (1971). *Phys. Chem. Glasses* **12**, 165.
Rich, T. C., and Pinnow, D. A. (1972). *Appl. Phys. Lett.* **20**, 264.
Richter, H., Breitling, G., and Herre, F. (1954). *Z. Naturforsch.* **9a**, 390.
Riebling, E. F. (1963). *J. Chem. Phys.* **39**, 1889, 3022.
Rowe, J. E. (1974). *Appl. Phys. Lett.* **25**, 576.
Rowe, J. E., and Ibach, H. (1973). *Phys. Rev. Lett.* **31**, 102.
Ruffa, A. R. (1968). *Phys. Status Solidi* **29**, 605.
Sasaki, T., Fukutani, H., Ishiguro, K., and Isumitani, T. (1965). *Jap. J. Appl. Phys. Suppl.* I, Vol. 4, 527.
Saskena, B. D. (1961). *Trans. Faraday Soc.* **57**, 242.
Savage, J. A., and Nielson, S. (1964). *Phys. Chem. Glasses* **5**, 82.
Savage, J. A., and Nielson, S. (1965a). *Infrared Phys.* **5**, 195.
Savage, J. A., and Nielson, S. (1965b). *Phys. Chem. Glasses* **6**, 90.
Scholze, H. (1959). *Glastech. Ber.* **32**, 81, 142.
Scott, J. F., and Porto, S. P. S. (1967). *Phys. Rev.* **161**, 903.
Schroeder, H. (1965). *Proc. Int. Congr. Glass, 7th, Brussels* Vol. I, paper 7.
Schultz, P. C. (1974). *J. Am. Cer. Soc.* **57**, 309.
Schultz, P. C., and Smyth, H. T. (1972). In "Amorphous Materials" (R. W. Douglas and B. Ellis, eds.), pp. 453–461. Wiley (Interscience), New York.
Sharma, T. N., Sakaino, T., and Horiya, T. (1956). *Bull. Tokyo Inst. Tech. Ser. B.* **3**, 155.
Shih, C. K., and Su, G. J. (1965). *Proc. Int. Congr. Glass, 7th, Brussels* Vol. I, paper 48.
Sigel, G. H. Jr. (1968). Ph.D. Thesis, Georgetown Univ., Washington, D.C. (unpublished).
Sigel, G. H. Jr. (1971). *J. Phys. Chem. Solids* **32**, 2373.

Sigel, G. H. Jr. (1973/74). *J. Non-Cryst. Solids* **13**, 372.

Sigel, G. H. Jr. (1975). Unpublished results.

Sigel, G. H. Jr., and Ginther, R. J. (1968). *Glass Tech.* **9**, 66.

Silver, A. H., and Bray, P. J. (1958). *J. Chem. Phys.* **29**, 984.

Simon, I. (1957). *J. Am. Ceram. Soc.* **40**, 150.

Simon, I. (1960). *In* "Modern Aspects of the Vitreous State" (J. D. Mackenzie, ed.), Vol. 1, pp. 120–151. Butterworths, London.

Simon, I., and McMahon, H. O. (1952). *J. Chem. Phys.* **20**, 905.

Simon, I., and McMahon, H. O. (1953a). *J. Am. Ceram. Soc.* **36**, 160.

Simon, I., and McMahon, H. O. (1953b). *J. Chem. Phys.* **21**, 23.

Skolnik, L. (1975). *In* "Optical Properties of Highly Transparent Solids" (S. S. Mitra and B. Bendow, eds.), pp. 405–443. Plenum Press, New York.

Slater, J. C., and Johnson, K. H. (1972). *Phys. Rev. B* **5**, 844.

Smirnova, E. V. (1965). *Sov. Phys. Dokl.* **10**, 247.

Smith, H. L., and Cohen, A. J. (1963). *Phys. Chem. Glasses* **4**, 173.

Snitzer, E. (1966). *Appl. Opt.* **5**, 1487.

Sobolev, N. N., and Seederov, T. A. (1957). *Opt. Spek.* **3**, 560.

Spitzer, W. G., and Kleinman, D. A. (1961). *Phys. Rev.* **121**, 1324.

Steele, F. N., and Douglas, R. W. (1965). *Phys. Chem. Glasses* **6**, 246.

Stepanov, B. I., and Prima, A. M. (1958). *Opt. Spek.* **4**, 774; **5**, 15.

Stevels, J. M. (1947). *Proc. Int. Congr. Pure Appl. Chem., 11th* **5**, 519.

Stookey, S. D. (1949a). *J. Ind. Engl. Chem.* **41**, 856.

Stookey, S. D. (1949b). *J. Am. Ceram. Soc.* **32**, 246.

Stolen, R. H. (1970). *Phys. Chem. Glasses* **11**, 83.

Stroud, J. S. (1961). *J. Chem. Phys.* **35**, 844.

Stroud, J. S. (1962). *J. Chem. Phys.* **37**, 836.

Stroud, J. S. (1964). *Phys. Chem. Glasses* **5**, 71.

Stroud, J. S. (1971). *J. Am. Ceram. Soc.* **54**, 401.

Su, G. J., Borrelli, N. F., and Miller, A. R. (1962). *Phys. Chem. Glasses* **3**, 167.

Swarts, E. L., and Cook, L. M. (1965). *Proc. Int. Congr. Glass, 7th, Brussels* Vol. I, paper 23.

Sweet, J. R., and White, W. B. (1969). *Phys. Chem. Glasses* **10**, 246.

Takahashi, K., and Goto, Y. (1970). *Yogyo-Kyokai-Shi* **78**, 11. (In Japanese.)

Takahashi, S., Miyashita, T., Edahiro, T., Horiguchi, M., and Masuno, K. (1974). *Proc. Int. Congr. Glass, 10th, Kyoto* **6**, 64.

Tanabe, Y., and Sugano, S. (1954). *J. Phys. Soc. Jpn.* **9**, 753, 766.

Tarte, P. (1964). *In* "Physics of Non-Crystalline Solids" (J. A. Prins, ed.), pp. 549–565. Wiley (Interscience), New York.

Tauc, J. (1975). *In* "Optical Properties of Highly Transparent Solids" (S. S. Mitra and B. Bendow, eds.), pp. 245–260. Plenum Press, New York.

Tenney, A. S., and Wong, J. (1972). *J. Chem. Phys.* **56**, 5516.

Tiedemann, J. (1966). *Angew. Chem. Int. Ed.* **5**, 85.

Tischer, R. E. (1968). *J. Chem. Phys.* **48**, 4291.

Tossell, J. A., Vaughan, D. J., and Johnson, K. H. (1973). *Chem. Phys. Lett.* **20**, 329.

Toyuki, H., and Akagi, S. (1972). *Phys. Chem. Glasses* **13**, 15.

Tucker, R. F. (1962). *Advan. Glass Technol.* **1**, 103–114.

Turner, W. E. S., and Weyl, W. (1935). *J. Soc. Glass Tech.* **19**, 208.

Urnes, S. (1960). *In* "Modern Aspects of the Vitreous State" (J. D. Mackenzie, ed.), Vol. I, pp. 10–37. Butterworths, London.

Urnes, S. (1969). *Phys. Chem. Glasses* **10**, 69.

Van Uitert, L. G., Pinnow, D. A., Williams, J. C., Rich, T. C., Jaeger, R. E., and Grodkiewicz, W. H. (1973). *Mater. Res. Bull.* **8**, 469.

Van Vleck, J. H. (1932). *Phys. Rev.* **41**, 208.
Van Wazer, J. R. (1958). *In* "Phosphorus and Its Compounds," Vol. I. Wiley (Interscience), New York.
Veinberg, H. I. 1962. *Zh. Fiz. Chem.* **36**, 81.
Wannier, G. (1937). *Phys. Rev.* **52**, 191.
Warren, B. E., and Biscoe, J. (1938). *J. Am. Ceram. Soc.* **21**, 49, 259.
Warren, B. E., and Mavel, G. (1965). *Rev. Sci. Instrum.* **36**, 196.
Warren, B. E., Kruter, H., and Morningstar, O. (1936). *J. Am. Ceram. Soc.* **19**, 202.
Weckerle, H. (1933). *Glastech. Ber.* **11**, 273, 314.
Wemple, S. H. (1973). *Solid State Commun.* **12**, 701.
Wemple, S. H., Pinnow, D. A., Rich, T. C., Jaeger, R. E., and Van Uitert, L. G. (1973). *J. Appl. Phys.* **44**, 5432.
Westman, A. E. R. (1960). *In* "Modern Aspects of the Vitreous State" (J. D. Mackenzie, ed.), Vol. I, pp. 63–90. Butterworths, London.
Weyl, W. (1945). *J. Soc. Glass Tech.* **29**, 291T.
Weyl, W. A. (1951). "Coloured Glasses." Soc. of Glass Technol., Sheffield.
Weyl, W., and Thumen, E. (1933). *Sprechsaal* **66**, 197.
White, K. I., and Midwinter, J. E. (1973). *Opto-electronics* **5**, 323.
Wiech, G. (1967). *Z. Phys.* **207**, 428.
Williams, D. J., Bradburg, B. T., and Maddocks, W. R. (1959). *J. Soc. Glass Tech.* **43**, 337T.
Wilmot, G. B. (1954). Ph.D. Thesis, Massachusetts Inst. of Technol., unpublished.
Wong, J., and Angell, C. A. (1971). *Appl. Spectrosc. Rev.* **4**, 97–232.
Wybourne, B. G. (1965). *In* "Spectroscopic Properties of Rare Earths," Wiley (Interscience), New York.
Yip, K. L., and Fowler, W. B. (1974). *Phys. Rev. B* **10**, 1400.
Yokota, R., and Shimizu, K. (1957). *J. Phys. Soc. Jpn.* **12**, 833.
Young, C. G. (1969). *Proc. IEEE* **57**, 1267.
Zarzycki, J. (1956). *Travaux du IVe Congr. Int. du Verre, Paris* **4**, 323.
Zarzycki, J. (1957). *Verres. Refract.* **11**, 3.
Zarzycki, J. (1964). *In* "Physics of Non-Crystalline Solids" (J. A. Prins, ed.), pp. 527–548. Wiley (Interscience), New York.
Zarzycki, J., and Naudin, F. (1960). *Verres Refract.* **14**, 113.
Zsigmondy, R. (1909). "Colloids and the Ultramicroscope" (J. Alexander, translator). Wiley, New York.

Photochromic Glass

ROGER J. ARAUJO

Research and Development Laboratories
Corning Glass Works
Corning, New York

I. Introduction

Photochromic materials change color when exposed to light and revert to their original color when the light is removed. Photochromism has been discussed recently in a book by that title (Brown, 1971). These materials can be used to control the intensity of light and to store images, at least temporarily, without chemical development. Furthermore, most photochromics exhibit high resolution capability in image storage. Their main disadvantage is that compared to ordinary photographic films they suffer a low darkening efficiency because they are not chemically developed. Also, many of these materials lose efficiency with use.

Photochromic glasses have several advantages over other photochromic materials simply because they are glasses. By use of standard glass-melting and glass-forming techniques they can be made in any desired size or shape. Their transparency and durability under chemical attack, scratching, or moderate heating are also advantages for some applications. The most important advantage is that most photochromic glasses show no fatigue.

The number of glasses showing photochromic behavior is small compared to the number of photochromic crystalline and liquid materials, but it is growing. Photochromic ophthalmic lenses are already commercially available, and it is reasonable to expect use of photochromic glass in architectural applications, as well as in the field of information storage and display. Photochromic glasses, then, form a group of materials too important to be ignored.

II. Homogeneous Glasses

There are several families of homogeneous glasses which darken reversibly in response to ultraviolet excitation.

A. Reduced Alkali Silicate Glasses

Cohen and Smith (1962) reported that certain strongly reduced silicate glasses are photochromic. The properties of various compositions from this class of glasses have been discussed by Swarts and Pressau (1965).

The peak position of the induced absorption band in these strongly reduced glasses depends upon the specific alkali included in the glass, but is not very sensitive to the exact concentration of the alkali. In soda–silica glasses the band peaks at 570 nm and the shape remains constant over a composition range of soda concentration from 8% to 30%. If the soda is totally replaced by lithia or potassia, the photochromic band peaks at 494 or 717 nm, respectively.

Although the peak position of the induced band does not strongly depend on impurities, the strength of the absorption can be either considerably enhanced or suppressed. Cerium and europium, in concentrations as low as 100 ppm, considerably enhance the darkening. In alkalirich compositions, small additions of zirconia (~ 100 ppm) greatly enhance darkening, whereas in silicarich compositions some degree of inhibition of darkening by zirconia impurities is observed. Photochromic coloring is generally inhibited by small concentrations of several transition or heavy metal cations. Additions of 200 ppm of titanium, vanadium, or iron completely suppressed coloring.

Normally, glasses doped with cerium, europium, or zirconium can be darkened by the 3660-Å mercury line, whereas undoped glasses are darkened only by higher-energy light, such as the 2537-Å line. It is interesting to note that exposure to the high-energy source sensitizes the glass to darkening by the longer wavelengths. This sensitizing effect is not permanent, however, being substantially diminished after 6 hr.

Swarts and Cook (1965) reported that when glasses of this type were melted in the absence of nitrogen, the photochromic response was greatly attenuated or missing altogether. The ultraviolet absorption of the glass increased with increasing amounts of nitrogen dissolved in the glass. Therefore, it was tentatively concluded that an absorption mechanism intrinsic to nitrogen enhanced photochromism in undoped glasses. Furthermore, the introduction of nitrogen led to more complete reduction of cerium and europium when these were included in the glass.

Swarts and Pressau (1965) studied the kinetics of darkening of a europium-doped, reduced soda–silica glass which was irradiated by 3600-Å light at −155°C. At this low temperature the fading rate was negligible. The investigators found that the logarithm of the induced absorption was linear with time, suggesting that darkening follows first-order kinetics. Analysis of the decay of the induced absorption band which occurred upon cessation of irradiation showed that the fading kinetics could be described approximately by a second-order rate equation. Glasses which had been activated by 2537-Å light showed very little deviation from a second-order decay law, whereas glasses activated by 3660 Å frequently deviated in the direction of first-order kinetics during fading.

Muller and Milbey (1967) also studied the kinetics of the fading process in reduced alkali–silicate glasses. They report that undoped glasses follow a simple second-order decay law, whereas the fading process in glasses containing cerium or europium is best described by an expression containing two second-order terms. They suggest that the slow-decaying component is related to the basic glass, whereas the fast-decaying one is related to the sensitizer ions.

The dependence of the rate of growth of the absorption on temperature was not studied, but the fading rate is a very sensitive function of temperature. One composition colored by 3660 Å showed a rate constant thirty times greater at room temperature than at −155°C. Optical bleaching of the induced absorption was observed, but no attempt was made to study the kinetics.

The equilibrium value of the induced absorption at 570 nm was found to decrease under continued ultraviolet exposure with fatigue occurring most rapidly in the rare-earth-doped glasses. The original photochromic response could be restored by heating the samples to the annealing point. Heating repeatedly to temperatures near the softening point destroyed the photochromic response. Coincident with prolonged ultraviolet exposure was the growth of a new absorption band at about 250 nm which did not fade at room temperature.

Comparisons of the darkening rate constants after various lengths of time of exposure was not made, but the fading rate constant showed a substantial increase after the glass had been subjected to prolonged exposure.

An 8-mm thick sample of a europium-doped soda silica glass attained a maximum darkening of 6 dB after about 20 sec of irradiation by a 3660-Å light source with an intensity at the sample of 0.04 W/cm^2. This corresponds to a sensitivity of about 100 $mJ/cm^2/dB$. The fading was essentially complete in 5 min after the cessation of irradiation.

Although the darkening is approximately two orders of magnitude lower than the limit one would expect if each incident photon gives rise to one color center having an oscillator strength of one, it is probably adequate for many applications. The fading rate is certainly fast enough for ophthalmic or architectural applications; however, the tendency to fatigue places limitations on the usefulness of these glasses. Furthermore, the reducing conditions required during melting make it difficult to make samples of these photochromic glasses larger than a few hundred grams.

A similar effect is described by Stroud (1966) in a patent which teaches that a photochromic glass can be made by adding from 0.005 to 1.0 wt % cerium oxide and 0.005 to 1.0 wt % manganese oxide to reduced sodium silicate glasses. Reduction can be achieved by merely adding small amounts of carbohydrate to the batch to be melted.

Such glasses are transparent and either colorless or only slightly yellow when formed. Upon exposure to ultraviolet the glasses darken to a purple color. When the ultraviolet light is removed, the color fades and the glasses return rapidly to their original appearance.

Stroud speculates that an excited cerous ion catalyzes the activation of a manganous ion which can readily lose an electron to produce a manganic ion and a free electron. The reversibility of this darkening phenomenon is due to the fact that the glass is free of polyvalent cations which act as electron traps. Therefore, the free electrons can recombine with the manganic ions to destroy the color center.

The present author (unpublished results) was unable to reproduce Stroud's results without use of stronger reducing conditions. With sufficiently strong reducing conditions, the manganese appeared to be inconsequential. It appears that Stroud's results are very similar to those of Cohen and Smith.

B. CdO–B_2O_3–SiO_2 Glasses

Meiling (1971) described a family of homogeneous glasses based on the CdO–B_2O_3–SiO_2 system that are photochromic and show no fatigue. Stable glasses can be made with a wide range of ratios of silica to boric oxide, if the mole percent of cadmium oxide is between 45 and 65. Many of the glasses in this composition area are photochromic with or without dopants.

At a fixed ratio of silica to boric oxide, the degree of darkening that could be achieved was found to depend strongly on the composition, with the maximum darkening being obtained when the cadmium oxide concentration was 60 mole %. At constant cadmium oxide percentage, the photochromic effects depend only weakly on composition. Heat treatment was not found to alter the photochromic effects.

The shape of the induced spectrum was very nearly flat throughout the visible spectrum with slightly more darkening in the blue region than in the red. This is to be contrasted with the reduced alkali–silicate glasses where, as already noted, the induced absorption is characterized by a distinct peak whose position depends on the composition. The undoped samples are not very sensitive. Typically, it required energies of the order of 1 J/cm^2 to obtain 2 dB of darkening in a 2-mm-thick sample. Doping with copper increases the sensitivity by approximately a factor of 2, and doping with copper and silver chloride increases the sensitivity by a factor of 3. Several hours are required for complete bleaching of any of these glasses.

The darkening sensitivity of the doped glasses is comparable to that of the reduced alkali silicates, but their fading characteristics are inferior. The cadmium borosilicates are superior to the reduced alkali silicate in that they do not fatigue and that they are easily made in large quantities.

C. $Na_2O \cdot Al_2O_3 \cdot 2SiO_2 \cdot \frac{1}{2}NaX$ Glasses

Araujo et al. (1975) observed that many glasses having a stoichiometry similar to nepheline would, if doped with sodium chloride and sodium sulfide, darken when irradiated with short wavelength ultraviolet (2537 Å). No deterioration of the photochromic effect is observed if up to half of the sodium oxide is replaced by potassium oxide. The silica can be partly or wholly replaced by germania and the alumina by gallium oxide. The sodium sulfide can be replaced by sodium selenide or sodium telluride. Sodium fluoride is actually preferable to sodium chloride because of the lower loss during melting and because of the more efficient action of the fluoride as a flux.

Unfortunately these glasses are fairly deep brown when made and the induced photochromic darkening does not lead to very high contrast ratios. Addition of ions such as cadmium which complex the sulfur, diminish the residual color of the glass, but apparently the complexing is so complete that photochromism is also diminished.

Exposure of a previously darkened glass to 3660-Å irradiation causes rather bright fluorescence. A unique application of this effect may be in display systems. If the glass is darkened in a pattern by exposure to 2537-Å irradiation through a mask, only the darkened areas fluoresce upon subsequent exposure to the longer wavelengths and the pattern is observed in a fluorescence mode.

It is believed that the mechanism of darkening in these glasses is similar to the mechanism which is responsible for photochromism in the mineral hackmanite which will be discussed in the next section.

III. Suspension of Crystallites in Glasses

Many inorganic materials are soluble in molten glasses but precipitate out when the glasses are cooled or held at some intermediate temperature for a long time. In some cases the precipitate is in a fine enough suspension so that the glass retains a high degree of transparency in reasonable thicknesses. Photochromism has been observed in several systems where crystallites are suspended in a glassy matrix.

The advantage of glasses containing crystallites over the homogeneous glasses is that their physical properties can often be varied independently of their photochromic properties. For example, adjustment of refractive index was necessary for ophthalmic use, as it was for an information display device (Megla, 1970). Color of the undarkened glass was adjusted in the development of the Photosun® ophthalmic lenses. Heat absorption was obtained in the glasses containing copper–cadmium halides by adding reduced iron ions; then the thermal expansion coefficient had to be reduced to prevent thermal breakage under the abrupt temperature gradients caused by shadows. Such control would be very difficult, if not impossible, in the homogeneous photochromic glasses.

A. Hackmanite

Lee (1936) reported that natural hackmanite can be darkened by long-wavelength ultraviolet light and optically bleached by visible light. Medved (1954) reported that natural hackmanite can be darkened by x-rays and then optically bleached or thermally bleached at elevated temperatures (450°C). He postulated that the color center responsible for the induced absorption is an F center; i.e., an electron trapped at a halogen ion vacancy. Hodgeson *et al.* (1967) confirmed this by ESR measurement.

Radler (1964) attempted to make a photochromic glass-like material containing hackmanite. This was done by firing sodium carbonate, silicic acid, and alumina in the correct proportions to produce a mass of material having a stoichiometry corresponding to the chemical formula $Na_2O \cdot Al_2O_3 \cdot 2SiO_2$. To this base material various amounts of boric acid, alkali halides, sodium sulfate, and sodium silicofluoride were added and the batch refired.

No photochromism was ever observed unless the material was fired in a hydrogen atmosphere in graphite crucibles. As the boric oxide content of the batch was increased, the glasslike character of the material increased, but the photochromic character was decreased. The induced color varied somewhat with the halide used. Materials containing chlorides and iodides tended to be blue, whereas materials containing bromides tended to be red-violet.

The investigators pointed out that the photosensitive phase was separated from the glass phase. The considerable haze or opacity was ascribed to the fact that the two phases differed in refractive index.

One of the better materials transmitted only about 30% of the visible light in the undarkened state when it was ground and polished to 0.9 mm. Upon exposure to ultraviolet light, the sample darkened to 3.0% transmission at 540 nm.

These glasses did not thermally bleach at room temperature and the optical bleaching was reported to be inefficient, although no quantitative measure of the efficiency was given. No report was given about fatigue or resolution capability in these glasses.

Araujo *et al.* (1975) describe a photochromic glass–ceramic in which hackmanite is grown from the glass. As previously indicated, many of the glasses are photochromic prior to the growth of the hackmanite phase. In the early low-temperature ($\sim 650°C$) stages of heat treatment, carnegeite, nepheline, and sodium fluoride are formed. When the temperature is raised ($\sim 850°C$), the nepheline lines and sodium fluoride lines are diminished and sodalite lines appear in the x-ray traces.

Unlike the crystalline materials, neither glass nor glass–ceramic is much affected by reducing conditions either in the melt or in the final annealing steps. Unfortunately neither the glasses nor the glass–ceramics show the cathodochromic properties found in crystalline materials by Phillips (1970). Therefore, the glasses are not attractive for use in display systems of the type considered by Taylor *et al.* (1970). Synthetic sodalites are discussed by Williams *et al.* (1969).

B. Silver Molybdate

Sawchuk and Stookey (1966) reported that photochromic materials were made by precipitating silver molybdate or silver tungstate from a sodium aluminoborosilicate base glass. These materials are characterized by high haze levels and low darkening efficiencies. Bleaching was not discussed.

C. Thallium Chloride

Sakka and Mackenzie (1972) report that thallium halide can be precipitated by heat treatment from a potassium, barium aluminophosphate glass and the resulting material can be darkened by exposure to near uv. The induced absorption is similar to that found in irradiated TlCl crystals by the same authors (1973) and x-ray diffraction patterns of these glasses verify the presence of TlCl crystallites.

Darkening is considerably enhanced by doping with copper or silver in both the glasses and the crystals. Half fading times for thermal bleaching at room temperature are in the tens of hours, but optical bleaching under a 250-W tungsten light at a distance of 30 cm is essentially complete in less than five minutes.

Thermal fading is reasonably well described by a second-order rate equation, and an activation energy of 1.3 eV is found for the fading rate constant.

D. J. Kerko (unpublished results) reported darkening of TlCl in a sodium lead phosphate glass only when the glass had been sufficiently heat treated to cause opacity. When the glass was doped with silver, photochromism was observed in the transparent glass.

P. A. Tick (unpublished results) was unable to produce photochromism without silver in the base glass described by Sakka.

D. Copper Cadmium Halides

Araujo (1967) has described transparent photochromic glasses containing crystallites of mixed copper and cadmium halides. The induced coloration consisted of a very broad absorption band diminishing somewhat in strength with increasing wavelength throughout the visible and a sharp band superimposed at 625 nm (see Fig. 1). In 2-nm thicknesses the transmission in the clear state of a typical glass is about 90%. Two minutes of irradiation with a 200-W mercury lamp from a distance of 6 in reduced the transmission of a

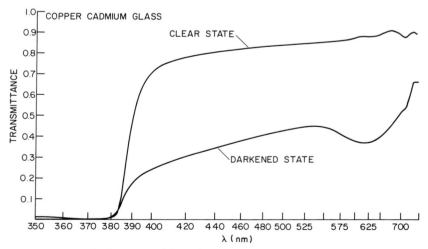

Fig. 1. Spectrum of Cu–Cd glass in clear and darkened states.

typical glass to about 40% at the 625-nm absorption peak. The degree of thermal bleaching in the first 5 min is comparable to that observed in Photogray® lenses when they are darkened under identical conditions. The rate of change of thermal bleaching with respect to time is, in general, smaller for glasses containing copper cadmium halides than it is for glasses containing silver halides. That is, the initial fade rate of a copper cadmium glass may be lower than that of a particular silver halide glass, but the time required for complete fading is normally shorter.

No optical bleaching has been observed for any of the copper cadmium glasses. This characteristic excludes these glasses from use in some information storage devices, but it may be an advantage in architectural or ophthalmic applications. The absolute darkening sensitivity has not been measured, but it appears to be comparable to the more sensitive silver halide glasses. The induced absorption coefficient is considerably higher than that found in silver halide glasses. It ranges from 35 to 75 cm^{-1} in the copper cadmium glasses and from 5 to 20 cm^{-1} in the silver halides. Hence very thin layers can be very dark. This author (unpublished results) found that a very thin (~ 50 μm) photochromic layer can be obtained by melting a copper-free glass, and subsequently introducing the copper by an ion-exchange technique. The transmission in the visible was reduced from 90% to about 50% by the absorption induced in this thin layer when the glass was irradiated.

Although it appears to this author that the family of glasses which can act as suitable hosts for copper cadmium halides is not as extensive as that for the silver halides, it is large enough to satisfy requirements for many applications. Host glasses which are easily formed and as of good optical quality and which fill a considerable range in refractive index of expansion coefficient are relatively plentiful.

These glasses can be made under moderately strong reducing conditions, and so one can easily incorporate reduced iron in the base glass to achieve infrared absorption without deleterious effects on the photochromic properties. This is, of course, impossible in the silver halide glasses because the silver is so easily reduced to colloidal silver.

These glasses have not yet been used for commercial applications, but it is expected that the Corning Glass Works, who introduced photochromic glasses to the ophthalmic market, will in 1977 make available photochromic lenses containing copper cadmium halides instead of silver halides. Their objective in doing so is to make available to the user a larger choice of colors by adding this glass which darkens to a green color to the presently available group of glasses which darken to a gray or brown color.

The darkening mechanism of these glasses is not well understood, but it is believed that the active crystal phase is a solution of cuprous ions in a crystal of cadmium chloride. Photolysis is accompanied by the production of cupric ions in nonequilibrium sites thus giving the glass a green color. This belief is based on a presumed analogy between these glasses and the copper-doped cadmium chloride crystals studied by Kan'no *et al.* (1973).

Tick (unpublished results) reported an induction time of tens of microseconds when darkening photochromic glasses containing silver halide but no induction whatsoever in glasses containing copper cadmium halide. This seems to indicate that colloid formation is not a dominant mechanism in these glasses and this fact is consistent with the proposed mechanism. The thermal fading seems to be very well described by the diffusion model (Araujo and Borrelli, 1976), the details of which will be discussed later in this chapter. The darkening rate, when corrected for simultaneous fading, is very well approximated by a first-order expression. This, too, seems to be consistent with the idea that the absorption is primarily due to a single simple color center such as a cupric ion. It is also important to note that the time dependence of the absorption during fading is only weakly dependent on the monitored wavelength for these glasses but very strongly so for the silver halides.

IV. Glasses Containing Silver Halides

A. Introduction

Silver halides can be precipitated to produce transparent photochromic glasses in a wide range of base glasses. The alkali boroaluminosilicates developed by Armistead and Stookey (1965) are perhaps the most thoroughly studied. Alkali and alkaline earth borates have been described by Suzuki (1971), alkali and alkaline earth phosphates by Kerko (unpublished results), heavy metal aluminoborates by Araujo *et al.* (1972), and modifier-free borophosphosilicates by R. F. Reade (unpublished results).

Ordinarily, the silver halide crystallites and the attendant photochromism are developed by holding the homogeneous glass containing dissolved silver and halide ions at temperatures between 400 and 800°C for periods of 15 min to 4 hr. Electron microscopic examination of the photochromic glasses indicated a useful range of the crystallite size to be between 80 and 150 Å. When the particles are smaller than the lower limit, the glass does not darken well; when larger than the upper limit, the glass is hazy. The average spacing between particles is of the order of 1000 Å.

Both composition and thermal history are important in determining photochromic properties. A small amount of copper increases the darkening sensitivity by several orders of magnitude and is included in almost all transparent glasses. Moser *et al.* (1959) explained the effect of copper in single-crystal silver halide in terms of hole trapping. Fanderlik (1968) indicated that doping with silver sulfide increases sensitivity. Mitchell (1957) has indicated that this is also due to hole trapping. Araujo *et al.* (1967) found that cadmium enhances the darkening sensitivity, but only in the presence of copper.

The choice of halogen strongly affects photochromic properties. The fundamental absorption edge of the silver halide, and therefore the darkening sensitivity, extends to longer wavelength as the atomic number of the halogen increases. Furthermore, since the solubilities of the various silver halides differ considerably from each other in most base glasses, the choice of halogen affects the crystallization kinetics and, indirectly, photochromic properties.

The base glass composition, of course, affects the physical properties of the glass, as well as the photochromic properties. The levels of lead, alkali, and boric acid markedly affect darkening and fading rates. Perhaps the most important function of the base glass composition is the temperature dependence of the silver halide solubility. A good host glass is one in which the solubility of the silver halide is high at the high temperatures used to melt the glass but low at intermediate temperatures. In such a case, the glass can be supersaturated with silver halide, and the excess solute can be precipitated out precisely by heating at intermediate temperatures. The effect of heat treatment, of course, depends on composition; but, in general, higher temperatures or longer times tend to produce larger particles, darker color after exposure, and slower thermal bleaching. The rates at which glasses are cooled sometimes have a marked influence on photochromic properties. No general statement about the effects of cooling rate is possible because these effects vary with the composition of the glass. Examples in which slow cooling prevents darkening, as well as examples in which slow cooling slows the fade rate and increases darkening, have been seen by this author.

The large variations in composition and heat treatment which are possible provide a wide range of properties (Armistead and Stookey, 1964; Megla, 1966; Smith, 1967; Gliemeroth and Mador, 1970; Araujo, 1973). The induced absorption extends throughout the whole visible spectrum in every glass in this family, but some glasses absorb more strongly in the red giving them a bluish tint. Other glasses show a distinct pink color when darkened. The activation wavelengths are ordinarily in the interval from 320 to 380 nm but can be longer. Activation by red light (610 nm) has been observed as an extreme example. The time required for complete thermal bleaching at room temperature varies from one minute to many months. Optical bleaching efficiency varies from slightly more than zero to the very high value of 100 mJ/cm^2/dB. The speed of thermal bleaching and the efficiency of optical bleaching can be controlled independently. Within the interval between 50 and 90°F, the temperature dependence of equilibrium darkening varies from virtually zero to almost 1% transmission change in a 2-mm sample for each degree temperature change. Fast-fading glasses almost always show a large temperature dependence. The darkening sensitivity varies from zero to a high value of 1 mJ/cm^2/dB in a 2-mm sample. This is the highest sensitivity available in any photochromic glass. There are some glasses which will neither darken nor bleach significantly in response to some narrow interval of "neutral" wavelengths in the green part of the spectrum.

The speed of darkening and the equilibrium darkening are closely related to the intensity of the activating light. Deviations from a linear relationship between the equilibrium absorption and the intensity are very large in some glasses and only slight in others. In general, linearity is more closely approximated as the thermal fading rate increases (Araujo, 1968a).

Megla (1966) indicated that photochromic glasses containing silver halides show no signs of fatigue after being subjected to more than 300,000 darkening and bleaching cycles. The same investigator indicated that the resolution capability of this class of glasses exceeds 2000 line pairs per millimeter.

If a piece of photochromic glass is elongated under a sufficiently high stress, it polarizes light in the darkened state (Araujo *et al.*, 1970). This is because the silver halide particles become elongated and the material therefore exhibits anisotropy in its refractive index.

Seward (1975) reported that thermally darkenable photochromic (TDPC) glasses have been produced by precipitating copper-doped silver halide crystallites in certain lanthanum borate base glasses. A TDPC glass is characterized by a darkened state which is stable at room temperature but which can be bleached by visible light. The glass will return to the darkened state with the passage of time. The rate of darkening and the color of the darkened state are affected by temperature. These TDPC glasses are further characterized by the vanishing of the darkened color above 430°C and redarkening of the glass below 310°C. These temperatures are very similar to the melting and crystallization temperatures of silver chloride droplets in glass which were reported by Hammel and McGary (1969).

B. Formation

As already indicated, the photochromic properties of a glass containing silver halides depends on composition and heat treatment. The composition of the glass is not, however, determined exclusively by the composition of the batch from which it is melted. Melting conditions can be very important because they strongly influence retention levels and oxidation states.

The factors that influence retention may differ for almost each individual glass. For example, Reade (unpublished results) observed the surprising result that about 10% chloride was retained in certain phosphate glasses, but about 50% chloride was retained when 5% B_2O_3 was added to the batch and melted at the same temperature. Obviously a complete discussion of this topic is impossible and only a few observations that are reasonably general will be made. Silver retention is decreased by increased halides in the batch. Halogen retention is decreased as the halide content is increased in the batch. Bromide retention is seriously reduced by increased chloride in the batch but the reverse is not true. Bromide retention is reduced by nitrates in the batch. T. R. Kozlowski (unpublished results) explained this as the result of a reaction at low temperatures (~ 600–$800°C$) in the early stages of melting to form nitrosyl bromide. Sulfate has an even stronger effect on the retention of bromide than does nitrate.

Pavlovskiy *et al.* (1971) point out the dramatic influence of redox reactions on the retention of iodide ions. They melted sodium aluminoborosilicates at $1500°C$, barium aluminoborates at $1250°C$, and sodium aluminoborogerminates at $1100°C$. In all cases the direct addition of iodine to a batch containing silver leads to its complete loss. However, if iodine was added to a batch containing no silver, chlorine, or bromine its volatilization from the borate and germanate glasses was only about 25% and from the silicate glass about 50%.

Maki and Tashiro (1972) studied the melting of an alkali boroaluminosilicate glass as a function of various atmospheres and also reached the conclusion that loss of chloride was increased by oxidation in an air atmosphere.

The concept of oxidation state of the total glass is of limited value in photochromic glasses, and it is more meaningful to consider the oxidation state of both the copper and silver. The oxidation state of the copper is important because of the hole trapping capability of the cuprous ion and the electron trapping capability of the cupric ion (Moser and Burnham, 1964). The oxidation state of the silver is, of course, important because ionic silver is required for the formation of silver halide crystals. Furthermore, the present author has seen indications that the presence of small amounts of neutral silver can increase the darkening sensitivity of some glasses and can affect the shape of the induced absorption spectrum.

J. H. Schreurs (unpublished results) using ESR techniques studied the factors which affect the oxidation state of the copper in a variety of silicate glasses. When the glasses were melted without silver or halogen the copper was always oxidized. When silver was included in the batch, the copper was somewhat more reduced. Chloride in the batch has a profound effect on reducing the copper. In certain high-alumina glasses there is no trace of the cupric ion spin resonance signal when there was 2% chloride in the batch and a very strong signal when there was no chloride.

Using indirect evidence as to the state of the silver (neither Ag^+ nor Ag^0 shows a spin signal). J. H. Cowan and H. D. Smith (unpublished results) concluded that a fast initial heating rate during melting was the single most important factor in the reduction of silver (other than composition changes).

If the glass is cooled sufficiently rapidly from the molten state to room temperature, no photochromism is observed because no appreciable quantity of silver halide has been precipitated. For development of photochromism, the glass is heated to a temperature range high enough so that mobilities are high and precipitation can be achieved in practical times. In most glasses temperatures can exist such that the glass is not supersaturated with respect to the silver halide and precipitation cannot occur. Normally the temperature used for heat treatment is considerably lower than this critical temperature because the average size of the silver halide particles formed increases as the temperature increases, and this can lead to objectionable haze levels.

Seward (unpublished results) attempted to study the precipitation kinetics of the silver halides in photochromic glasses without actually measuring particle sizes. He assumed (with good justification) that silver halide crystallites will not darken at sizes much less than 80 Å and that as the crystallite gets larger the fade rate decreases, while the induced absorption spectrum extends to longer wavelengths. He also assumed that the amount of ultraviolet absorption is approximately proportional to the total volume of silver halide precipitated; i.e., independent of the crystallite size. Using these assumptions he inferred details of the precipitation from a study of photochromic properties as a function of heat treatment. Figure 2 indicates the decided separation of temperatures at which the nucleation rate and growth rate of the silver halide liquid droplets are maximized for one particular glass. In other glasses, the separation of maximum nucleation rate and growth rate temperatures were not distinct. Whenever the separation was distinct, it seemed that the nucleation rate was increased by increasing the bromide concentration or by increasing the silver concentration. The growth rate, on

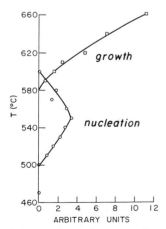

Fig. 2. Relative nucleation and growth rates as a function of temperature.

the other hand, was not much affected by bromide concentration, but was increased by increasing silver or chloride concentrations.

Moriya *et al.* (1972) studied the growth rate of silver chloride and silver bromide in an alkali aluminoborosilicate glass using x-ray line broadening techniques. In this particular glass the maximum precipitation temperature was 940°C for silver chloride and 1080°C for silver bromide. They report the following rates ($\bar{v}$ is the mean volume of silver halide liquid droplets in cm^3; t is time in seconds):

$$\bar{v}/t = 4.99 \times 10^2 \exp(-92500/RT) \quad \text{AgCl} \quad (510\text{--}600°C)$$

$$\bar{v}/t = 1.74 \times 10^{-3} \exp(-80000/RT) \quad \text{AgBr} \quad (585\text{--}705°C).$$

Moriya also argues that since the mean diameter of liquid particles of silver chloride and silver bromide increased in proportion to the cube root of the heating time, the growth of large droplets is accompanied by the dissolution of small ones. This is consistent with the findings of G. B. Carrier and D. Weinberg (unpublished results) who, by the use of electron microscopy and small angle x-ray scattering, respectively, indicated that the mean particle size increased but the number density of particles decreased as the heat-treatment temperature was increased. Hammel and MacGary (1969) also indicate that at least in one high-boron glass precipitation is more than 80% complete in 15 min and growth, which continues for more than 48 hr, proceeds by a ripening mechanism.

The factors which influence the temperature dependence of the solubility of the silver halides in various host glasses are not well understood. G. B. Hares (unpublished results) was unable to establish a relationship between the temperature dependence of the solubility and a change in boron coordination as a function of temperature.

Gliemeroth and Bach (1971) suggest that immiscibility of the glass other than the silver halide immiscibility may be important. Their electron micrographs show inhomogeneous droplets, parts of which are denser and parts of which are less dense than the surrounding glass. They propose that the droplet is silver halide (more dense) accompanied by a boron-rich material (less dense). Carrier (unpublished results) has seen similar structures in virtually all the photochromic glasses he has examined. Gliemeroth's interpretation is consistent with the work of S. Tong (unpublished results), who immersed finely ground photochromic glasses in boiling water for one minute. The dissolved material was found to be boron rich. If the same composition was subjected to this test without first being heat treated to produce photochromism, no boron was found in the water.

J. B. Chodak (unpublished results) challenged this interpretation because he had observed similar microstructures in glasses that did not contain boron. The present author believes that Chodak's results may have an analogous explanation since his (Chodak's) glasses contained tantalum oxide. Hence, it is conceivable that immiscibility would lead to a phase of lower electron density than the major phase.

While Gliemeroth's arguments are appealing, the issue is not conclusively decided. Carrier (unpublished results) reports that under continued exposure to the electron beam the more dense and less dense portions of the precipitated droplet move considerably. Since this observation is not understood, caution must be used in interpreting the micrographs.

The need for caution is emphasized by the results of Fanderlich and Prod'homme (1973) who obtained larger values for the mean diameter of the droplet when using light-scattering techniques than when using electron micrographs. They suggest that this discrepancy is due to an evacuated region formed by the droplet shrinking away from the host glass during cooling.

As indicated in the introductory remarks, the photochromic properties are not uniquely determined by the composition of the glass and by the distribution of particle sizes attained during precipitation. Treatment of the glass at low temperatures, after precipitation has been completed, can have a profound influence on photochromic properties. Sakka *et al.* (1974) demonstrate that, in certain alkali boroaluminosilicate glasses containing silver chloride, reheating raises or lowers the degree of darkening depending on the composition of the base glass, the copper level, the temperature of precipitation of the silver chloride as well as, of course, the temperature of reheating. J. Poitras (unpublished results) indicated that the influence of reheating was stronger in glasses containing pure silver chloride than in glasses containing mixed silver chloride and bromide. Seward (unpublished results) suggests that the effect of reheating is less pronounced in glasses that were highly nucleated than in slightly nucleated glasses. All workers agree that the effect is strongest between 300 and 400°C.

Sakka *et al.* (1974) indicate that the silver chloride crystals are not pure but rather solid solutions of sodium chloride and silver chloride. They also present x-ray evidence showing that separation into two phases occurs when the glass is heated at 380°C.

This author knows no evidence indicating the effect of heat treatments on the relationship between the ratio of halogens in the crystal and in the glass. For glasses given long high-temperature heat treatments to produce larger particles, Moriya (1973), and later H. J. Holland and H. P. Oberlander (unpublished results) found that the crystal is richer in bromide than the glass. (See Fig. 3.)

C. Mechanism

The darkening and bleaching mechanisms of photochromic glass are believed to be quite analogous to the photolytic effects found in bulk silver

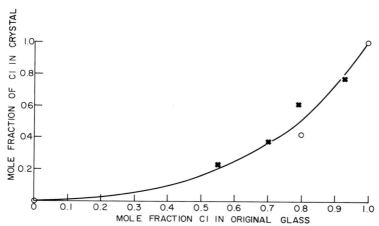

Fig. 3. Concentration of chloride in solid solution versus chloride concentration of original glass. × = data from Moriya (1973); ○ = data from Holland and Oberlander.

halide crystals. The volume darkening of such crystals has been discussed in terms of a colloidal silver color center by Seitz (1951), Brown and Wainfan (1957), and Moser *et al.* (1959). This last author credited Duboc with suggesting that the absorption bands observed in bulk silver halide crystals could be produced by random distribution of small aggregates of silver which are not spheres but ellipsoids.

Moriya (1974) suggested that the absorption band observed in many photochromic glasses around 500 nm can be attributed to small colloids of silver formed by irradiation. Seward (1974) suggested that ellipsoidal colloids of silver might be responsible for absorption throughout the visible. Using the Mie Theory as applied to colloids in the alkali halides by Doyle (1960), he calculated that the wavelength of maximum absorption of a spherical silver particle in silver halide should be about 4800 Å. If the particle is not spherical, the absorption band splits into two components. The absorption, which occurs when the electric vector of the light is parallel to the long axis of an oblate colloid, moves toward longer wavelengths as the particle becomes more highly elongated and the perpendicular absorption moves slightly to shorter wavelengths (see Fig. 4). Stookey and Araujo (1968) showed that this is actually the case for prolate colloids of silver in glass.

When bleached with light 607 nm, a photochromic glass showed a large change in absorption around 600 nm with smaller changes at longer and shorter wavelengths. At 500 nm there was virtually no change, but a larger change in absorption was observed at still shorter wavelengths (see Fig. 5).

The behavior is consistent with the elongated colloid model. A colloid whose elongation ratio is appropriate absorbs the 607-nm light and is

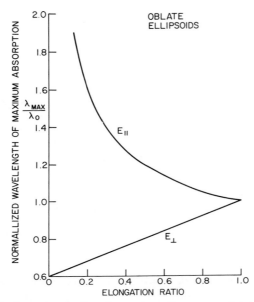

Fig. 4. Normalized wavelength of maximum absorptions versus elongation ratio for oblate ellipsoids.

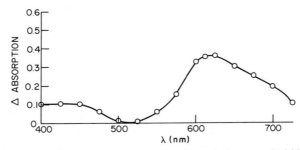

Fig. 5. Change in absorption as a function of wavelength due to optical bleaching. $\lambda_{BL} =$ 607 mμm; 0.05 J.

bleached, thereby decreasing the absorption at 607 nm. Moreover, since the same colloid also absorbed light at shorter wavelengths than 500 nm when the electric vector was perpendicular to the long axes, the absorption is decreased there also (see Fig. 6.)

Seward's observations are, in the opinion of this author, fairly compelling evidence not only for the induced absorption being due in large part to the existence of silver colloids but also for the colloid absorption being responsible for much, if not all, of the optical bleaching.

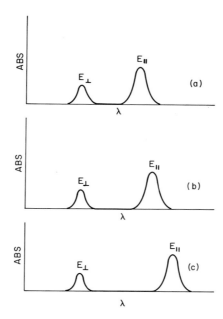

Fig. 6. Double peaked absorption of silver colloids having three different elongation ratios.

This interpretation of optical bleaching is widely accepted in silver halide physics. Moser and VanHeyningen (1966) cite the work of VanHeyningen and Rosencranz showing that there is no photoresponse in undarkened silver halide crystals but a large response after the crystal is darkened by exposure to blue light. They (Moser and VanHeyningen) interpret these results as clear evidence that electrons can be photoliberated from photolytically produced silver aggregates and that this is the mechanism by which exposure in the photoproduct absorption band results in bleaching of the band.

As indicated in the introductory remarks, optical bleaching efficiencies vary widely. This is not inconsistent with the idea that optical bleaching is due to absorption of the colloid. Gilleo (1953) showed that the work function of silver varies considerably with the nature of the insulator with which it is in contact. It is 4.3 eV for silver on NaCl and only 1.1 eV for silver on AgCl. Mott and Gurney (1948) explain the reduction of the work function of a metal by an insulator by assuming the vacuum levels match. Thus the difference in energy between the Fermi level of the metal and the bottom of the conduction band of the insulator is merely the difference between the work function of the metal in vacuum and the electron affinity of the insulator.

P. L. Young (unpublished results) suggested that the Fermi level of the insulator might be important (see Fig. 7). With the Fermi levels matched,

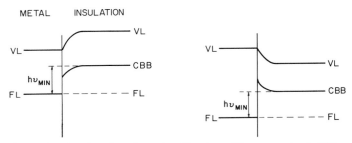

Fig. 7. Energy bands at interface of metal and insulator. VL, vacuum level; CBB, conduction band bottom; FL, fermi level; $h\nu_{min}$, lowest frequency light which can lead to optical bleaching.

since the vacuum level must be continuous across the interface, both the vacuum level and the conduction band bottom of the insulator bend. In such a case not only does the probability of photochromism depend on the wavelength maximum and the density of states as in Mott's proposal, but also it can depend on the tunnelling through a barrier in some cases and "resonant" energy levels in other cases.

Moser and VanHeyningen (1966) report that free holes are liberated by absorption in the cupric ion absorption band. Marquardt *et al.* (1971) also indicate that holes are released from the cupric ion by irradiation with blue or green light at low temperatures.

Hole liberation might also be a mechanism for optical bleaching but probably not an efficient one because of the competitive darkening processes at the short wavelengths where the cupric ion in silver halides absorb light.

As previously indicated, the hole trapping by the cuprous ion is essential to darkening. The cupric ion is an electron trap which can limit the amount of darkening because at sufficiently high concentrations it competes successfully with silver colloids for the photoelectrons. Furthermore, it is likely that cupric ions which are incorporated in the silver halide crystal during formation are indistinguishable from those formed by photolysis. As a result, cupric ions originally in the crystal ought to increase the fade rate somewhat, all other things being equal, because statistically they increase the probability of a hole being available to attack the silver colloid.

Dotsenko *et al.* (1974) studied the effect of the total copper concentration in the glass on the parameters in the rate expressions for thermal bleaching. They expressed the fade rate as the sum of two first-order terms. They found that for glasses containing silver chloride, increased copper increased both fade rate constants and increased the weighting factor of the faster fading term. For glass containing mixed halides, the rate constants and the weighting factors of the terms all went through extremes. They suggest that the different behavior of the two types of glass may be because the mixed halide crystal contains more cupric ions originally than does the simple silver chloride crystal. While the kinetic equation used by these workers may not be physically meaningful, their suggestion about solubilities is probably correct.

Fournier and Shaw (1974) also studied the influence of copper on darkening level and initial fade rates at various temperatures. They observed that at room temperature, increased copper decreased the equilibrium darkening level and increased the initial fade rate. In studying the temperature dependence they assumed that the fading allowed an nth order expression, where n was a small number. No attempt was made at verification of such an expression. For the sake of obtaining numerical results they assumed $n = 1.5$ and also $n = 2.0$. With each assumption they determined the temperature dependence of the "rate constant" by measuring the temperature dependence of the initial fade rate and "correcting" it for the value of the absorption at equilibrium. For example, if by assumption

$$-dA/dt = kA^{1.5} \qquad (1)$$

where A is absorption and k is a rate constant, then the "rate constant" is equal to the initial fade rate divided by the 1.5 power of the equilibrium value of A.

Using this technique, they found that both the pre-exponential and the activation energy of the "rate constant" increased with increasing copper concentration.

These authors were able to explain this effect only by assuming that during fading holes were stimulated into the valence band and by assuming further that as the total copper increased the copper ions interacted increasing the trap depths. They argued that the increased copper ion concentration naturally gives rise to a larger pre-exponential.

In the opinion of this author, thermal excitation of the holes into the valence band is not very probable. Marquardt et al. (1971) indicated that for optical transitions the trap depth of a hole on a cupric ion is 2.2 eV. Although the thermal trap depth may be smaller than this, it seems unlikely to this author that it could be shallow enough to make thermal activation significant at room temperature. Furthermore, it is unnecessary to make such assumptions, since all the effects of copper are more simply explained on the basis of the already known fact that the cupric ion is a competitive electron trap.

Consider the rate of change of some absorbing species concentration A formed by the photolysis of the photochromic glass,

$$\frac{dA}{dt} = k_d \frac{Cu^+}{1 + fCu^{2+}} - k_r ACu^{2+} \qquad (2)$$

where Cu^+ and Cu^{2+} are concentration of the ions, k_d is a darkening rate constant, k_r is a fading rate constant, and f is a number which indicates the extent of the competitive electron trapping by the cupric ion. Let η be the fraction of copper initially in the crystal as cupric ions. Then

$$Cu^{2+} = A + \eta T \tag{3}$$

and

$$Cu^+ = (1 - \eta)T - A \tag{4}$$

where T is the total copper in the crystal. At equilibrium

$$0 = \frac{k_d[(1 - \eta)T - A]}{1 + f(A + \eta T)} - k_r A(A + \eta T) \tag{5}$$

If f is not zero and η is a constant, then the darkening term in Eq. (5) will initially increase with increasing T but cannot do so indefinitely because of the T in the denominator. The fading term must, however, continue to increase with increasing T unless A decreases. Since A is determined by the competition between the darkening and fading times, it must show a maximum as a function of T. Numerical calculations show this maximum very clearly and show further that the initial fade rate predicted by Eq. (2) reaches an asymptotic value with ever increasing T.

The effect of total copper on the equilibrium absorption and initial fade rate was studied by Hares, Tick, and Seward (unpublished results) in a large number of different photochromic glasses and in every case these effects were observed.

The apparent interaction of the copper ions invoked by Fournier and Shaw may be merely an artifact of using an inadequate rate equation. Using Eq. (2) and assuming an Arrhenius temperature dependence for k_r and a constant k_d, Araujo (unpublished results) was able to calculate a temperature dependence of the equilibrium absorption similar to that found by Shaw (see Fig. 8).

Schreurs (unpublished results), using spin resonance techniques, observed that as the total copper in the glass increases, the ratio of the cupric ion to total copper increases (See Fig. 9). If one assumes that the same effect occurs to the ratio of cupric ion to total copper in the crystal, then Eq. (2) can be used to give a better description of Fournier and Shaw's results (see Fig. 10). Furthermore, if Eq. (2) is used to predict initial fade rates as well as equilibrium absorptions, then the results show the effect of temperature quoted by Shaw (see Table I). The temperature dependence of the "rate constant" appears to depend on the copper concentration even though no such dependence was introduced into the generation of the numbers.

Several authors attempted more detailed studies of the kinetics of bleaching of photochromic glasses. King's (1968) data indicated that optical bleaching and thermal bleaching were two separate mechanisms. Araujo (1968b, 1971) showed that the diffusion model was in qualitative agreement with King's

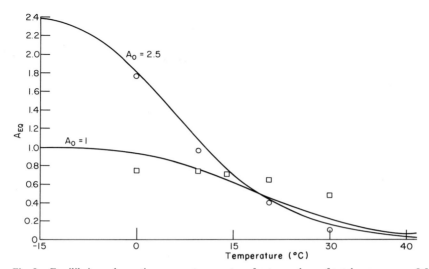

Fig. 8. Equilibrium absorption versus temperature for two values of total copper. $\eta = 0.5$. A_0 is the absorbance corresponding to complete conversion of Cu^+ to Cu^{2+}. It is the largest possible value of the absorbance. A_{eq} is the absorbance at equilibrium. $\bigcirc$ = high copper; $\square$ = low copper. (After Fournier and Shaw, 1974.)

data and, furthermore, that the diffusion model correctly predicted the behavior of the thermal fade rate with the darkening conditions. Dotsenko *et al.* (1973) asserted that their data do not show a dependence of the fade rate on darkening conditions. However, they only varied their exposure times from 7.5 to 30 sec. Hence, they failed to generate the conditions under which the effect becomes obvious. It is precisely this effect which makes performance characteristics of production glasses difficult to predict.

The correct predictions of the diffusion model prompted confidence on the part of its author in its essential correctness and inspired the decision to further develop the model to quantitatively describe practical photochromic glasses. The quantitative application to glasses containing silver halides was complicated by the fact that the model only deals with the number of electrons or holes, while the absorption induced in the silver halides is not proportional to the number of trapped electrons. In a colloid of silver the absorption per electron changes with the size and shape of the colloid.

Because, as previously indicated, the absorption induced in photochromic glasses containing copper cadmium halide did not seem to be strongly related to the formation of colloids, glasses of that type were chosen for study by Araujo and Borrelli (1976). They verified that in photochromic

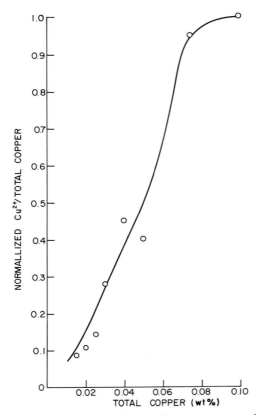

Fig. 9. Normalized ratio of cupric to total copper versus total copper.

glasses containing copper cadmium halides the fading depends on darkening conditions. The diffusion model described was shown to be consistent with the observed intensity dependence of equilibrium darkening, the fading kinetics, the darkening kinetics, and the dependence of the initial fade rate on the darkening time.

D. The Diffusion Model

In the present review, some aspects of the general diffusion model will be discussed and some unpublished results presented. Data already appearing in the literature will not be reported here.

The photolysis of the active phase of a photochromic glass results in the generation of electron–hole pairs which can either be trapped or recombine. Color centers result from the trapping of these charge carriers. Optical

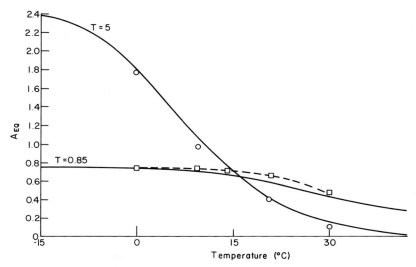

Fig. 10. Equilibrium absorption versus temperature for two values of total copper. $\eta = 1 - e^{-14T}$. (After Fournier and Shaw, 1974.)

TABLE I

TEMPERATURE DEPENDENCE OF VARIOUS ABSORPTION PARAMETERS

Total copper	Temperature	A	$\dfrac{dA}{dt}$ initial	$\dfrac{dA}{dt}\bigg/A^{1.5}$	$k^0 \exp(-\Delta E/T)$
5	273.2	4.678	0.018968	1.875×10^{-3}	$1.046 \times 10^8 \exp(-6.766.7 \times 10^3/T)$
	283.2	4.372	0.039577	0.00433	
	293.2	3.926	0.075101	0.00965	
	313.2	2.813	0.20824	0.04414	
10	273.2	8.873	0.035978	0.00136	$1.462 \times 10^8 \exp(-6.946 \times 10^3/T)$
	283.2	7.971	0.072151	0.00321	
	293.2	6.835	0.13075	0.00732	
	313.2	4.526	0.33505	0.03480	
20	273.2	16.272	0.065976	0.00101	$1.989 \times 10^8 \exp(-7.112 \times 10^3/T)$
	283.2	13.912	0.12593	0.00243	
	293.2	11.372	0.21754	0.00567	
	313.2	7.0369	0.52086	0.02790	

bleaching, if it is present, occurs by the optical stimulation of a trapped carrier to an allowed band followed by recombination of the free carrier with a trapped carrier of opposite sign. By contrast, thermal bleaching never involves free carriers. The trapped carriers diffuse until they are close enough for recombination to occur by tunnelling.

1. COORDINATES

Since the charge carriers are executing a random walk in a small particle, one expects their motion to be described by the diffusion equation. However, it must be emphasized that the fading rate of a glass is determined in this model by the distance between trapped electrons and trapped holes and not by the position of either in a particle. The choice of coordinate systems must reflect this fact. Araujo (1968b) described the technique of considering all the carriers of one sign to be contained in a small spherical region of radius r_0 with fading resulting from carriers of opposite sign crossing the boundary.

This is a reasonable model if only one colloid of silver is formed in each crystal and the number of holes trapped in any one crystal is large enough so that the spatial distribution of the holes around the colloid can be considered representative of the distribution in any other crystal. Chodak's (unpublished results) studies of optical bleaching make it clear that more than one colloid can grow in a single silver halide particle. Nevertheless, it is a reasonable approximation to analyze the carrier diffusion in photochromic glasses containing silver halides by solving the three-dimensional diffusion equation for the density of holes ρ at a distance r from the colloid and setting the number of trapped electrons or trapped holes equal to the integral of the density over the whole volume of a crystal; i.e.,

$$N = 4\pi \int \rho r^2 \, dr \qquad (6)$$

Araujo (1968b) was not strictly correct in trying to use this kind of mathematics to treat an ensemble of pairs where the distance between pairs is overwhelmingly larger than the separation of the two members of a pair. In such a case it can be shown that the three-dimensional diffusion equation is appropriate, but the total number of trapped carriers of one sign is equal to the integral over all distances; i.e.,

$$N = \int \rho \, dr. \qquad (7)$$

where ρ is the number of holes per unit length at the distance r from the colloid. This approach, however, leads to solutions which converge only very slowly. Since numerical calculations have shown that one-dimensional equations give results very similar to these, the one-dimensional equations were used to analyze the copper cadmium model.

2. THE GENERATING FUNCTION

The diffusion equation to be considered is

$$\frac{\partial \rho}{\partial t} = D \nabla^2 \rho + G(I, t, q) + R(t, q) \tag{8}$$

$R(t, q)$ represents the recombination of trapped holes and trapped electrons and is, of course, a function of the separation between them; i.e., it is a function of the coordinates, q. It is also a function of time, since it depends on the density of trapped carriers which is a function of time.

$G(I, t, q)$ represents the generation of trapped holes and electrons and its form will be discussed immediately.

The generation of trapped hole and electron pairs is the result of two processes, the generation of free electrons and holes by the absorption of uv and the trapping of those carriers. If an overwhelming fraction of the free carriers generated by the absorption of uv is trapped, the rate of generation of trapped carriers is proportional to the intensity, I, of the uv source. On the other hand, if most carriers simply recombine, the rate of generation of trapped carriers or color centers is proportional to the square root of the intensity of the light. So far it seems best results are obtained by using the linear intensity dependence.

In previous discussions, Araujo (1968b, 1971) assumed the trapping rate to be constant for the sake of mathematical simplicity. This simplification must be abandoned if quantitative results are desired. Since the number of hole trapping sites is limited, the generating function must decrease with time.

For simplicity of discussion, the number of hole trapping sites is considered to be limited and indeed it is likely that this is always the case, since the hole traps are impurities such as Cu^+. Thus, the generating function is assumed to be proportional to the number of cuprous ions. The most general expression for the generating function, within these constraints, is

$$G(I, q, t) = k_d(q) I \xi(q, t) \tag{9}$$

where $k_d(q)I$ is a rate constant which depends on the coordinates and is proportional to the intensity of exciting light I, and ξ is the density of cuprous ions which is a function of the coordinates and, of course, the time. If the diffusion coefficient of the cuprous ion is very large compared to the diffusion coefficient of the cupric ion, it can be shown that ξ will no longer depend on the coordinates and can be written as

$$\xi = k \left[A_0 - \int \rho \, dv \right] \tag{10}$$

where A_0 is the initial number of cuprous ions, and $\int \rho \, dv$ is the general expression for the number of cupric ions formed (also the number of trapped electrons).

For convenience, one can separate out explicitly the coordinate dependence of $k_d(q)$. Thus, the form of the generating function most extensively used is

$$G(I, q, t) = k_d I \left(A_0 - \int \rho \, dv \right) f(q) \tag{11}$$

where $f(q)$ is some function of the coordinates. It is the probability that an electron and hole are initially trapped at a separation q.

Araujo (1968b) suggested that in order that the fade rate depend on darkening conditions in the manner observed, the ρ function must exhibit a maximum with respect to coordinates during finite darkening times. This is possible only if the generating function has a maximum or is a monotonically decreasing function of the coordinates. The very short lifetimes of free carriers makes it clear that the generating function must decrease for large values of the coordinates. Furthermore, it seems likely that the generating function must be small for very small values of the coordinates because the probability that a cuprous ion exists in a given volume shrinks as the volume does. Therefore, it is reasonable to assume that the generating function has a maximum. For mathematical simplicity the specific form assumed is

$$f(q) = \delta(q - q_1) \tag{12}$$

where δ is the delta function. q_1 is the separation of electrons and holes at the instant of initial trapping. A study of subtle effects during the early stages of darkening requires a more realistic $f(q)$, and one must resort to numerical solutions.

3. THE RECOMBINATION FUNCTION

Since the basic assumption is that neither electrons nor holes are stimulated into the allowed bands in thermal fading, recombination, by definition, must occur by tunneling. For the silver halide model, the probability for recombination at a distance of q must be proportional to the number of trapped electrons and to the density of trapped electrons at q.

$$R(t, q) = k_r(q)\rho \int \rho \, dv \tag{13}$$

D. A. Nolan (unpublished results) has shown that for the colloidal silver model in silver halides, $k_r(q)$ is approximately exponential. With such a form for $k_r(q)$, the diffusion equation is nonlinear and insoluble except by numerical techniques. Therefore, Nolan's form is abandoned and instead

$$k_r(q) = \delta(q - q_0) \tag{14}$$

is considered, where q_0 is the separation of electrons and holes when they recombine.

For the copper cadmium model (isolated pairs), the probability for recombination at a distance of q must be proportional to the number of pairs having a separation of q.

$$R(t, q) = k_r(q)\rho \tag{15}$$

For both the copper cadmium halide system and the silver halide, Eq. (14) is a satisfactory representation for some purposes.

To evaluate the fading rate one simply integrates Eq. (8) after having set $G(I, t, q)$ equal to zero; i.e., the sample is no longer exposed to activating irradiation. This gives simply (in one dimension)

$$-d(A)/dt = D(\partial\rho/\partial x)_{x=0} + \left[\int k_r(x)\rho(x)\,dx\right] \tag{16}$$

where A is proportional to the absorbance defined by $-\ln(I/I_0)$.

If one uses Eq. (14) for k_r, Eq. (16) becomes simply

$$-d(A)/dt = D(\partial\rho/\partial x)_{x=0} \tag{17}$$

if one imposes the boundary condition $\rho = 0$ at $x = 0$.

Clearly this is the same result obtained by ignoring the recombination term in Eq. (8) and indeed is the justification for doing so.

For subtle effects in initial stages of fading, a more realistic form of k_r must be used, and the equations solved numerically. For example, Nolan (unpublished results), using an exponential function, predicted that glasses with a small diffusion coefficient can be expected to get darker at equilibrium as the temperature increases over certain ranges. The explanation for this apparent anomaly is that the equilibrium absorption is determined by the darkening rate and the instantaneous fade rate. As the temperature is increased, the density of pairs within the range in which tunnelling is significant decreases, slowing the instantaneous fade rate and causing the glass to get darker. Of course, the average fade rate over a finite fading time increases with temperature due to the higher diffusion coefficient. Borrelli (unpublished results) verified this result experimentally (see Fig. 11).

To apply the diffusion model quantitatively to the silver halides, a property proportional to the total number of trapped electrons or holes must be measured. The spin signal of the cupric ion seemed to this author to be well suited to this purpose. Therefore, a glass was made under slightly reducing

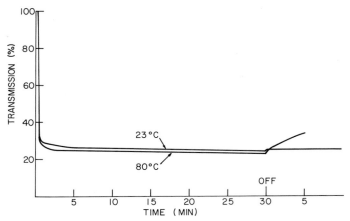

Fig. 11. Darkening and fading curves at low temperatures.

conditions so that there was no cupric ion spin signal in the glass before irradiation, and this glass was used for a kinetic study. Schreurs (unpublished results) observed a spin resonance signal when the glass darkened and measured the signal peak to peak height as a function of time during the fading. Figure 12a indicates that Schreurs' data is well described by the three-dimensional diffusion equation, and Fig. 12b indicates a rather poor fit to the one-dimensional equation.

These results tend to substantiate the correctness of considering all the electrons to be in one spherical region in the silver halides but considering the color centers in the copper cadmium halides to be essentially isolated pairs.

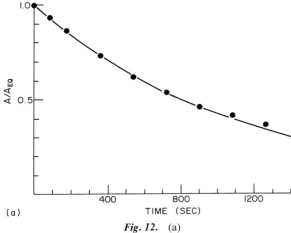

(a)

Fig. 12. (a)

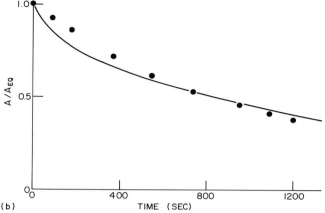

Fig. 12. Comparison of EPR data with calculated by (a) three-dimensional model, (b) one-dimensional model. Ordinate is the ratio of absorbance to equilibrium absorbance.

We can now see the beginnings of understanding of photochromic glasses. It is to be hoped that as this understanding increases, glasses can be produced having properties suitable to a wider range of applications.

References

Araujo, R. J. (1967). U.S. Patent 3,325,299.

Araujo, R. J. (1968a). Recent Advances in Display Media, NASA SP 159.

Araujo, R. J. (1968b). *Appl. Opt.* **7**, 781.

Araujo, R. J. (1971). *In* "Photochromism" (G. H. Brown, ed.), Chapter VIII, p. 680. Wiley, New York.

Araujo, R. J. (1973). *Feinwerktech. Micron.* **77**, 52.

Araujo, R. J., and Borrelli, N. F. (1976). *J. Appl. Phys.* **47**, 1370.

Araujo, R. J., Eppler, R. A., and Krause, E. F. (1967). U.S. Patent 3,328,182.

Araujo, R. J., Cramer, W. H., and Stookey, S. D. (1970). U.S. Patent 3,540,793.

Aranjo, R. J., Sawchuk, L. G., and Seward, T. P. (1972). U.S. Patent 3,703,388.

Araujo, R. J., Beall, G. H., and Sawchuk, L. G. (1975). U.S. Patent 3,923,529.

Armistead, W. H., and Stookey, S. D. (1964). *Science* **144**, 150.

Armistead, W. H., and Stookey, S. D. (1965). U. S. Patent 3,208,860.

Brown, G. H. (1971). "Photochromism." Wiley, New York.

Brown, F. C., and Wainfan, N. (1957). *Phys. Rev.* **105**, 93.

Cohen, A. J., and Smith, H. L. (1962). *Science* **137**, 981.

Dotsenko, A. V., Zakharov, V. K., and Tsekhomskiy, V. A. (1973). *Opt. Technol.* **40**, 687.

Dotsenko, A. V., Papunashvili, N. A., and Tsekhomskiy, V. A. (1974). *Sov. J. Opt. Technol.* **41**, 395.

Doyle, W. T. (1960). *Proc. Phys. Soc. (London)* **75**, 649.

Fanderlik, I. (1968). *Int. Congr. Glass, 7th, London.*

Fanderlich, I., and Prod'homme, L. (1973). *Verres Re'fract.* **27**, 97.

Fournier, J. T., and Shaw, R. R. (1974). *Meeting Am. Ceram. Soc., 76th, Chicago*. Abstract Only, *Bull. Am. Ceram. Soc.* **53**, 355.

Gilleo, M. A. (1953). *Phys. Rev.* **91**, 534.

Gliemeroth, G., and Mador, K. H. (1970). *Angew. Chem. Int. Ed.* **9**, 434.

Gliemeroth, G., and Bach, H. (1971). *Glastech. Ber.* **44**, 305.

Hammel, J. J., and McGary, T. (1969). *In* "Reactivity of Solids" (J. W. Mitchell, ed.), p. 695. Wiley, New York.

Hodgeson, W. G., Brinen, J. S., and Williams, E. F. (1967). *J. Chem. Phys.* **47**, 3719.

Kan'no, K., Naoe, S., Mukai, S., Nakai, Y., and Miyanaga, T. (1973). *Solid State Commun.* **13**, 1325.

King, C. B. (1968). Cited in Araujo (1968b).

Lee, I. (1936). *Am. Min.* **21**, 764.

Maki, T., and Tashiro, M. (1972). *Inst. Chem. Res., Kyoto Univ.* **80**, Chem. Abstr.

Marquardt, C. L., Williams, R. T., and Kabler, M. N. (1971). *Solid State Commun.* **9**, 2285.

Medved, I. B. (1954). *Am. Min.* **39**, 615.

Megla, G. K. (1966). *Appl. Opt.* **5**, 945.

Megla, G. K. (1970). *IEEE Int. Conv. Digest*, 166.

Meiling, G. S. (1971). U.S. Patent 3,615,771.

Mitchell, J. W. (1957). *Sci. Ind. Photogr.* **28**, 368.

Moriya, Y. (1973). *Yogyo-Kyokai-Shi* **81**, 259.

Moriya, Y. (1974). *Int. Congr. Glass, Kyoto*, **10**.

Moriya, Y., Tokunaga, S., and Kawai, T. (1972). *Yokio-Kyoksi-Shi* **80**, 35. Only abstract in English.

Moser, F., and Burnham, D. C. (1964). *Phys. Rev.* **136**, A744.

Moser, F., and VanHeyningen, R. S. (1966). *In* "Theory of Photographic Process" (C. E. K. Mees and T. J. James, eds.), 3rd ed., Chapter 1, p. 26. Macmillan, New York.

Moser, F., Nail, N. R., and Urbach, F. (1959). *Phys. Chem. Solids* **9**, 217.

Mott, N. F., and Gurney, R. W. (1948). "Electronic Processes in Ionic Crystals," 2nd ed. Dover, New York.

Muller, A., and Milbey, M. E. (1967). *Am. Phys. Soc., March Meeting, Chicago, Illinois*.

Pavlovskiy, V. K., Tunimanova, I. V., and Tsekhomskiy, V. A. (1971). *Opt. Technol.* **38**, 37.

Phillips, W. (1970). *J. Electrochem. Soc.: Solid State Sci.* **117**, 1557.

Radler, R. (1964). Synthesis of Inorganic Materials for High Density Computer Memory Applications. Tech. Doc. Rep. AL TDR 64–170.

Sakka, S., and MacKenzie, J. D. (1972). *J. Am. Ceram. Soc.* **35**, 553.

Sakka, S., and Mackenzie, J. D. (1973). *Bull. Chem. Soc. Jpn.* **46**, 848.

Sakka, S., Matsusita, K., and Kamiya, K. (1974). *Int. Congr. Glass, 10th, Kyoto*.

Sawchuk, L. G., and Stookey, S. D. (1966). U.S. Patent 3,293,052.

Seitz, F. (1951). *Rev. Mod. Phys.* **23**, 328.

Seward, T. P. (1974). Symposium on Photochromic Glasses, *Am. Ceram. Soc. Meeting, 76th, Chicago*, Illinois.

Seward, T. P. (1975). *J. Appl. Phys.* **46**, 689.

Smith, G. P. (1967). *J. Mater. Sci.* **2**, 139.

Stookey, S. D., and Araujo, R. J. (1968). *Appl. Opt.* **7**, 777.

Stroud, J. S. (1966). U.S. Patent 3,255,026.

Suzuki, J., and Kume, M. (1971). U.S. Patent 3,617,316.

Swarts, E. L., and Cook, L. M. (1965). *Int. Congr. Glass, 7th, Brussels*.

Swarts, E. L., and Pressau, J. P. (1965). *J. Am. Ceram. Soc.* **48**, 333.

Taylor, M. J., Marshall, D. J., Forrester, P. A., and McLaughlen, S. D. (1970). *Radio Electron. Eng.* **40**, 17.

Williams, E. F., Hodgeson, W. G., and Brinen, J. S. (1969). *J. Am. Ceram. Soc.* **52**, 139.

Anomalous Birefringence in Oxide Glasses

TAKESHI TAKAMORI

IBM Thomas J. Watson Research Center
Yorktown Heights, New York

and

MINORU TOMOZAWA

Materials Engineering Department
Rensselaer Polytechnic Institute
Troy, New York

I. Introduction

Under certain conditions any glasses can become anisotropic. Stress is the most common cause of the optical anisotropy in glasses, and can be observed as birefringence. The stress birefringence in a solid glass is proportional to the applied stress. This phenomenon, known as the stress optical effect, has been of technological importance in evaluating the "permanent strains" (Kohl, 1967) caused by disannealing, tempering, sealing, ion-exchange, etc. The birefringence induced by these thermal or chemical processes is accompanied by a compensating stress birefringence of the opposite sign. Therefore, any disturbance of the stress balance in a single piece of glass (e.g., machining or etching) will change its birefringence pattern.

A different type of optical anisotropy can sometimes be induced in glass by thermomechanical processing. When a glass is cooled under uniaxial load from its viscoelastic temperature range, for example, the glass often remains optically anisotropic even after the load is removed. If the cooling is slow enough to avoid thermal strain, the birefringence observed will be uniform throughout a specimen without any compensating birefringence of the opposite sign. This birefringence is sensitive to the atomic structure or microstructure of the glass, and its pattern of distribution is not disturbed by subsequent machining. This type of birefringence is often referred to as "anomalous birefringence" as opposed to the birefringence caused by the stress-optical effect. This anomalous birefringence is a potential tool for the structural study of glasses. In this chapter, the current status of our understanding of this anomalous birefringence will be reviewed. Although the phenomenon itself has been known for some time, only recently has its origin been clarified; consequently, no previous review article has been written on this subject.

II. Definition and Experimental Method

Several terms used in this chapter are now defined, mostly phenomenologically. For further details, see the comprehensive treatment of birefringence by Ernsberger (1970).

If light propagates through a material with different velocities depending upon the plane of polarization, the material is said to be "birefringent." The schematic illustration of Fig. 1 was reproduced from the book by Ernsberger (1970) for a tetragonal crystal which is uniaxially birefringent. This figure can also be interpreted as a piece of glass stressed in the vertical direction. The velocity of light propagating parallel to the c axis of the crystal is identical in all planes of polarization (Fig. 1a). The c axis of the material in this case coincides with its "optic axis." The velocity of light propagating perpendicular to the c axis, on the other hand, varies depending upon the plane of polarization. Light polarized to vibrate in the plane perpendicular to the c axis (Fig. 1b) will have the same velocity as the light propagating parallel to the c axis (Fig. 1a), and is called "ordinary" ray. Light polarized to vibrate in the plane parallel to the c axis (Fig. 1c) will have a different velocity from the ordinary ray and is called "extraordinary"

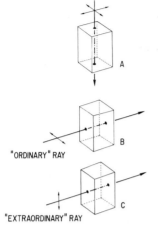

Fig. 1. Interaction of light with a tetragonal crystal. (After Ernsberger, 1970.)

"ORDINARY" RAY

"EXTRAORDINARY" RAY

ray. The refractive index of a material is a nondimensional number and is equal to the ratio of the velocity of light in vacuum to that in the material. Thus, to light propagating in the direction perpendicular to the c axis, the optically uniaxial material in Fig. 1 exhibits two different refractive indices depending upon the planes of polarization.

Birefringence B in a material is defined by

$$B = n_e - n_o \tag{1}$$

where n_e and n_o are, respectively, refractive indices for the extraordinary and ordinary rays. Depending upon the sign of B in Eq. (1), the birefringence is referred to positive or negative. The relative retardation R by this birefringence is

$$R = D(n_e - n_o) \tag{2}$$

where D is the thickness of the material along the light path. Usually the retardation per unit thickness, R/D (nm/cm), is used as a measure of the birefringence.

Experimental techniques to measure the birefringence, especially optical instruments involved, have been adequately covered by numerous textbooks on stress optical effect (e.g., Partridge, 1949; Espe, 1968; Longhurst, 1973). An ASTM designation (C148-59T) was also established designating details of the operation. In this section, therefore, only the specimen preparation particular to the anomalous birefringence observation will be described.

In almost all of the reported works on the anomalous birefringence, the sample glasses have been prepared by stretching rather than compressing. Stretching will give a uniaxial anisotropy more easily than compressing.

In many earlier works, a glass rod was simply stretched manually over the gas torch. The rod was first heated in a gas flame, then while cooling freely in the atmosphere, it was pulled with increasing force until it could no longer be stretched. For a little more quantitative work, a vertical or horizontal electric furnace has been used together with a mechanical load. Figure 2 is an example of the pulling furnace used by Thomas (1964). An end of the glass rod is connected to a dynamometer which is fixed on the wall, and the other end to a cable which is pulled hydraulically. Each end of the rod is embedded in a low melting alloy to avoid stress concentration at the joints. At the middle of the furnace, a polarimeter with Babinet compensator is set up to measure the birefringence of the glass rod. The arrangement in Fig. 2 is essentially the same as those of other workers, and is adequate for the measurement of the birefringence as a function of temperature.

Alternatively, the specimen which was subjected to the thermomechanical treatment is often brought to the room temperature with or without the mechanical load and the birefringence is measured at the room temperature. In this procedure it is important to eliminate the thermal strains by cooling the specimen slowly. The variation of the birefringence as well as the specimen dimension with the subsequent heat treatment without external stress can give valuable information regarding the structural origin of the birefringence.

Phenomenologically, two types of anomalous birefringence can be distinguished. One type, which is often attributed to "frozen-in strain," is observed

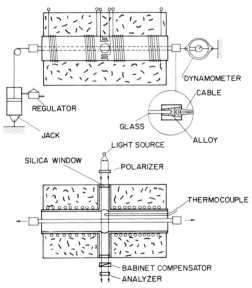

Fig. 2. Furnace for stretching. (After Thomas, 1964.)

in both homogeneous and heterogeneous glasses and is annealable by an ordinary procedure. The other type is related to the anisotropic micro-structure of glasses and persists even after annealing. The latter will be called "microstructural birefringence" to distinguish it from the former. These two types of anomalous birefringence will be discussed separately.

III. Single-Phase Glass and Birefringence by Frozen-in Strain

Filon and Harris (1923) noted unusual birefringence in a flint glass specimen which was subjected to a longitudinal pressure of 130.6 bars at 400°C for 42 hr and cooled slowly under this pressure. This was one of the earliest observations of "anisotropy" in glasses. The "frozen-in elastic deformation" observed here was explained by assuming the "diphasic" nature of glass; the structure which consists of elastic and fluid units in glass. Stirling (1955) has conducted more systematic work on frozen-in elastic deformation of Pyrex-type borosilicate and soda–lime silicate fibers which were cooled under tensile load. Optical retardation of positive sign was noticed, and on heating the fibers anisotropic contraction was observed. The result was explained by supposing the glasses to behave as a collection of viscoelastic elements with a distribution of relaxation time. Merker (1960, 1969) studied similar optical anisotropy of fine glass fibers and sheet glasses, and correlated the magnitude of the anisotropy to polarizability and mobility of the cations in the network. Magnetic anisotropy of glass fibers was studied by Toor (Selwood, 1956), who measured the magnetic susceptibility of two types of glass fibers, one a diamagnetic sodium magnesium silicate ($Na_2O \cdot MgO \cdot 4SiO_2$) and the other a paramagnetic glass of the composition $Na_2O \cdot NiO \cdot 4SiO_2$. The former remained isotropic within the limit of experimental error, but the latter became anisotropic when cooled under tensile load. Most of the glasses described above (except Pyrex) are not phase separated. A similar optical anisotropy and also an anisotropic contraction on heating can be observed even in fused silica after cooling under load (Takamori, 1964). The "diphasic" structure in fused silica is no more than a mathematical model. When the specimens are heat treated at the annealing temperature, the birefringence of this type disappears completely and the length of the specimen decreases.

IV. Glass-in-Glass Phase Separation and Birefringence by Frozen-in Strain

The frozen-in strains of single phase can also exist in phase-separated glasses cooled under load. In addition, there will be frozen-in strain due to the two-phase structure. Suppose that the higher viscosity phase of the phase-separated glass is continuous. Then at the temperature at which the glass rod is stretched, the pulling load must be borne essentially by the higher viscosity phase, since the lower viscosity phase is too fluid at this temperature to hold the load (Takamori, 1976). The glass is then cooled under load to the temperature at which the lower viscosity phase is solidified. The contraction of the elastically deformed higher viscosity phase upon removal of the load causes an axially compressive stress in the lower viscosity phase and tensile stress in the higher viscosity phase. The total stress-optical effect of the microphases yields either negative or positive birefringence depending on the volume fraction and the relative magnitudes of the stress-optical coefficients of each phase, hence on the composition of each phase. A definitive observation of the birefringence of this type has not been reported. We will call this type II frozen-in strain. This birefringence, if ever observed, should also be annealed out by an ordinary annealing procedure, just as with other birefringences caused by strains.

Figures 3 and 4 illustrate the results on the frozen-in strain in commercial Pyrex glass. In this experiment, the as-received glass rods were (1) heat-treated at 633°C for varied lengths of time (hereafter called prepulling heat-treatment) to induce phase-separation, (2) pulled at the same temperature with a constant load (maximum stress ≈ 6 kg/mm^2) to attain twice the original length, and (3) cooled quickly under the same load to near the nominal annealing point by air-blow and then slowly to below 300°C in the furnace to avoid thermal strain. The sample rod prepared in this way looks like a uniaxial crystal with optical anisotropy. Using a polarimeter, a positive birefringence was observed, uniform throughout the sample, with the light traveling perpendicular to the pulling axis, but no birefringence with the light parallel to the axis. Curve A in Fig. 3 shows this positive birefringence. The sample was then annealed at the annealing point (560°C) for 1 hr followed by furnace cooling. With polarized light, the sample rod still appeared like a uniaxial crystal but now with a negative birefringence. Curve B in Fig. 3 shows this negative birefringence. The positive birefringence observed before the annealing has been caused by the strains frozen-in by the thermomechanical processing, and it was released by annealing as indicated the arrows in the figure. The negative birefringence observed after the annealing procedure is due to the microstructure, as will be discussed in the next section.

In addition to the fading of the positive birefringence the release of the frozen-in strain in the sample glass cooled under tensile load is accompanied by a significant change in thermal expansion curves, as shown in Fig. 4. In this figure the thermal expansion curve of the Pyrex glass stretched and cooled under load (curve C) is compared with those of the as-received

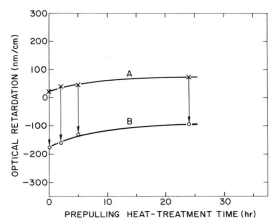

Fig. 3. Birefringence in Pyrex glass cooled under tensile load. (a) before and (b) after annealing.

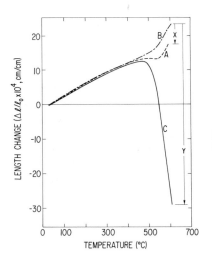

Fig. 4. Thermal expansion curves of Pyrex glass. (a) as-received, (b) annealed, and (c) stretched and cooled under load.

(curve A) and the annealed samples (curve B). The curves were obtained at a heating rate of 3°C/min. Compared with the 0.05% contraction of the as-received sample indicated by the arrow X in the figure, the overall contraction of the stretched sample was about 0.6% (arrow Y). After this contraction, the sample revealed significant negative birefringence, and the repeat measurement gave the expansion curve almost identical to the curve B in Fig. 4.

The amount of the frozen-in strain in the stretched glass, measured by both thermal expansion (Takamori, 1963) and constant temperature contraction (Stirling, 1955) always appeared greater in borosilicates (mostly phase separable) than in other technical silicate glasses (such as soda–lime silicates). At present, however, it is not clear what fraction of the measured frozen-in strain in the borosilicate glasses is due to the two-phase structure (type II) that is superimposed on the frozen-in strain of single phase. As discussed before, both types of frozen-in strain are expected to be released by an ordinary annealing. More elaborate experiments are needed to clarify this point.

It is possible that there is birefringence in phase-separated glasses caused by strains but it is not annealable. This birefringence could be caused by the thermal contraction difference of anisotropically deformed different phases, and would not be eliminated by annealing. Its occurrence should be always possible especially if the higher viscosity phase is continuous and does not flow at the annealing temperature as is the case in most technical borosilicate glasses, but such birefringence apparently has not been reported. Even after the birefringence has disappeared and the glass becomes isotropic, the local stress from differential contraction should remain in the glass. The birefringence from this local stress cannot be observed by the ordinary polarimeter. Such internal stress in a phase-separated glass has been observed by Kerwawycz and Tomozawa (1974) by light scattering. It may not be possible to observe this stress birefringence due to differential contraction separately from the birefringence due to microstructure, which will be described in the next section. However, our experiments indicate that the fraction of the observed birefringence which is attributable to this differential contraction effect is much smaller than that due to microstructure. One indication along this line will be described toward the end of this chapter in connection with the microstructural birefringence in microporous glasses (cf. Section VIII).

V. Glass-in-Glass Phase Separation and Microstructural Birefringence

A. *Review*

Several papers report "structural" birefringence in borosilicate glasses, and some of the compositions studied are listed in Table I. Many of the results reported in the literature are confusing at two major points. First is the confusion of the microstructural birefringence with the superimposed birefringence caused by frozen-in strains, as discussed in previous sections. The second problem comes from the conflict in predicted and measured sign of the birefringence. In this subsection, literature on these two problems

TABLE I

COMPOSITION OF BOROSILICATE GLASSES APPEARING IN THE LITERATURE

| Glass compositions (wt%) | | | | | | | | |
SiO_2	B_2O_3	Li_2O	Na_2O	K_2O	Al_2O_3	PbO	As_2O_3	References
66	25	—	7	—	2	—	—	Long (1954)
73	16.5	—	4.5		—	6	—	Indenbom (1953)[a]
70	23	—	7	—	—	—	—	Botvinkin and Ananich (1963)
66.6	27.5	—	5.8	—	—	—	0.1	Takamori (1963)
66.85	22.25	—	7.60	—	2.30	0.20	—	Thomas (1964)[b]
57.10	33.08	—	9.82	—	—	—	—	Schill (1967)
80.5	12.9	—	3.8	0.4	2.2	—	—	Takamori (1974)[c]
70.0	28.0	1.2	—	0.5	1.1	—	—	Takamori and Tomozawa (1976)[d]

[a] Nonex. Composition taken from Espe (1968).
[b] MO, Sovirel.
[c] Pyrex. Composition taken from Shand (1958).
[d] Code 7070, Corning. Composition taken from Shand (1958).

is reviewed. The sign of birefringence will be discussed further in the next subsection.

Since Botvinkin and Ananich's work (1959, 1963) has been referred to often, it will be discussed in some detail. Sample rods of composition (by weight) 70% SiO_2, 23% B_2O_3, and 7% Na_2O were stretched over a gas torch and cooled under load, and birefringence was measured as a function of sample temperature. Because of this method of sample preparation, the magnitude and even the sign of birefringence varied from sample to sample. In all cases, however, at the first heating of the sample glass (which had not been annealed after stretching) up to the viscoelastic temperature range, the birefringence changed in such a direction that the magnitude of the negative birefringence became greater. Only the negative birefringence reached after the first heating and cooling was found to be reversible on subsequent thermal cycling between the room temperature and the annealing temperature range ($\approx 600°C$). The negative birefringence observed was probably structural. The positive birefringence superimposed on the structural one in the initial sample can probably be attributed to the effect of the frozen-in strain, just as curve A in Fig. 3. Botvinkin and Ananich (1963), however, attributed it to the chain orientation in sodium borate phase on the basis of additional experiments on B_2O_3 and two sodium borate glasses, which had about the same composition of the lower viscosity phase of the phase-separated borosilicate glass. Only positive birefringence was observed in these glasses after they were cooled under load. Upon heating, the positive birefringence simply decreased to zero, and no negative birefringence appeared. The authors considered that the observation of the positive birefringence in these simple glasses and the x-ray study for the oriented B_2O_3 glass fibers in the literature (Goldstein and Davies, 1955) confirmed their suggestion of chain orientation. However, the positive birefringence observed for the B_2O_3 glass disappeared below 280°C (Botvinkin and Ananich, 1959) and for sodium borosilicates (composition not reported) below 500°C (Botvinkin and Ananich, 1963), all corresponding to the reported annealing temperatures of these glasses (Takamori, 1976). Therefore, the positive birefringence observed in B_2O_3 and the borate glasses is probably also the one caused by frozen-in strain rather than microstructural in origin. Furthermore, the x-ray proof for the oriented B_2O_3 glass fibers reported in the literature itself is questionable (Takamori, 1965).

The release of the frozen-in strain causes a significant contraction of the sample glass, as seen in Fig. 4. Such stress-release observed on thermal expansion curves has been presented as evidence for structural anisotropy in some papers (Indenbom, 1953; Ananich and Botvinkin, 1966), and must be similarly questioned.

Stirling (1955), on the other hand, in the study of the frozen-in strains attributed the positive birefringence observed in a Pyrex-type borosilicate glass to the frozen-in effect only. However, the data reported by Stirling, which show the birefringence as a function of effective stress (tensile), can be extrapolated to negative birefringence at zero effective stress (Takamori, 1974). This result suggests the existence of the superimposed microstructural birefringence as seen in Fig. 3, since the birefringence by frozen-in strain depends on the applied stress while the microstrutural birefringence does not.

After sample glasses, cooled under tensile load, were annealed either intentionally (Thomas, 1964) or during the first heating of thermal cycling (Botvinkin and Ananich, 1963) almost all of the microstructural birefringence reported in sodium borosilicate glasses were negative. In earlier work this negative birefringence has been explained by strain. Indenbom (1953) and Long (1954) explained their observations on the basis of type II frozen-in strain discussed in the previous section. This thermomechanical strain is probably annealed out, however. Long *et al.* (1956), interpreted their observations on a microstructural basis.

Botvinkin and Ananich (1963) proposed form birefringence as the cause of the negative birefringence remaining in stretched samples after annealing. Form birefringence is observed in a heterogeneous material consisting of a matrix of one material and nonspherical particles of another material of different refractive index with some degree of preferential alignment, and is well known in colloid chemistry (Chamot and Mason, 1958). As experimental support, Botvinkin and Ananich (1963) measured birefringence produced by immersing elongated samples of porous Vycor type in a series of liquids having various indices of refraction. By impregnating the elongated porous glass with benzene, whose index of refraction is considered close to that of the sodium borate phase, they observed negative birefringence almost corresponding to that of the initial unleached glass. This experiment seemed to be the most direct proof for form birefringence as the cause of the negative birefringence and other workers, including one of the authors, (Thomas, 1964; Schill, 1967; Takamori, 1974) have considered negative birefringence to be form birefringence until recently.

However, the basic theory of form birefringence worked out by Ambronn and Wiener (Coker and Filon, 1957) predicts positive birefringence, instead of negative, when a medium contains a number of suspended prolate spheroids such as in a stretched phase-separated glass. Thus from the theoretical point of view negative birefringence observed in stretched glass cannot be form birefringence (Ernsberger, 1975). This conflict was resolved recently and is discussed in the next subsection.

B. Form Birefringence and Distribution Birefringence

1. Experimental Observations

The conflict concerning the sign of structural birefringence was resolved by the correlation of the observed birefringence with microstructure (Takamori and Tomozawa, 1976). A commercial lithium borosilicate glass, Corning Code 7070 (see Table I) has a dispersed microstructure, rather than connective, when heat treated at around 600°C, as shown in Fig. 5 (Tomozawa and Takamori, 1977a). Such a dispersed microstructure simplifies the interpretation of the experimental results.

As-received glass rods were (1) heat-treated at 595°C for varied times to induce phase separation (prepulling heat treatment). Then, (2) each sample was uniaxially stretched at the same temperature with a constant load (maximum stress ≈ 6 kg/mm²). At twofold elongation the load was removed and the glass rod was rapidly quenched in air. Subsequently, (3) each rod was annealed to remove thermal strain by heating at the annealing point for 1 hour and cooling in the furnace at 2°C/min. In spite of the drastic increase in the measured viscosity of the glasses with progressive phase separation (Simmons et al., 1974), the lower viscosity phase plays a major role in thermally straining the phase separated borosilicates (Takamori, 1976), and the nominal annealing point is adequate to release any annealable strain related to the lower viscosity phase. After annealing, the micro-

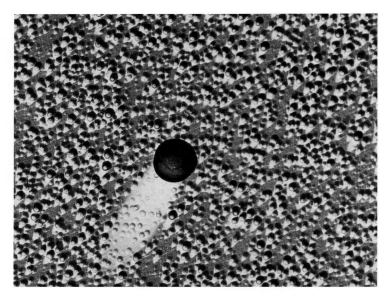

Fig. 5. Two-stage replica electron micrograph of fractured and etched surface of a phase-separated commercial borosilicate glass (Corning code 7070). Glass was heat-treated at 600°C for 100 hr. Latex sphere shows 0.5 μm in diameter.

structural birefringence in the sample glass was measured with a polarimeter at room temperature. Then, (4) the samples were subjected to another heat treatment at 564°C for various lengths of time, hereafter called postpulling heat treatment, and furnace cooled (2°C/min). Any change of the birefringence in the samples by this heat treatment was measured again at room temperature.

Figure 6 shows the thermal relaxation behavior of the microstructural birefringence in the glass as a function of the postpulling heat treatment time at 564°C. The time indicated for each curve shows the prepulling heat treatment time given to the specimen to induce phase separation prior to stretching. The positive birefringence observed in Fig. 6 was not caused by frozen-in strains, as discussed in Fig. 3, because the load was removed before the sample rod was taken out of the furnace, and the sample was annealed at the annealing point before the measurement. It can be seen from the figure that positive birefringence, which is observed right after stretching, relaxed quickly through zero to a negative value, then asymptotically to zero. The magnitude of the maximum positive birefringence is greater for the specimen subjected to a longer prepulling heat treatment, while the magnitude of the negative birefringence is approximately the same (≈ -200 nm/cm) for all specimens.

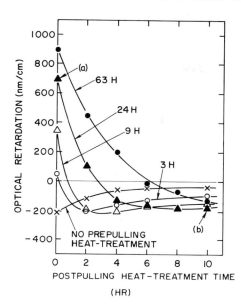

FIG. 6. Birefringence observed as a function of the postpulling heat-treatment time at 564°C. Times indicated on curves are the prepulling heat-treatment times at 595°C. (After Takamori and Tomozawa, 1976.)

Figures 7a and 7b represent microstructures of the glass specimens with positive and negative birefringence, respectively. These microstructures, corresponding to points (a) and (b) in Fig. 6, were obtained from a fractured and etched surface taken parallel to the tensile axis. Figure 7a, which corresponds to positive birefringence, reveals prolate spheroidal shapes, while Fig. 7b, which corresponds to the negative birefringence, shows almost spherical particles. A comparison of Figs. 7a and 7b shows that the size of the minor phase particle appears to be similar in both figures, indicating that the shape change is the major process taking place during the postpulling heat treatment. This observation was further confirmed by an experiment of extended conditions. A glass rod was subjected to the prepulling heat treatment of 100 hr at 595°C and stretched to elongate 2.5 times its original length. The positive birefringence was measured as 1100 nm/cm. As shown in Fig. 8a, the prolate spheroidal shapes of the minor phase were more clearly seen in this case. To attain negative birefringence in this sample by the postpulling heat treatment, it was necessary to heat-treat it at 600°C for 16 hr. Figure 8b shows the microstructure of this sample with a negative birefringence.

Apparently there are two different mechanisms contributing to the microstructural anisotropy; one is contributing to a positive birefringence term, and the other is contributing to a negative birefringence term. The positive birefringence depends on the time of the prepulling heat treatment and

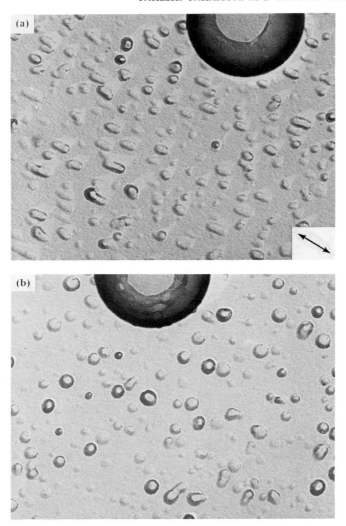

Fig. 7. Two-stage replica electron micrographs of the fractured surface taken parallel to the tensile axis. (a) before and (b) after a postpulling heat treatment at 564°C for 10 hr. These micrographs (a) and (b) correspond to (a) and (b) of Fig. 6, respectively. The arrow in (a) indicates the tensile axis. Latex sphere shows 0.5 μm in diameter. (After Takamori and Tomozawa, 1976.)

relaxes faster than the negative birefringence. The current series of experiments shows clearly that it is the positive birefringence which is caused by the form birefringence. This is consistent with Ambronn-Wiener's theory (Coker and Filon, 1957). A question still remains: What is the cause of the

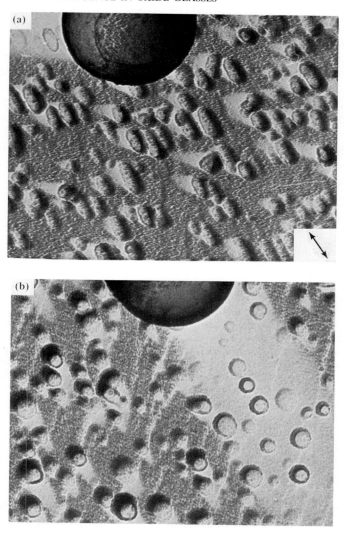

Fig. 8. Two-stage replica electron micrographs of the fractured surface taken parallel to the tensile axis. The glass was subjected to 100 hr of prepulling heat treatment at 595°C and stretched. (a) before and (b) after the postpulling heat treatment at 600°C for 16 hr. The arrow in (a) indicates the tensile axis. Latex sphere shows 0.5 μm in diameter.

negative birefringence observed after the spheroidal deformation of the minor phase particles returned to spherical as shown in Figs. 7 and 8?

When a glass with an isotropic distribution of minor phase particles is stretched, the originally spherical particles elongate into prolate spheroids.

At the same time, the distribution of the dispersed phase becomes anisotropic. After the postpulling heat treatment at 564°C, the prolate spheroids return to their initial spherical shape but stay at the same position and the overall specimen dimensions remain almost unchanged.[†] The growth rate of the minor phase particles at the postpulling heat treatment temperature is extremely slow and the rearrangement of the particles requires a much longer time at this temperature. The anisotropic distribution of the spherical minor phase gives a component of interparticle distance (on the average) greater in the direction parallel to the tensile axis than in directions perpendicular to the axis. This sequence of events is schematically shown in Fig. 9. This

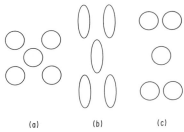

Fig. 9. Schematic representation of microstructures. (a) After the prepulling heat-treatment, the sample is phase separated but still isotropic. (b) After pulling, the sample is anisotropic with a uniaxially elongated minor phase and gives positive birefringence. (c) After the postpulling heat treatment, the sample is still anisotropic due to the anisotropic distribution of the spherical minor phase and gives negative birefringence. (After Takamori and Tomozawa, 1976.)

anisotropic distribution of the minor phase particles apparently causes the observed negative birefringence. The authors have proposed that this type of birefringence be called "distribution birefringence" analogous to form birefringence (Takamori and Tomozawa, 1976). The specimens showing form birefringence also have a superimposed distribution birefringence.

There is a close analogy between the birefringence caused by viscous flow of the minor phase in phase separated glasses and the birefringence caused by elastic strain in homogeneous glasses (stress-optical effect). The stress-optical birefringence is caused by the superposition of two effects (Mueller, 1938). One effect is the anisotropic distribution of atoms produced by the elastic deformation of the specimen and this effect produces negative birefringence under a uniaxial tension. The second effect is due to a deformation of atoms and this effect produces positive birefringence under a uniaxial

---

[†] For example, a sample glass subjected to 63 hr of prepulling heat treatment showed 0.4% shrinkage after 10 hr of postpulling heat treatment. This magnitude should be compared to 200% of the overall elongation by pulling.

tension. In most glasses the second effect dominates. The distribution birefringence observed in the above study corresponds to the negative birefringence caused by the anisotropic distribution of atoms and the form
birefringence corresponds to the effect due to the deformation of atoms. The
difference between the two phenomena is that in the phenomenon observed
as microstructural birefringence it was possible to separate the distribution
effect from the form effect, while in the stress-optical effect the two superimposed effects are closely related, and it is impossible to observe them
separately.

The variation of the magnitude of positive birefringence with prepulling
heat-treatment time can be explained by the shape variation of the prolate
spheroidal particles. With this glass, a short heat treatment at 595°C is
sufficient to nearly complete phase separation (Tomozawa and Takamori,
1977a), and the volume fraction of the minor phase can be regarded to be
almost constant for all specimens. However, interfacial energy will have a
different effect on spheroidal particles of different sizes (Seward, 1974). The
smaller particle has a smaller radius of curvature, giving rise to a greater
driving force toward a spherical shape. Therefore, a specimen which is subjected to a shorter prepulling heat treatment gives a prolate spheroid closer
to the spherical form, thus the positive form birefringence for this specimen
is smaller. The specimen which was given no prepulling heat treatment and
was stretched at 595°C is probably phase separated during the stretching,
and the minor phase particles remained spherical since they are so small,
giving only negative birefringence due to their anisotropic distribution.

As discussed in Section V,A, only negative birefringence was observed
as a microstructural birefringence of stretched sodium borosilicates in the
literature. This also can be explained in line with the current discussion,
although many of the sodium borosilicate glasses in Table I reveal interconnected structures upon phase separation (Tomozawa and Takamori,
1977a) in contrast to that shown in Fig. 5 of a lithium borosilicate glass.
In most of the works on those glasses, as-prepared glasses were stretched
without extensive prepulling heat treatment, and therefore the relaxation
of the concurrent form effect was so fast even during stretching that only
the effect of the anisotropic distribution was left in the samples. In fact, an
experiment by the authors indicated that a sodium borosilicate glass (70%
SiO_2, 23% B_2O_3, and 7% Na_2O by weight), for which Botvinkin and Ananich
(1959, 1963) found the negative birefringence only, can show the positive
birefringence before the postpulling heat treatment, when it was subjected to
such an extensive prepulling heat treatment as 100 hr at 600°C.

If a phase-separated glass consists of a higher viscosity minor phase dispersed in a lower viscosity matrix, on the other hand, the minor phase will
remain spherical through the stretching procedure, and its distribution can

be made anisotropic by the flow of the lower viscosity matrix. According to the present theory, such a glass will show negative birefringence only, even with an extensive prepulling heat treatment. This prediction was also confirmed on a glass of SiO_2 40, B_2O_3 50, and Na_2O 10 wt %.

2. THEORETICAL ANALYSES

a. Form Birefringence. We noted already that the Ambronn–Wiener theory is available for the form birefringence caused by fibrous and lamellar structures (Coker and Filon, 1957). More general cases of spheroidal particles are considered in the following.

The high-frequency dielectric constant ε of a system containing spheroidal particles with three principal radii a, b, b, is given by (Sillars, 1937)

$$\varepsilon = \varepsilon_1 \left[1 + \frac{q(\varepsilon_2 - \varepsilon_1)}{\varepsilon_1 + (\varepsilon_2 - \varepsilon_1)L} \right] \tag{3}$$

where ε_1 and ε_2 are the dielectric constants of the matrix phase and the dispersed phase, respectively, q the volume fraction of the dispersed phase, L a form factor which depends upon the radius ratio a/b and takes L_a or L_b depending upon whether the direction of the electric field is along the a or b axis. The magnitudes of L_a and L_b range from 1 to 0 and 0 to 0.5, respectively, when a/b ratio ranges from 0 (lamellar structure) to ∞ (fibrous structure). $L_a + L_b \leq 1$ for all cases. For a prolate spheroid ($a > b$);

$$L_a = \frac{1 - e^2}{e^2} \left(-1 + \frac{1}{2e} \ln \frac{1 + e}{1 - e} \right) \tag{4}$$

$$L_b = \frac{1}{2}(1 - L_a) \tag{5}$$

where

$$e = 1 - (b^2/a^2) \tag{6}$$

For an oblate spheroid ($a < b$),

$$L_a = \frac{1 + f^2}{f^2} \left(1 - \frac{1}{f} \arctan f \right) \tag{7}$$

$$L_b = \frac{1}{2}(1 - L_a) \tag{8}$$

where

$$f^2 = (b^2/a^2) - 1 \tag{9}$$

At high frequencies, the dielectric constant is equal to the square of the refractive index, i.e., $\varepsilon_1 = n_1{}^2$, $\varepsilon_2 = n_2{}^2$, and $\varepsilon = n^2$.

For a glass which is subjected to a uniaxial stress, the refractive indices of the extraordinary ray n_e and of the ordinary ray n_o are given by

$$n_e^2 = n_1^2 \left[1 + \frac{q(n_2^2 - n_1^2)}{n_1^2 + (n_2^2 - n_1^2)L_a} \right] \tag{10}$$

$$n_o^2 = n_1^2 \left[1 + \frac{q(n_2^2 - n_1^2)}{n_1^2 + (n_2^2 - n_1^2)L_b} \right] \tag{11}$$

Therefore

$$n_e^2 - n_o^2 = \frac{q n_1^2 (n_2^2 - n_1^2)^2 (L_b - L_a)}{[n_1^2 + (n_2^2 - n_1^2)L_a][n_1^2 + (n_2^2 - n_1^2)L_b]} \tag{12}$$

The denominator in Eq. (12) is positive regardless of the magnitudes of n_1, n_2, or a/b. Therefore, the sign of birefringence is determined only by $(L_b - L_a)$, i.e., the shape of the particles, being positive for prolate (or rod shape) and negative for oblate (or disk shape).

When a phase-separated borosilicate glass is stretched, the glass contains prolate particles as shown in Figs. 7 and 8, and thus a positive form birefringence is expected. An example of calculation of the form birefringence by prolate particles as a function of n_2 is shown in Fig. 10, in which $n_1 = 1.458$, the value for Vycor glass, and $q = 0.4$ were assumed. The numbers in the figure are a/b ratios. A similar formulation by Thomas (1964) gave opposite results since he used a wrong expression for the electric field in the particle.

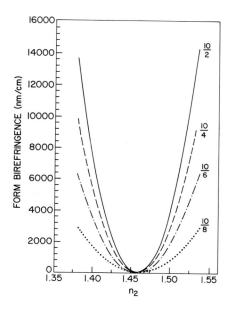

Fig. 10. Form birefringence versus n_2 based on Eqs. (10) and (11), when $n_1 = 1.458$ and $q = 0.4$.

b.   *Distribution Birefringence by Spheres.*   When the second phase consists of spheres (Sillars, 1937)

$$L_a = L_b = \tfrac{1}{3} \tag{13}$$

thus, from Eq. (3),

$$n^2 = n_1{}^2 \left[1 + \frac{3q(n_2{}^2 - n_1{}^2)}{2n_1{}^2 + n_2{}^2} \right] \tag{14}$$

The mean polarization P (moment per unit volume) of this material due to the second phase particles is given by

$$\frac{4\pi P}{E_0} = \frac{n^2}{n_1{}^2} - 1 = \frac{3q(n_2{}^2 - n_1{}^2)}{2n_1{}^2 + n_2{}^2} \tag{15}$$

where E_0 is the uniform field in the absence of the minor phase particle. In the derivation of this equation, it is assumed that the polarization of one particle is not influenced by the polarization of other particles. In reality, the polarization is slightly modified by the influence of other particles. When the distribution of the particles is random and isotropic, and thus the equiprobability surface of finding the next neighbors is spherical, the polarization will become

$$P' = \frac{1}{4\pi} \frac{3q(n_2{}^2 - n_1{}^2)}{2n_1{}^2 + n_2{}^2} \left(E_0 + \frac{4\pi}{3} P' \right) \tag{16}$$

Therefore,

$$P' = \left(\frac{1}{4\pi} \frac{3q(n_2{}^2 - n_1{}^2)}{2n_1{}^2 + n_2{}^2} E_0 \right) \left(1 - \frac{q(n_2{}^2 - n_1{}^2)}{2n_1{}^2 + n_2{}^2} \right)^{-1} \tag{17}$$

and the corresponding refractive index is given by

$$n^2 = n_1{}^2 \left\{ 1 + \left[\left(\frac{3q(n_2{}^2 - n_1{}^2)}{2n_1{}^2 + n_2{}^2} \right) \left(1 - \frac{q(n_2{}^2 - n_1{}^2)}{2n_1{}^2 + n_2{}^2} \right)^{-1} \right] \right\}$$

$$\approx n_1{}^2 \left[1 + \frac{3q(n_2{}^2 - n_1{}^2)}{2n_1{}^2 + n_2{}^2} + \frac{3q^2(n_2{}^2 - n_1{}^2)^2}{(2n_1{}^2 + n_2{}^2)^2} \right] \tag{18}$$

The approximation is valid since the term containing q is small compared with unity. The third term in the bracket of Eq. (18) is small compared with the second term and this equation is approximately equal to Eq. (14).

When the distribution of spherical particle is not isotropic, the coefficient of P' in the right-hand side of Eq. (16) has to be modified. When the specimen is stretched uniaxially, the originally spherical equiprobability surface for the distribution of particles becomes spheroidal. Thus the polarization in directions parallel and perpendicular to the tensile axis, P_a' and P_b', respectively, become

$$P_a' = \frac{1}{4\pi} \frac{3q(n_2^2 - n_1^2)}{2n_1^2 + n_1^2} (E_0 + 4\pi L_a P_a') \tag{19}$$

$$P_b' = \frac{1}{4\pi} \frac{3q(n_2^2 - n_1^2)}{2n_1^2 + n_2^2} (E_0 + 4\pi L_b P_b') \tag{20}$$

The corresponding refractive indices are given by

$$n_e^2 \approx n_1^2 \left[1 + \frac{3q(n_2^2 - n_1^2)}{2n_1^2 + n_2^2} + \frac{9q^2 L_a(n_2^2 - n_1^2)^2}{(2n_1^2 + n_2^2)^2} \right] \tag{21}$$

$$n_0^2 \approx n_1^2 \left[1 + \frac{3q(n_2^2 - n_1^2)}{2n_1^2 + n_2^2} + \frac{9q^2 L_b(n_2^2 - n_1^2)^2}{(2n_1^2 + n_2^2)^2} \right] \tag{22}$$

Therefore

$$n_e^2 - n_0^2 \approx 9q^2 n_1^2 (n_2^2 - n_1^2)^2 (L_a - L_b)/(2n_1^2 + n_2^2)^2 \tag{23}$$

The sign of this distribution birefringence is determined by $(L_a - L_b)$ only, i.e., the distribution of spherical particles, being positive for oblate distribution and negative for prolate distribution. This is opposite to that of the form birefringence. An example of calculation of the distribution birefringence is shown in Fig. 11, in which again $n_1 = 1.458$ and $q = 0.4$ were assumed. The birefringence actually observed right after the stretching is the sum of those in Figs. 10 and 11. This total birefringence is illustrated in Fig. 12, which indicates the positive birefringence. In the calculations in Figs. 10–12, the magnitude of L_a and L_b in the form effect and the distribution effect are considered the same. In practice, however, these factors do not have to be the same even for the same material, since the ratio a/b of the spheroidal particle

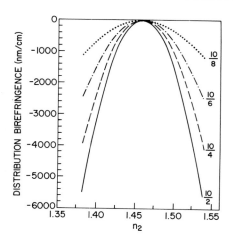

Fig. 11. Distribution birefringence versus n_2 based on Eqs. (21) and (22), when $n_1 = 1.458$ and $q = 0.4$.

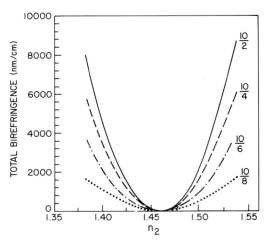

Fig. 12. Total birefringence versus n_2 (sum of birefringences in Figs. 10 and 11). No relaxation was assumed.

is not necessarily the same as that of distribution. We have seen in Section V,B,1 that the form birefringence relaxes much faster than the distribution birefringence. This observation indicates that the a/b ratio in the actual form effect is smaller than that in the distribution effect, resulting contribution of the latter in the total birefringence greater than in Fig. 12. In an extreme case of this trend we observed only negative birefringence before the postpulling heat treatment.

VI.   Crystallization and Optical Anisotropy

A.   Crystallizable Glass and Microstructural Birefringence

Russian workers have reported a series of results on the microstructural birefringence observed in glasses containing TiO_2 as a nucleating agent. Ananich et al. (1965) observed positive birefringence in a glass of spodumene composition ($Li_2O \cdot Al_2O_3 \cdot 4SiO_2$) plus 10% TiO_2. The magnitude of the birefringence, observed right after stretching and annealing, was in the range of 100–2000 nm/cm, depending on the stretching conditions. The birefringence change during heating the sample glass was measured up to 800°C. The birefringence started to increase after 700°C at which the glass starts to crystallize. After heating up to 800°C, the thermal expansion characteristics showed a drastic change, indicating a significant amount of crystallization. From these observations, it was inferred that the increase of the birefringence by the heat-treatment is due to the growth of the crystals with orientation in the direction of the stretching.

The microstructural birefringence in a SiO_2–Al_2O_3–BaO–TiO_2 glass was also observed at the same laboratory (Ananich *et al.*, 1967). After the sample glasses were stretched at 900°C for 5 min and annealed, they report that the structural birefringence often reached the order of 20,000 nm/cm. The birefringence change of these samples was measured as a function of heat treatment temperatures. Samples were held for 2 hr at selected temperatures. The birefringence measured at room temperature started to increase after the heat treatment at 800°C. The maximum was attained at 950°C. However, in contrast to the glass of spodumene composition, the crystallization of this glass could not be detected by x-ray diffraction until about 900°C, and even at the maximum of the microstructural birefringence (950°C) the crystallinity of the sample was 40–50%. It was inferred from these observations that the microstructural birefringence in this particular glass sensitivity responded to a kind of transition at the "precrystallization period" to bring the glass structure close to that of the future crystalline phase. Ananich *et al.* (1967) did not give the composition of the glass studied in the above nor the sign of birefringence observed. According to their more recent paper (Botvinkin *et al.*, 1968), however, they state that either positive or negative birefringence can be observed in the barium aluminosilicate glasses depending on their compositions.

The effect of TiO_2 on the structural birefringence of various glasses was also studied by those workers (Botvinkin *et al.*, 1968). Sample glasses were prepared by adding various amounts of TiO_2 to the three base glass compositions: spodumene ($Li_2O \cdot Al_2O_3 \cdot 4SiO_2$), celsian ($BaO \cdot Al_2O_3 \cdot 2SiO_2$) and cordierite ($2MgO \cdot 2Al_2O_3 \cdot 5SiO_2$). The amount of elongation by stretching was kept constant for all samples. The microstructural birefringence was observed only when the glass contained sufficient amount of TiO_2 to cause bulk crystallization, and the magnitude of the birefringence increased with the increase of TiO_2 content. When the samples were heated up to 800°C the magnitude of birefringence increased in all cases, either positive or negative, as we have already seen in their earlier work. They suggest that the observed birefringence and its dependence on TiO_2 content and temperature results either from the form birefringence or the presence of oriented anisotropic elements.

Recently an experiment on a lithium silicate glass was conducted by the authors. The composition of the glass is based on the well-studied binary $26Li_2O \cdot 74SiO_2$ (Doremus, 1973) with a 0.5 mole % addition of Al_2O_3 to lower the immiscibility boundary (Tomozawa, 1972) in an attempt to minimize phase separation during sample rod preparation. As-prepared sample rods were stretched at 510°C with a constant load (maximum stress ≈ 2.5 kg/mm^2), without any prepulling heat treatment. The phase separation of the glass must have proceeded during stretching, however, since the stretching temperature (510°C) was well within the immiscibility gap of this glass, and it took approximately 30 min to attain the elongation of 2.3 times original length. At this elongation, the load was removed and the glass was taken out of the furnace. After a simple annealing procedure to remove thermal strain (500°C for 30 min, followed by furnace-cooling), the birefringence of the sample rod was -340 nm/cm. By the postpulling heat treatment at 565°C for various lengths of time, the magnitude of this negative birefringence increased as shown in Fig. 13. This tendency is qualitatively the same as the observations by the Russian workers. In samples which were subjected to longer postpulling heat treatments, the occurrence of the crystallization gradually became apparent even with the polarimeter observations. Figure 14 shows the polarizing microscopic picture of a sample which was heat treated at 565°C for 8 hr after stretching, and ground and polished to leave a 1-mm-thick platelet with faces parallel to the stretching axis. The spheroidal particle in Fig. 14 shows positive birefringence with the major axis of the spheroid as the optic axis. Having the superimposed negative birefringence of the matrix glass, therefore, the spheroidal particles oriented to the stretching axis of the sample appeared darker than those oriented perpendicular to the stretching axis. Any preferred orientation of the particles to the stretching axis is hardly noticeable in this figure. Apparently, therefore, the increase of the negative birefringence with the postpulling heat treatment in Fig. 13 is not directly related with the crystals in this glass.

When the $26Li_2O\cdot74SiO_2$ glass undergoes phase separation, its microstructure has been shown to consist of silica-rich particles dispersed in the lithium-rich matrix (Doremus, 1973). The silica-rich particles should be rigid enough to remain undeformed when the sample rods are stretched at 510°C,

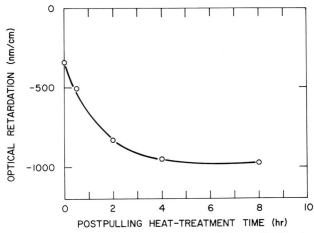

Fig. 13. Structural birefringence in a stretched and annealed crystallizable glass as a function of the postpulling heat-treatment time.

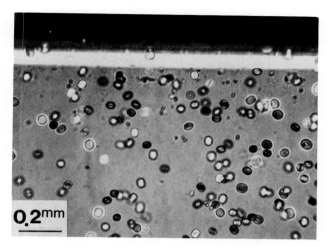

Fig. 14. Polarizing microscope picture of 26 $Li_2O\cdot74\ SiO_2\cdot0.5\ Al_2O_3$ glass with crystals, stretched and heat treated. The tensile axis was horizontal on the paper.

and there should be no form birefringence due to glass-in-glass phase separation in the as-stretched sample, superimposed on the observed negative birefringence. The increase of the negative birefringence in Fig. 13, therefore, was not caused by the relaxation of the form birefringence just as discussed on Fig. 6.

There will be distribution birefringence due to phase-separated particles in the stretched sample. The progressive phase separation and/or crystallization through the postpulling heat treatment may cause the change of the refractive indices of separated phases and/or their volume fraction, which will change the distribution birefringence in the matrix glass [see Eqs. (21)–(23)]. There may be additional effects of the anisotropic distribution of crystalline particles, however, and any definite conclusions await a detailed study of microstructure.

B. *Glass with Nonoxide Precipitation and Dichroism*

1. METAL-PRECIPITATING GLASS

When small metallic particles are dispersed in a glass, they absorb and scatter light. If the particles are small enough, the glass remains transparent, but colored. By stretching such a glass, Land (1943) and Stookey and Araujo (1968) obtained optically anisotropic glasses. The anisotropy was observed not as birefringence in this case but as dichroism. The procedure of the sample preparation was, however, essentially the same as those of the other birefringent glasses so far discussed. Here is an example which shows that,

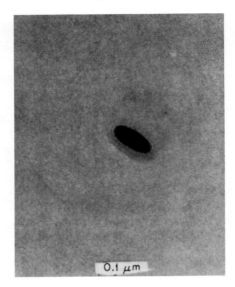

Fig. 15. Extraction replica electron micrograph of an elongated silver metal particle in glass. (After Stookey and Araujo, 1968.)

through the same processing, one can obtain a different kind of optical anisotropy in glass which has a different kind of microstructure.

A small amount of silver was added to various base glass compositions including soda-lime, lithium aluminosilicates, alkali borosilicates, etc. Depending on the base glass compositions, the prepulling heat-treatment ranged from 600 to 800°C for $\frac{1}{2}$ hr to obtain silver particle precipitation. The heat-treated glass was stretched from 50 to 500 times original length at the viscosity of about 10^7 P. The elongation of silver particles by this stretching was from 1.5- to 3-fold[†] as observed by electron micrographs, as shown in Fig. 15. In spite of the broad distribution of particle sizes in one sample, the elongation ratio of silver particles was found fairly uniform, indicating that the effect of different interfacial energies on elongated particles of different size (Seward, 1974) is small in this particular case. Visible spectra of the sample glasses were measured using polarized light. Figure 16 illustrates the dichroism observed.

2. PHOTOCHROMIC GLASS

Photochromic glasses are reviewed in detail in Chapter 3. Araujo *et al.* (1970) obtained photochromic polarizing glasses by stretching photochromic

[†] The viscosity of 10^7 P may be too low for the effective stretching of the silver particles in the glass. At higher viscosity, the glass will exert greater shear forces on the particles during stretching which will cause greater particles elongation. In many works on borosilicates, for example, the samples were stretched at the viscosity of 10^{10} to 10^{11} P.

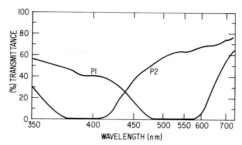

Fig. 16. Transmittance curves of a polarizing glass with elongated silver particles. P1: light polarized parallel to longitudinal axis of sample. P2: light polarized perpendicular to axis of sample. (After Stookey and Araujo, 1968.)

glasses. Silver halide containing glasses were first heat treated to make photo-chromism by developing silver halide crystals in matrix glass. The glasses were then stretched to attain elongation of the silver halide particles. Figure 17 shows elongated silver halide particles in a typical photochromic glass after stretching with a stress of ≈ 3.5 kg/mm^2 (Seward, 1974). The glass prepared

Fig. 17. Replica electron micrograph of elongated silver halide particles in a stretched photochromic glass. (After Seward, 1974.)

in this way could change reversibly from the clear unpolarized state to the darkened polarized state upon exposure to sunlight. At the darkened state, dichroism is apparently the dominant cause of the observed anisotropy of these glasses. Even at the clear unpolarized state, however, the glass should be optically anisotropic because of the possible microstructural birefringence (form and/or distribution) as discussed in the previous section.

VII. Chain Structure in Phosphate Glass and Birefringence

Phosphate glasses exhibit an unusually large amount of birefringence when stretched and cooled. These glasses have molecular structures unique among inorganic glasses, with chains, rings or branches in which the PO_4 tetrahedra are held together by the sharing of corner oxygen atoms (Westman, 1960). When the ratio of cation to phosphorus approaches unity (i.e., metaphosphate), the theoretical length of the resultant linear chain molecules becomes almost infinite. The actual metaphosphate glass, however, consists of various lengths of chains with a certain distribution. Glasses with such a chain structure should be oriented by uniaxial stretching at the viscoelastic temperature range. Goldstein and Davies (1955) observed an anisotropy in the $NaPO_3$ glass fibers prepared by stretching the glass during cooling. Without annealing the fibers, the anisotropy was revealed by an arc structure of x-ray diffraction patterns and a positive birefringence. The magnitude of the birefringence, they report, was 40,000 nm/cm. Their work was extended by Milberg and Daly (1963), who observed birefringences between 40,000 and 60,000 nm/cm in the $NaPO_3$ fibers similarly prepared. By analysis of the x-ray scattering pattern, they confirmed the strong orientation of chains along the fiber axis. The best agreement with the experimental distribution functions was obtained by assuming zigzag configurations in chains. Miller *et al.* (1973) prepared $NaPO_3$ fibers which showed a strong birefringence under a polarizing microscope. Preferential orientation of these samples was shown by laser Raman and infrared reflectance study.

The structure of the anisotropic phosphate glasses, however, appear to be different from that of anisotropic organic polymers in which oriented long chains exist and the anisotropy persists even after annealing at the glass transition temperature. In anisotropic $NaPO_3$ glass the effect of annealing is not clear, because crystallization occurs above about 300°C (Van Wazer and Callis, 1962). However, there is an indication that the anisotropy disappears when the specimens are annealed. Milberg and Daly (1963) state that disorientation can be achieved by heating the fibers to about 200°C, which is lower than the glass transition temperature of the $NaPO_3$ glass. Strangely, they observed no contraction accompanying the disorientation. It is likely, however, that an appreciable fraction of the observed positive birefringence in phosphate glasses comes from the frozen-in strain, since a zigzag chain structure can be strained greatly, for example, by a slight widening of chain angles.

VIII.   Microporous Glass and Microstructural Birefringence

So far we have discussed birefringence induced in various oxide glasses by thermomechanical processings. In this section we describe a totally different type of microstructural birefringence which was encountered during leaching of phase-separated glasses. The origin of this birefringence is not completely clear, but the phenomenon is of considerable interest.

A borosilicate glass of composition SiO_2 70, B_2O_3 23, and Na_2O 7 wt % was heat treated at 600°C for 100 hr to induce phase separation and quenched in air. After the surface layer of the heat-treated glass (Tomozawa and Takamori, 1977b) was removed by 20% HF etch, the glass was cut into various shapes. Then the glass specimens were soaked in 3NHCl solution saturated with NH_4Cl at $\approx 75°C$ to leach out the minor phase constituents. The resultant specimens, consisting of SiO_2-rich microporous skeletons, were immersed in the various "index of refraction liquids" and observed by polariscope.

Figure 18 illustrates typical polariscope observations. A cube of about 3 × 3 × 3 mm was cut from the phase-separated glass and leached. Then the cube was sliced and a thin platelet (< 1-mm thick) was cut out from the middle of the cube. When the platelet was immersed in the liquid with the refractive index of 1.460, which is close to that of Vycor glass (1.458), no birefringence was observed, as shown in Fig. 18a. Although the sample glass was quenched in air after the phase-separation heat treatment, the stress birefringence was not observed, since the minor phase had been leached out. This observation further confirms our previous contention that the thermal strain in phase-separated glasses is induced dominantly in the lower viscosity phase (Takamori, 1976). When the same platelet was immersed in a liquid with the refractive index of 1.320, strong negative birefringence was observed along the leaching direction, as shown in Fig. 18b. In the figure, yellow and blue colors correspond to the negative birefringence along horizontal and vertical axis, respectively. Figure 18c shows the same result obtained when the platelet was immersed in water of which the refractive index is 1.333.

Zhdanov (1958) has shown that a part of the structure in microporous glasses is of a secondary character, being created by decomposed deposits from the soluble minor phase within the cavity of the silica skeleton. Such colloidal deposits in the microporous glass was shown by scanning electron microscopy by Elmer *et al.* (1970). Figure 18b possibly indicates that these colloidal deposits have preferential orientation with respect to the leaching direction, resulting in a microstructural birefringence similar to those discussed in Section V. In the present case, the n_2 value of Eq. (12) or (23) comes from the liquid. Whether it is caused by the form effect or by the distribution effect will be determined by light scattering and/or small angle x-ray scattering in the near future. Although quantitative experiments are yet to be done, the magnitude of the birefringence appears to depend both on the heat-treatment conditions and leaching conditions. The study of this microstructural birefringence in the leached specimens should help in clarifying the leaching mechanism of borosilicate glasses.

    Apparently, the birefringence observed in Fig. 18 is different from the birefringence caused by the stress build-up during leaching borosilicate glasses. Hayashi et al. (1965) and Botvinkin et al. (1973) report the observations of strong compressive strain in the leached portion surrounding the unleached portion of the glass during HCl leaching. Botvinkin et al. (1973) have reported that the tensile stress in the unleached portion of the sample was as much as 1300 kg/cm². This stress was attributed to the volume change during the leaching procedure, which released the internal stress caused by the differential contraction. If there is a significant amount of differential contraction upon cooling the phase-separated glass, the glass should expand after leaching, since the higher expansion phase is leached out. If this were really the case, we should observe the birefringence of a fairly strong compressive strain in the leached portion surrounding the unleached portion. The following observation, however, indicates that the stress birefringence by differential contraction of anisotropic phases is much smaller than the microstructural one (see Section IV). A glass rod, 6 mm in diameter, of the same composition as in the preceding paragraph was heat-treated and leached by the same conditions. But this time, the leaching was terminated on half way through the sample. This sample rod was sliced parallel to the rod axis, and a thin platelet (< 1-mm thick) including the rod axis was obtained. When this platelet was immersed in the index of refraction liquid with the refractive index of 1.460, it was observed, as shown in Fig. 19a, that a slight tensile strain (positive birefringence, blue horizontally) in the leached porous portion of the glass surrounded the unleached portion, which is under a slight compression along the rod axis (horizontal in the figure). The sign of stress in this figure is opposite to what was expected from differential contraction. When the same platelet was immersed in the index of refraction liquid with the index of 1.320, the microstructural birefringence dominated over the stress birefringence, as shown in Fig. 19b. When the same platelet was immersed in water, as shown in Fig. 19c, a strong tensile strain was observed along the rod

---

Fig. 18. Polariscopic pictures of a microporous glass, immersed in the index of refraction liquid with the refractive index of (a) 1.460 and (b) 1.320, and (c) in water.

Fig. 19. Polariscopic pictures of a partially leached microporous glass. Sample was immersed in the index of refraction liquid with the refractive index of (a) 1.460 and (b) 1.320, and (c) in water.

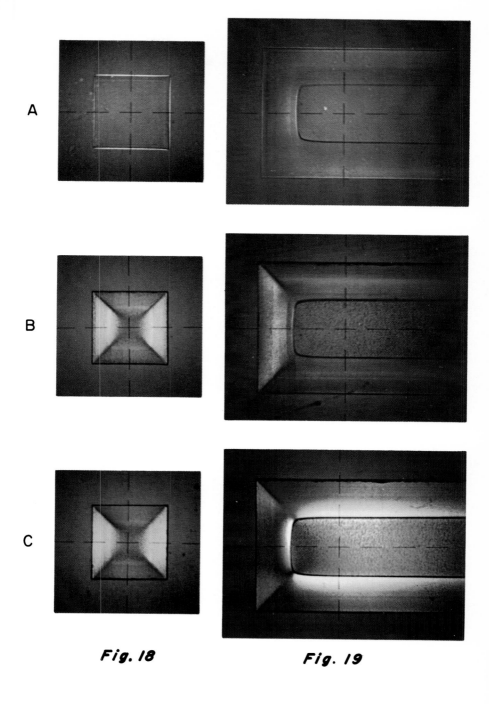

A

B

C

Fig. 18   *Fig. 19*

axis in the unleached portion of the glass, and also a strong compressive strain in the surrounding leached portion with a superimposed microstructural birefringence. The results in Fig. 19, therefore, indicate that the volume change observed by other workers during leaching, at least in certain cases, might not be caused by any dimensional change of the leached glass itself, but was caused by the swelling of the microporous glass by its reaction with water. In contrast to water, the index of refraction liquids commercially available are not reactive to glasses.

IX.   Summary and Conclusion

In this chapter, we have discussed anomalous birefringence in glasses as reported in the literature and also some recent observations at the authors' laboratories. Most of the information discussed is summarized in Table II. We believe that this area of study is a promising one, since "birefringence tells us things about structure at all levels from the atoms up to those that scatter

TABLE II

SUMMARY OF ANOMALOUS BIREFRINGENCE IN GLASSES

Classification	Structural requirement	Direction of inducing stress	Sign of birefringence	Typical glass reported
Frozen-in strain		Tensile	$+^a$	Soda–lime
		Compressive	−	Flint
Frozen-in strain (type II)	Phase separation		?	None[b]
Differential contraction of anisotropic phases	Phase separation		?	None[b]
Form birefringence	Phase separation	Tensile (elongation)	+	Borosilicate
		Compressive	−	
Distribution	Phase separation	Tensile (elongation)	−	Borosilicate
		Compressive	+	
Chain orientation	Molecular structure	Tensile (elongation)	+	$NaPO_3$
Anisotropic array of micropores	Microporous		$-^c$	Vycor-type microporous glass

[a] Most of the technical glasses, except dense flint type, have positive stress-optical coefficients.

[b] These birefringences are quite feasible when a phase-separated glass is processed thermomechanically. No definite observation has been reported yet, however.

[c] Along the direction of leaching.

light too severely to maintain coherence of the probe beam" (Ernsberger, 1975). As we have seen, there have been only a limited number of workers in this area and, consequently, the information available at present is limited and uncertain. Many questions remain to be answered in future work.

ACKNOWLEDGMENTS

The authors thank J. B. Landermann for experimental assistance and C. F. Aliotta for electron microscopy. Thanks are also due to F. M. Ernsberger of PPG Industries for criticism and encouragement in part of the work in this chapter.

References

Ananich, N. I., and Botvinkin, O. K. (1966). In "The Structure of Glass" (E. A. Porai-Koshits, ed.), Vol. 6, pp. 132–134. Consultants Bureau, New York.

Ananich, N. I., Botvinkin, O. K., and Dyatlova, L. V. (1965). Glass Ceram. **22**, 435–436.

Ananich, N. I., Botvinkin, O. K., Bogdanova, G. S., and Orlova, E. M. (1967). Inorg. Mater. 1144–1145.

Araujo, R. J., Cramer, W. H., and Stookey, S. D. (1970). U.S. Patent 3540793.

Botvinkin, O. K., and Ananich, N. I. (1959). Steklo i Keramika **16**, 6–11.

Botvinkin, O. K., and Ananich, N. I. (1963). Advan. Glass Technol., Part 2, 86–94.

Botvinkin, O. K., Ananich, N. I., Bogdanova, G. S., and Orlova, E. M. (1968). Inorg. Mater. **4**, 1159–1161.

Botvinkin, O. K., Mironova, N. L., and Shpilevskaya, G. L. (1973). In "The Structure of Glass" (E. A. Porai-Koshits, ed.), Vol. 8, pp. 122–125. Consultants Bureau, New York.

Chamot, E. M., and Mason, C. W. (1958). "Handbook of Chemical Microscopy," 3rd ed., pp. 335–383. Wiley, New York.

Coker, E. G., and Filon, L. N. G. (1957). "Treatise on Photo-Elasticity," 2nd ed., pp. 276–84. Cambridge Univ. Press, London and New York.

Doremus, R. H. (1973). "Glass Science." Wiley, New York.

Elmer, T. H., Nordberg, M. E., Carrier, G. B., and Korda, E. J. (1970). J. Am. Ceram. Soc. **53**, 171–175.

Ernsberger, F. M. (1970). "Polarized Light in Glass Research" PPG Ind. Pittsburgh, Pennsylvania.

Ernsberger, F. M. (1975). Private communications.

Espe, W. (1968). "Materials of High Vacuum Technology," Vol. II. Pergamon, Oxford.

Filon, L. N. G., and Harris, F. C. (1923). Proc. Roy. Soc. A **103**, 561–571.

Goldstein, M., and Davies, T. H. (1955). J. Am. Ceram. Soc. **38**, 223–226.

Hayashi, J., Nishiraku, T., and Ohtani, K. (1965). Mitsubishi Denki Giho **39**, 1092–1096.

Indenbom, V. L. (1953). Dokl. Akad. Nauk SSSR **89**, 509–511.

Kerwawycz, J., and Tomozawa, M. (1974). J. Am. Ceram. Soc. **57**, 467–470.

Kohl, W. H. (1967). "Handbook of Materials and Techniques for Vacuum Devices." Van Nostrand–Reinhold, Princeton, New Jersey.

Land, E. H. (1943). U.S. Patent 2319816.

Long, B. (1954). Brit. Patent 705212.

Long, B., Kantzer, M., and Orlu, M. (1956). Travaux du IVe Congres Int. du Verre, Paris pp. 292–295.

Longhurst, R. S. (1973). "Geometrical and Physical Optics," 3rd ed. Longman, London.
Merker, L. (1960). *Naturwissenschaften* **47**, 105.
Merker, L. (1969). *Glastech. Ber.* **42**, 419–424.
Milberg, M. E., and Daly, M. C. (1963). *J. Chem. Phys.* **39**, 2966–2973.
Miller, P. J., Exarhos, G. J., and Risen, Jr., W. M. (1973). *J. Chem. Phys.* **59**, 2796–2802.
Mueller, H. (1938). *J. Am. Ceram. Soc.* **21**, 27–33.
Patridge, J. H. (1949). "Glass-To-Metal Seals." Soc. Glass Technol., Sheffield.
Schill, F. (1967). *Veda Vyzk. Prum. sklarskem.* (9) 7–16.
Selwood, P. W. (1956). "Magnetochemistry," 2nd ed., pp. 189–192. Wiley (Interscience), New York.
Seward, T. P. III (1974). *J. Non-Cryst. Solids* **15**, 487–504.
Shand, E. B. (1958). "Glass Engineering Handbook," 2nd ed., p. 4. McGraw-Hill, New York.
Sillars, R. W. (1937). *J. Inst. Elec. Eng. (London)* **80**, 378–394.
Simmons, J. H., Mills, S. A., and Napolitano, A. (1974). *J. Am. Ceram. Soc.* **57**, 109–117.
Stirling, J. F. (1955). *J. Soc. Glass Technol.* **39**, 134T–144T.
Stookey, S. D., and Araujo, R. J. (1968). *Appl. Opt.* **7**, 777–779.
Takamori, T. (1963). *J. Am. Ceram. Soc.* **46**, 366–370.
Takamori, T. (1964). Unpublished.
Takamori, T. (1965). *J. Am. Ceram. Soc.* **48**, 170–171.
Takamori, T. (1974). *J. Am. Ceram. Soc.* **57**, 277.
Takamori, T. (1976). *J. Am. Ceram. Soc.* **59**, 121–123.
Takamori, T., and Tomozawa, M. (1976). *J. Am. Ceram. Soc.* **59**, 377–379.
Thomas, R. (1964). *Verres Refract.* **18**, 299–307.
Tomozawa, M. (1972). *In* "Advances in Nucleation and Crystallization in Glasses" (L. L. Hench and S. W. Freiman, eds.), pp. 41–50. Am. Ceram. Soc. Columbus, Ohio.
Tomozawa, M., and Takamori, T. (1977a). *J. Am. Ceram. Soc.* **60** (in press).
Tomozawa, M., and Takamori, T. (1977b). Unpublished.
Van Wazer, J. R., and Callis, C. F. (1962). *In* "Inorganic Polymers" (F. G. A. Stone and W. A. G. Graham, eds.), pp. 28–97. Academic Press, New York.
Westman, A. E. R. (1960). *In* "Modern Aspects of the Vitreous State" (J. D. Mackenzie, ed.), Vol. I. pp. 63–91. Butterworths, London.
Zhdanov, S. P. (1958). *In* "The Structure of Glass," pp. 125–134. Consultants Bureau, New York.

Light Scattering of Glass

JOHN SCHROEDER

Department of Chemistry
School of Chemical Sciences

and

Materials Research Laboratory
University of Illinois
Urbana, Illinois

I. Introduction

The use of optical scattering to study amorphous solids (glasses) dates from the work of Lord Rayleigh (1919) and continues to the present day. Raman (1927) examined fourteen different optical glasses, and was the first to conclude that true molecular light scattering was observable in glass. Others, namely, Krishnan (1936a), Krishnan and Rao (1944), Rank and Douglas (1948), Debye and Bueche (1949), Levin (1955), Maurer (1956), and Prod'home (1957), reinforced this conclusion of Raman. Maurer (1956) and Goldstein (1959) concluded that insofar as certain aspects of light scattering are concerned, a glass is indistinguishable from a liquid. Mauer (1956) has also observed an annealing effect on the scattered intensity. Fabelinskii (1968) states that a similar effect was observed by Levin (1955), Kolyadin (1956, 1960), Voishvillo (1962), and Andreev *et al.* (1960). All of the above researchers found that the measured intensity of the scattered light of each sample exceeded that calculated on the basis of equilibrium density fluctuations by at least one order of magnitude.

To explain this discrepancy, Raman (1927) proposed the "freezing" of concentration fluctuations and Mueller (1938) later generalized this to "frozen-in" (thermally arrested) density fluctuations. That the excess scattered light observed should be explainable in this fashion is not at all unreasonable, if the nonequilibrium process whereby glass is formed by supercooling the melt past its freezing point through its "glass transition" range is considered. All of the above measurements suffer from the same limitations: the complex composition of all the optical grade glasses (usually five or more components in each sample) are often not well defined so that a separation of the various scattering mechanisms is essentially impossible.

Velichkina (1953, 1958) has studied the light scattering from some relatively simple liquids that upon cooling form organic glasses. We give the results of her work as summarized by Fabelinskii (1968). The liquids that were studied were triacetin and chloroethyl ether (β-ether) in the temperature range from 200 to 300 K. Scattering in fused silica was also included for this temperature range. At the higher temperature end (>250 K) both the organic liquids were above their glass transition temperatures and the experimental results agreed rather well with simple theoretical calculations. At lower temperatures (<220 K) both are glasses and theory and measurements are in disagreement. For fused silica the disagreement between theory and experiment holds over the entire temperature range. Neither a liquid theory nor a crystal theory satisfactorily describes the scattering in any of the glasses.

The traditional techniques (before the advent of the laser as a source of high photon flux) employed in the above referenced works were mainly restricted to the measurement of scattered light intensity from which one can obtain the magnitude of density and concentration fluctuations. Newer techniques permit the measurement of the spectrum of the scattered light as well, whereby the dynamics of these fluctuations now become accessible.

The availability of the spectrum allows the separation of the basic processes causing light scattering. The scattering by propagating fluctuations in the dielectric constant (Brillouin scattering) and scattering by nonpropagating diffusive fluctuations (Rayleigh scattering) are the two processes. The former was first predicted by Brillouin (1922) who suggested that light should be inelastically scattered by the acoustic modes (Debye modes) of an insulating media. Mandelstamm (1926) independently predicted the same effect. Brillouin scattering can be simply viewed as the diffraction of the incident light by the "gratings" created by sound waves (phonons). But these gratings are moving with the speed of sound, causing the scattered light to be Doppler shifted. In an isotropic solid (glass) only two acoustic modes exist, and the Brillouin lines will consist of two sets of doublets symmetrically placed about the incident frequency. The observed Brillouin splitting can simply be related to the sound velocity of the particular mode responsible for the scattering (Brillouin, 1922) and the widths of the shifted lines are related to acoustic attenuation (phonon lifetimes) (Leontovich, 1931). Brillouin scattering was first observed by Gross (1930a, 1932) in liquids and later (Gross, 1930b) in crystalline quartz. Subsequent searches (Gross, 1930b; (Ramm, 1934); (Raman and Rao, 1937, 1938); (Venkateswaran, 1942); (Rank and Douglas, 1948) and (Velichkina, 1953, 1958) for Brillouin components in glass were unsuccessful, until Krishnan (1950) first observed these lines in fused quartz, followed by a number of other observations (Pesin and Fabelinskii, 1960, 1961; Flubacher *et al.*, 1960; Shapiro *et al.*, 1966).

Rayleigh scattering caused by nonpropagating diffusive fluctuations was, of course, observed much earlier. Here the scattered line is centered at the same frequency as the incident light. This scattering results from nonpropagating density, temperature, and, for mixtures, concentration fluctuations. The intense scattering observed near critical points (opalescence) was first explained by von Smoluchowski (1908) and Einstein (1910) in terms of the divergence of the compressibility at the liquid–vapor critical point. In mixtures, the analogous effect results from the vanishing of the chemical potential derivative.

For a given selected scattering angle the Rayleigh line generally results from more than one type of nonpropagating fluctuation. In principle, these different sources can be separated on the basis of their temporal behavior. The temporal behavior is reflected in the spectrum of the scattered light. Since the kinetics of the composition fluctuations are considerably slower than those of density fluctuations, the former should lead to a narrower Rayleigh component and could thus be distinguished from the latter. In practice, however, both contributions are too narrow to be resolved and, consequently, it is necessary to resort to other means of separation. Specifically, in a glass density fluctuation are effectively frozen in at a temperature corresponding to a viscosity of about $10^{13.5}$ P (Tool, 1946; Pinnow *et al.*, 1968; Schroeder *et al.*, 1973), while composition fluctuations are thermally arrested at a viscosity of about 10^7-10^8 P. Hence by performing series of experiments at different temperatures and measuring the intensity as a function of time, the quasielastic spectrum can be taken apart to show the different contributions, and the growth of composition fluctuations as the system goes from a nonequilibrium to equilibrium state can be studied. For example, a liquid–liquid miscibility gap below the glass transition temperature can be explored by light scattering.

It is the purpose of this chapter to describe the application of present-day light-scattering technology to studies of the glassy state. For theoretical background, existing liquid theory will be used, and modifications described that are necessary for its application to glasses. Examples of experiments will be considered that exhibit certain aspects of theory most conveniently.

II.   Theoretical Background

A.   General Concepts

When a molecule or atom is subjected to a plane electromagnetic wave, the oscillating electric dipole so created will emit a secondary wave resulting in the phenomenon of scattering by a single-scattering center. In condensed media the distances between scatterers are usually small compared to the wavelength of the light (i.e., the medium is essentially a continuum), thus the secondary waves emitted by neighboring scatterers are almost in phase such that one may simply add their amplitudes. For homogeneous condensed media in all directions except forward, the scattered waves cancel and the forward scattering which is coherent with the incident light leads to the concept of the refractive index for the media. Hence a perfect homogeneous medium will not scatter; however, in an inhomogeneous medium, the secondary waves no longer cancel completely and scattering is obtained. We are primarily interested in inhomogeneities arising from thermodynamic fluctuations, i.e., local density and concentration fluctuations. These variations give rise to random fluctuations in the local dielectric constant of a number of volume elements, each of which is large enough to contain many molecules but whose linear dimensions are small compared to the wavelength of light. Using these assumptions the classical theory of light scattering in dense media was developed by von Smoluchowski (1908) and Einstein (1910).

The classical theory does not allow for the simple inclusion of relaxation effects, nor does it properly include the stochastic nature of the thermodynamic fluctuations which produce the light scattering. To handle these effects, Komarov and Fisher (1963) and Pecora (1964) have shown that the spectrum of the light scattered from condensed media is determined by the space and time Fourier transform of the Van Hove space–time correlation function (Van Hove, 1954). Many authors have treated the electrodynamic problem necessary to obtain the scattered field and the intensity of the scattered light (Benedek and Fritsch, 1966; Benedek, 1966; Cummins, 1969; Stevens et al., 1972; Chu, 1974). Here we will follow the developments given by Van Kampen (1969), Cummins (1969), Cummins and Swinney (1970) and Benedek (1966) and present only the pertinent equations important in the rest of this chapter to allow predictions of effects and interpretation of measured results.

For an isotropic medium (liquid, glass), where the intermolecular spacing is small compared to the wavelength of light, it is possible to treat the fluid or solid optically as a continuous medium in which thermally excited collective excitations perturb the local dielectric constant. In the above approximation one finds that a scattering experiment selects normal modes of the scattering system characterized by wavevectors $\mathbf{k} = \mathbf{K}_0 - \mathbf{K}_s$, where $\mathbf{K}_0$ is an incident wave vector and $\mathbf{K}_s$ the scattered wave vector. If this particular mode perturbs the dielectric constant ε_k, causing a fluctuation in it, then the cross section for scattering is given by Cummins and Swinney (1970) in the form

$$\frac{d^2\sigma}{d\Omega\,d\omega} = \left(\frac{\omega_0}{c_0}\right)^4 \frac{V^2}{(4\pi)^2}$$

$$\times \sin^2\varphi \left\{ \frac{1}{2\pi} \int_{-\infty}^{\infty} e^{-i(\omega-\omega_0)\tau} \langle \varepsilon_k^*(t) \cdot \varepsilon_k(t+\tau) \rangle \, d\tau \cdot \delta(\mathbf{K}_s - \mathbf{K}_0 \pm \mathbf{k}) \right\}$$

$$\tag{1}$$

where $\langle \varepsilon_k^*(t) \cdot \varepsilon_k(t+\tau) \rangle$ is the autocorrelation function for the dielectric constant, V the scattering volume, ω_0 the laser frequency, c_0 the speed of light, and φ is the angle between $\mathbf{E}_0$ and $\mathbf{K}_s$. In pure liquids and multicomponent fluids one may have both density and/or concentration fluctuations which do not propagate, but relax by diffusive processes such as heat transport and mass transport. Consequently, these fluctuations give rise to unshifted components in the scattered light which make up the Rayleigh spectrum.

First one may consider only fluctuations in density at constant pressure, and if the density fluctuates from its equilibrium value by $\Delta\rho$, then the return to equilibrium is governed by thermal diffusion; namely,

$$(\partial\rho/\partial t)_P = \chi \, \nabla^2 \rho \tag{2}$$

where $\chi = \Lambda/\rho c_P$ is the thermal diffusivity (c_P the specific heat and Λ the thermal conductivity). The autocorrelation function for density fluctuations at constant pressure of wave-vector $\mathbf{k}$ is given by

$$\langle \rho_k^*(t) \cdot \rho_k(t + \tau) \rangle = \langle |\Delta\rho_k|^2 \rangle \exp(-\chi k^2 |\tau|) \tag{3}$$

Inserting this into Eq. (1) Cummins and Swinney (1970) obtain the differential scattering cross section in the form

$$\frac{d^2\sigma}{d\Omega \, d\omega} = \left(\frac{\omega_0}{c_0}\right)^4 \frac{V^2}{(4\pi)^2} \sin^2\varphi \left(\frac{\partial\varepsilon}{\partial\rho}\right)^2 \langle |\Delta\rho_k|^2 \rangle \frac{\chi k^2/\pi}{(\omega - \omega_0)^2 + (\chi k^2)^2} \tag{4}$$

Next one may consider a binary liquid (or multicomponent fluid of different refractive indices). If the concentration of one component fluctuates away from the equilibrium value by Δc, then the return to equilibrium may be described by the mass transport law:

$$\partial c/\partial t = D \, \nabla^2 c \tag{5}$$

here D is the diffusion coefficient. Assuming that at $t = 0$, $\Delta c(t = 0) = \Delta c_0 \exp(i\mathbf{k} \cdot \mathbf{r})$, Eq. (5) reduces to

$$\partial c/\partial t = -k^2 D \, \Delta c \tag{6}$$

The autocorrelation function for concentration fluctuations becomes

$$\langle c_k^*(t) \cdot c_k(t + \tau) \rangle = \langle |\Delta c_k|^2 \rangle \exp(-Dk^2 |\tau|) \tag{7}$$

for concentration fluctuations of wavevector $\mathbf{k}$. With the aid of Eq. (1) the cross section for scattering of coherent light by concentration fluctuations characterized by this autocorrelation function becomes (Cummins and Swinney, 1970)

$$\frac{d^2\sigma}{d\Omega \, d\omega} = \left(\frac{\omega_0}{c_0}\right)^4 \frac{V^2}{(4\pi)^2} \sin^2\varphi \left(\frac{\partial\varepsilon}{\partial c}\right)^2 \langle |\Delta c_k|^2 \rangle \frac{Dk^2/\pi}{(\omega - \omega_0)^2 + (Dk^2)^2} \tag{8}$$

Although the above equations were derived for liquids, they are still general enough to apply to amorphous solids.

The most important idea that is brought out by the above discussion is that the Rayleigh spectrum is made up of contributions from density fluctuations, temperature fluctuations, and concentration fluctuations and the cross section becomes proportional to

$$\frac{d^2\sigma}{d\Omega \, d\omega} \propto \left(\frac{\partial\varepsilon}{\partial P}\right)_{T,c}^2 \langle |\Delta P_k|^2 \rangle + \left(\frac{\partial\varepsilon}{\partial T}\right)_{P,c}^2 \langle |\Delta T_k|^2 \rangle$$

$$+ 2 \left(\frac{\partial\varepsilon}{\partial T}\right)_{P,c} \left(\frac{\partial\varepsilon}{\partial P}\right)_{T,c} \langle |\Delta P_k| \cdot |\Delta T_k| \rangle + \left(\frac{\partial\varepsilon}{\partial c}\right)_{P,T}^2 \langle |\Delta c_k|^2 \rangle \tag{9}$$

The contribution of isochoric temperature fluctuations to the variation of ε is negligible (Münster, 1974; Fabelinski, 1968) and it is straightforward to show (see Section II,B) that a change of variables $(P, T, c) \rightarrow (\rho, T, c)$ gives

$$\frac{d^2\sigma}{d\Omega \, d\omega} \propto \left(\frac{\partial \varepsilon}{\partial \rho}\right)_{T,c}^2 \langle|\Delta\rho_k|^2\rangle + \left(\frac{\partial \varepsilon}{\partial c}\right)_{P,T}^2 \langle|\Delta c_k|^2\rangle \qquad (10)$$

In the following sections we shall examine the forms of $\langle|\Delta\rho_k|^2\rangle$ and $\langle|\Delta c_k|^2\rangle$ more closely and derive expressions applicable to glass systems.

B.   Intensity of the Scattered Light and Landau–Placzek Ratio for Multicomponent Liquids

In the previous section we have given an expression for the scattering cross section applicable to liquids; we shall now obtain expressions that are valid for the glassy state. Again many derivations exist for the intensity of the scattered light of liquids; our derivation will parallel the works put forward by Lekkerkerker and Laidlaw (1972, 1973) or Miller (1967), but the approach will differ from their treatment in that we will apply it to an isotropic solid and not a liquid.

The fluctuations in the optical dielectric constant $\delta\varepsilon$ are given by

$$\delta\varepsilon = \left(\frac{\partial \varepsilon}{\partial \rho}\right)_{T,\{c\}} \delta\rho + \sum_{j=1}^{n-1} \left(\frac{\partial \varepsilon}{\partial c_j}\right)_{T,\rho,\{c'\}} \delta c_j \qquad (11)$$

In this equation the subscript $\{c\}$ denotes the $n-1$ independent mass fractions $c_1, c_2, c_3, c_4, \ldots, c_{n-1}$, whereas the prime in $\{c'\}$ means that the mass fraction appearing in the partial derivative is omitted from the set $\{c\}$. Neglecting the temperature dependence of the optical dielectric constant at constant density is reasonable only if (Münster, 1974); (Fabelinskii, 1968); (McIntyre and Sengers, 1968)

$$\left(\frac{\partial \varepsilon}{\partial T}\right)_{\rho,\{c\}} \frac{T\alpha_P}{c_v K_T \rho} \ll \left(\rho \frac{\partial \varepsilon}{\partial \rho}\right)_{T,\{c\}} \qquad (12)$$

where α_P is the coefficient of volume expansion, K_T the isothermal compressibility, and c_v the specific heat at constant volume.

Next we must choose a set of state variables. While any set of independent variables will suffice a choice of specific ones will make our calculations somewhat less cumbersome. The criterion for choosing such state variables is described by Landau–Lifshitz (1969) and Mountain and Deutch (1969). The probability of a simultaneous deviation of several thermodynamic quantities from their mean values is denoted by $x_1, x_2, \ldots, x_n$. We define the entropy $s(x_1, x_2, \ldots, x_n)$ as a function of the quantities $x_1, x_2, \ldots, x_n$ and write the probability distribution in the form $w \, dx_1 \cdots dx_n$. Now Δs_T is the change in entropy of the system plus the surrounding caused by the fluctuation and from the Boltzmann principle we have

$$w = \text{const} \exp(\Delta s_T/k_B)$$

where

$$\frac{\Delta s_T}{k_B} = -\frac{1}{2} \sum_{i,K=1}^{n} \beta_{iK} x_i x_K \tag{13}$$

hence if we choose $(p, s, c_1, c_2, \ldots, c_{n-1})$ it can be shown that $\langle \delta s \, \delta c_i \rangle \neq 0$; therefore $(p, s, c_1, c_2, \ldots, c_{n-1})$ does not satisfy our criterion of statistical independence. But if we use instead of δs a reduced entropy δs_{red} defined by

$$\delta s_{\text{red}} = \delta s - \sum_{j=1}^{n-1} (\partial s/\partial c_j)_{T,P,\{c'\}} \, \delta c_j \tag{14}$$

our set of state variables now are independent, since $\langle \delta s_{\text{red}} \, \delta c_i \rangle = 0$. This is done in analogy to the work of Mountain and Deutch (1969) where they used a reduced quantity substituted for T, the temperature. [DeGroot and Mazur (1962) and Cohen et al. (1971) also give analogous expressions and reasons for doing this step.] It can be shown that for the set of variables $(s_{\text{red}}, P, \{c\})$ that δs_{red} is statistically independent of the concentration fluctuations.

Now using Eq. (11) and the expression for reduced entropy, we may write the fluctuations in the optical dielectric constant as

$$\delta \varepsilon = (\partial \varepsilon/\partial \rho)_{T,\{c\}}(\partial \rho/\partial s)_{P,\{c\}} \, \delta s_{\text{red}} + (\partial \varepsilon/\partial \rho)_{T,\{c\}}(\partial \rho/\partial P)_{s,\{c\}} \, \delta P$$
$$+ \sum_{j=1}^{n-1} (\partial \varepsilon/\partial c_j)_{T,P,\{c'\}} \, \delta c_j \tag{15}$$

However, the scattered light is given simply by (Benedek, 1966)

$$dI(\theta, \Phi, \varphi) = \left(\frac{c_0 E_0{}^2}{8\pi}\right)\left(\frac{\omega_0}{c_0}\right)^4 \frac{\sin^2 \varphi}{(4\pi)^2} (2\pi)^3 \langle |\delta \varepsilon_k|^2 \rangle \, d\Omega \tag{16}$$

where c_0 is light velocity, E_0 the applied field, φ the angle between polarization and propagation direction, and $d\Omega = \sin \theta \, d\theta \, d\Phi$. Thus we really require $\langle |\delta \varepsilon_k|^2 \rangle$, where $\delta \varepsilon_k$ is the kth spatial Fourier component of the fluctuation in the dielectric susceptibility.

$$\langle \delta \varepsilon_k{}^2 \rangle = (\partial \varepsilon/\partial \rho)^2_{T,\{c\}}(\partial \rho/\partial s)^2_{P,\{c\}} \langle \delta s_{\text{red}}^2 \rangle + (\partial \varepsilon/\partial \rho)^2_{T,\{c\}}(\partial \rho/\partial P)^2_{s,\{c\}} \langle \delta P^2 \rangle$$
$$+ \sum_{j=1}^{n-1} \sum_{K=1}^{n-1} (\partial \varepsilon/\partial c_j)_{T,P,\{c'\}}(\partial \varepsilon/\partial c_K)_{T,P,\{c'\}} \langle \delta c_j \, \delta c_K \rangle \tag{17}$$

The brackets $\langle \ \rangle$ denote the average over an equilibrium ensemble and again, as stated above, all quantities in these brackets have to be interpreted as the kth Fourier components of reduced entropy, pressure, and concentration fluctuations.

The pressure fluctuations at constant entropy can be identified to a good approximation with sound waves. Brillouin observed (Benedek, 1966) that the time dependence of the correlation function for the pressure fluctuation is essentially an attenuating and propagating sound wave. Thus, this part of the above equation can be identified with the shifted lines or Brillouin doublets. We are now left with the entropy fluctuations at constant pressure and with concentration fluctuations; both may be identified with heat conduction, thermal diffusion, and diffusion (Münster, 1974). All of these are nonpropagating modes; thus they are identified with the central or Rayleigh line.

Using thermodynamic fluctuation theory (Fabelinskii, 1968) one obtains for the mean square fluctuations

$$\langle \delta s_{\mathrm{red}}^2 \rangle = k_{\mathrm{B}} c_P / \rho V' \tag{18}$$

$$\langle \delta P^2 \rangle = k_{\mathrm{B}} T / V' K_s \tag{19}$$

$$\langle \delta c_j\, \delta c_K \rangle = (k_{\mathrm{B}} T / \rho V')(B^{-1})_{jK} \tag{20}$$

Here k_{B} is the Boltzmann constant, V' the volume of fluctuation, K_s the adiabatic compressibility at high frequency, T the temperature, and $(B^{-1})_{jK}$ the inverse of a symmetric matrix B, the elements of which are given (Lekkerkerker and Laidlaw, 1973; Jordan and Jordan, 1966) by

$$B_{jK} = (\partial/\partial c_K [\mu_j - \mu_n])_{T,P,\{c'\}} = (\partial \tilde{\mu}/\partial c_K)_{T,P,\{c'\}} \tag{21}$$

By using the expressions

$$(\partial \rho/\partial s)_{P,\{c\}} = -(\rho/V'\alpha_P)(K_T - K_s) \tag{22}$$

$$(\partial \rho/\partial P)_{s,\{c\}} = \rho K_s \tag{23}$$

where $\alpha_P = -\rho^{-1}(\partial \rho/\partial T)_P$ is the coefficient of thermal expansion and K_T, K_s are isothermal and adiabatic compressibilities, one may rewrite Eq. (17) in the form

$$\langle \delta \varepsilon_k^2 \rangle = \left(\frac{\partial \varepsilon}{\partial \rho}\right)_{T,\{c\}}^2 \frac{k_B T}{V'} \rho^2 (K_T - K_s) + \left(\frac{\partial \varepsilon}{\partial \rho}\right)_{T,\{c\}}^2 \frac{\rho^2 k_B T}{V'} K_s$$

$$+ \sum_{j=1}^{n-1} \sum_{K=1}^{n-1} \left(\frac{\partial \varepsilon}{\partial c_j}\right)_{T,P,\{c'\}} \left(\frac{\partial \varepsilon}{\partial c_K}\right)_{T,P,\{c'\}} \frac{k_B T}{\rho V'} (B^{-1})_{jK} \tag{24}$$

With Eq. (24) one can now express the intensity of the scattered light for a multicomponent fluid system.

In the above equation both the Rayleigh and Brillouin contributing parts have been identified, and Eq. (24) shall be expressed as the Landau–Placzek ratio. The Landau–Placzek ratio by definition (McIntyre and Sengers, 1968) is the ratio of the intensities of the central components I_c to the total Brillouin components $2I_B$; or

$$R_{LP} \equiv \frac{I_c}{2I_B} = \frac{(\rho \, \partial\varepsilon/\partial\rho)^2_{T,\{c\}}(K_T - K_s) \cdot T}{(\rho \, \partial\varepsilon/\partial\rho)^2_{T,\{c\}} K_s \cdot T}$$

$$+ \frac{\sum_{j,K=1}^{n-1} (\partial\varepsilon/\partial c_j)_{T,P,\{c'\}}(\partial\varepsilon/\partial c_K)_{T,P,\{c'\}}(B^{-1})_{jK} \cdot T}{(\rho \, \partial\varepsilon/\partial\rho)^2_{T,\{c\}} \rho T K_s} \tag{25}$$

The first part of the above equation is usually reduced to a simpler form and for a liquid one obtains $(\gamma - 1)$ where $\gamma = K_T/K_s$; however, for a glass, it is more convenient to leave it in the expanded form.

Before proceeding with our main task, which is to derive from liquid theory expressions for the scattered intensity, Landau–Placzek ratio, and the spectrum of the scattered light which are valid for the case of a glass, we shall briefly discuss the spectrum of the scattered light of a multi-component fluid. In the past decade many publications (Benedek, 1966; Chu, 1974; Cummins and Swinney, 1970; Münster, 1974; Haus, 1974; Fabelinskii, 1968; Lekkerkerker, 1973) have dealt with this subject. A particularly concise form limited to a two-component fluid, which will suffice for our discussion, is given by Münster (1974)

$I(\mathbf{k}, \omega)$

$$= \frac{I_0 k_0^2}{16\pi^2 R^2} \sin^2 \varphi \left\{ \left(\rho \frac{\partial\varepsilon}{\partial\rho}\right)^2_{T,c_2} k_B T K_T \left[\frac{1}{\gamma} \left(\frac{\Gamma k^2}{(\Gamma k^2)^2 + (\omega + v_s k)^2} \right. \right. \right.$$

$$+ \frac{\Gamma k^2}{(\Gamma k^2)^2 + (\omega - v_s k)^2} \bigg) + \left(1 - \frac{1}{\gamma}\right) \frac{2}{z_2 - z_1} \left(\frac{(Dk^2 - z_1)z_1}{z_1^2 + \omega^2} - \frac{(Dk^2 - z_2)z_2}{z_2^2 + \omega^2} \right) \bigg]$$

$$+ \left(\frac{\partial\varepsilon}{\partial c_2}\right)^2_{T,P} \frac{k_B T M_2}{N_L d_1} \left(\frac{\partial\mu}{\partial c_2}\right)^{-1}_{T,P} \frac{2}{z_2 - z_1} \left[\frac{(z_2 - Dk^2)z_1}{z_1^2 + \omega^2} - \frac{(z_1 - Dk^2)z_2}{z_2^2 + \omega^2} \right]$$

$$+ \left(\frac{\partial\varepsilon}{\partial\rho}\right)_{T,c_2} \left(\frac{\partial\varepsilon}{\partial c_2}\right)_{T,P} \frac{2k_B T}{c_P} \frac{2Dk_T k^2}{z_1 - z_2} \left[\frac{z_1}{z_1^2 + \omega^2} - \frac{z_2}{z_2^2 + \omega^2} \right] \right\} \tag{26}$$

Here N_L is Avogadro's number, d_1 the mass component 1 (solvent) per unit volume, μ_2 the chemical potential of component 2, and $c_2 = m_2/m_1$;

$$z_1 = \tfrac{1}{2}(\chi k^2 + Hk^2) + \tfrac{1}{2}[(\chi k^2 + Hk^2)^2 - 4\chi Dk^4]^{1/2} \tag{27}$$

and

$$z_2 = \tfrac{1}{2}(\chi k^2 + Hk^2) - \tfrac{1}{2}[(\chi k^2 + Hk^2)^2 - 4\chi Dk^4]^{1/2} \tag{28}$$

and χ, H are defined by

$$H = D[1 + (k_T/Tc_P)(\partial\mu/\partial c)_{P,T}] \tag{29}$$

$$\chi = \Lambda/\rho_0 c_P, \qquad k_T = c_1 c_2 (D'/D)T_0 \tag{30}$$

where D'/D is the Soret coefficient; also,

$$\Gamma = \tfrac{1}{2}\left\{\frac{\tfrac{4}{3}\eta + \eta_v}{\rho_0} + \chi(\gamma - 1)\right.$$

$$\left. + \frac{Dv_s^2}{\rho_0^2(\partial\mu/\partial c)_{T,P}}\left[\left(\frac{\partial\rho}{\partial c}\right)_{P,T} + \frac{k_T}{c_P}\left(\frac{\partial\rho}{\partial T}\right)_{P,c}\left(\frac{\partial\mu}{\partial c}\right)_{T,P}\right]^2\right\} \tag{31}$$

In the above equations η and η_v are the shear viscosity and bulk viscosity, respectively; Λ the heat conductivity, k_T the thermal diffusion ratio, v_s the sound velocity, and D is the isothermic diffusion constant. The above equations are somewhat unwieldy to handle, and we shall just discuss the physical significance of $I(\mathbf{k}, \omega)$. The spectrum of the scattered light consists of the two Brillouin lines and the unshifted central Rayleigh line. Both of the Brillouins show the simple Lorentzian shape. The frequency shift is equal to $\pm v_s k$ and the half width is equal to (Γk^2). The unshifted central line is made up of two Lorentzian lines with half widths given by z_1 and z_2. Usually no simple physical interpretation of this line is possible, since both the half widths and the coefficients are determined by the superposition of a number of parameters, such as diffusion, thermal diffusion, and heat flow.

The physical information available from measurements of the spectral distribution of the scattered light, in the case of the Brillouin lines, is the sound velocity (hypersonic) from the Brillouin shifts and the damping for the sound waves (phonon lifetimes) from the Brillouin linewidths. The central peak usually consists of the superposition of two Lorentzians with amplitudes involving many parameters (Mountain and Deutch, 1969), and the contributions to this unshifted peak are generally more difficult to obtain. However, there are a number of binary systems that meet certain conditions resulting in a simplified expression for the central component line shape. As a result, it is often possible to obtain a specific transport coefficient from the width of the central line. For the case of a binary liquid where the heat conductivity is much greater than the diffusion ($\chi \gg D$), then $z_1 \approx \chi k^2$ and $z_2 \approx Dk^2$, and the equation for the spectral distribution of the central line becomes (Mountain and Deutch, 1969):

$$I(\mathbf{k}, \omega)_{\text{cent.}} \propto \left(\frac{\partial\varepsilon}{\partial c}\right)_{T,P}^2 \frac{k_B T}{\rho}\left(\frac{\partial\mu}{\partial c}\right)_{T,P}^{-1}\frac{2Dk^2}{(Dk^2)^2 + \omega^2}$$

$$+ \left(\rho\frac{\partial\varepsilon}{\partial\rho}\right)_{T,c}^2 k_B T K_T\left(1 - \frac{1}{\gamma}\right)\frac{2\chi k^2}{(\chi k^2)^2 + \omega^2} \tag{32}$$

The central line appears as a superposition of a sharp Lorentzian line (due to diffusion) and a highly broadened Lorentzian line (due to heat conduction). It is possible to separate the two components, and the diffusion constant can be calculated from a measurement of the central line shape. Are'fev et al. (1967) and Berge' et al. (1970) have measured the diffusion coefficient by using the central Rayleigh linewidth to be $D \approx 10^{-5}$ cm/sec. Berge' et al. (1970) were able to measure both Lorentzian components separately and they determined a value of $\chi \approx 10^{-3}$ cm/sec. Hence, the condition that $\chi \gg D$ seems to be well met in some binary liquids, and light scattering does provide a useful method of determining diffusion constants in a wide class of these liquids.

One last point to consider is in what characteristic way the spectrum of the scattered light will change as a critical point of phase separation is approached. From experimental observations of the central component for mixtures near the critical point it has been shown that the Rayleigh line becomes very intense and extremely narrow (White et al., 1966; Chu, 1967a,b). This means that the Rayleigh line is the carrier of the critical opalescence, whereas the Brillouin lines will remain relatively constant. This can be seen if Eq. (26) is rewritten in the form of the Landau–Placzek ratio and if $(\partial \mu / \partial c)$ approaches zero at the critical point. Since the critical opalescence of binary mixtures is caused by the long-range correlation of the local concentration fluctuations (Münster, 1974), any regression of these fluctuations can only proceed through nonpropagating diffusive modes, and thus will only be associated with the Rayleigh line.

C. Light Scattering Relations Applicable to Glass Systems

In the above section the necessay light scattering background for liquids was presented; now this treatment will be applied to glass. The expression for the Landau–Placzek ratio of a single-component glass, where only density fluctuations can exist, will be considered first. The first part of Eq. (25) is written in the form

$$R_{LP}(\rho) = \left[\left(\rho \frac{\partial \varepsilon}{\partial \rho} \right)_T^2 K_T \cdot T \middle/ \left(\rho \frac{\partial \varepsilon}{\partial \rho} \right)_T^2 K_s \cdot T \right] - 1 \qquad (33)$$

and it only contains that part of the Landau–Placzek ratio that is attributed to microscopic density fluctuations. The central intensity (Rayleigh intensity) is proportional to

$$I(\mathbf{k}, \omega) \propto \langle |\Delta \rho_k|^2 \rangle = \rho^2 (k_B T / V) K_{T,0} \qquad (34)$$

where the $K_{T,0}$ denotes the equilibrium isothermal compressibility of the liquid. According to Pinnow et al. (1968)

$$K_{T,0} = \delta K + K_{s,rel} + K_{s,\infty} \qquad (35)$$

where $\delta K = K_{T,0} - K_{s,0}$, the difference between the isothermal and adiabatic compressibilities; $K_{s,rel}$, the relaxational compressibility, describes the fluctuations associated with structural variations and is given by $K_{s,rel} = K_{s,0} - K_{s,\infty}$; and the last term is the high-frequency compressibility. In a viscoelastic material that supports high-frequency shear stress, $K_{s,\infty} \to [\rho v_{L,\infty}^2]^{-1}$ and $K_{s,rel} = K_{s,0} - (\rho v_{L,\infty}^2)^{-1}$, with $v_{L,\infty}$ being the hypersonic sound speed. Furthermore, Pinnow *et al.* (1968) show that δK can be decomposed into two parts, such that

$$\delta K = \delta K_{rel} + [(\alpha_\infty)^2 T/\rho c_{P,\infty}] \tag{36}$$

with α and c_P being the thermal expansion coefficient and specific heat, respectively. (α_∞ and $c_{P,\infty}$ are the high-frequency values of these parameters.) With $\alpha_\infty \ll \alpha$, $\delta K \approx \delta K_{rel}$ and Eq. (34) becomes

$$I(\mathbf{k}, \omega) \propto \langle |\Delta \rho_k|^2 \rangle = (k_B \rho^2/V)[T(\delta K_{rel}) + T K_{s,rel} + T(\rho v_{L,\infty}^2)^{-1}] \tag{37}$$

When $\omega_P \tau \gg 1$ (where τ is the average relaxation time) the oscillatory mode is not coupled to the relaxational process (Pinnow *et al.*, 1968) and the first two terms will appear in the central component (unshifted). The quantity $K_{s,rel} + \delta K_{rel}$ may simply be identified with $K_{T,rel}$, the relaxing part of the isothermal compressibility, and the nonpropagating density fluctuations in a liquid are just proportional to $k_B T K_{T,rel}$. The time scale for these density fluctuations is determined by the structural relaxation time of the liquid. As a glass is formed from the melt, by lowering the temperature, the relaxation time increases and can become very large; roughly, τ varies proportionally with the viscosity (η) as the temperature is changed (Herzfeld and Litovitz, 1959). If the rate of cooling is such that the equilibrium structure of the melt can no longer change, these density fluctuations are literally "preserved" or frozen into the newly formed solid. Hence, one can assume that the magnitude of these fluctuations is characterized by a new temperature T_f, the "fictive temperature" (Tool, 1946; Ritland, 1959). At T_f the structural relaxation time of the melt is so long that upon lowering the temperature further no structural rearrangement is possible, and T_f has become the governing temperature for the molecular structure of a glass. This represents only a first approximation, and the hypothesis is equivalent to assuming a single relaxation time instead of a spectrum of relaxation times for a system. Nevertheless, for light scattering it is a useful concept, and in going from a liquid to a glass $T K_{T,rel}(T) \to T_f K_{T,rel}(T_f)$. However, the Brillouin intensity comes only from lattice contributions, since structural contributions are inaccessible due to their long relaxation times. Consequently, the term describing the vibrational modes (phonons) of the random glass lattice need not be altered, since these phonon modes are in equilibrium at the lattice temperature T.

Following the above discussion and inserting the correct temperatures into Eq. (37) we obtain an expression of the mean square density fluctuations valid for a glass (Schroeder *et al.*, 1973)

$$\langle |\Delta \rho_k|^2 \rangle = (\rho^2 k_B / V)\{T_f(K_{T,0} - K_{s,0})$$
$$+ T_f[K_{s,0} - (\rho v_{L,\infty}^2)^{-1}] + T(\rho v_{L,\infty}^2)^{-1}\} \tag{38}$$

From an order of magnitude standpoint, in the above equation the second term dominates (Laberge *et al.*, 1973) while the remaining two terms each contribute about 8% of the total.

Consequently, from Eq. (38) the density fluctuation part of the Landau–Placzek ratio for an amorphous solid (glass) now takes the form

$$R_{LP}(\rho) = \frac{T_f}{T} \left[\frac{K_{T,0}(T_f) - (\rho v_{L,\infty}^2)^{-1}}{(\rho v_{L,\infty}^2)^{-1}} \right] \left[\left(\rho \frac{\partial \varepsilon}{\partial \rho}\right)_{T_f}^2 \bigg/ \left(\rho \frac{\partial \varepsilon}{\partial \rho}\right)_T^2 \right] \tag{39}$$

$[\rho(\partial \varepsilon / \partial \rho)]_{T_f}^2$ will differ from $[\rho(\partial \varepsilon / \partial \rho)]_T^2$ at most by a few per cent (Bucaro and Dardy, 1974b) and for most applications can be neglected from Eq. (39).

For local concentration fluctuations the second part of Eq. (25) must be modified in analogous fashion to the case of density fluctuations. Any temperature dependence in the Rayleigh part will be changed from $T \rightarrow T_f'$, where T_f' is the fictive temperature associated with thermally arrested concentration fluctuations of wave vector **k**, and in general $T_f' \neq T_f$. Thus for a n-component system in general terms the Landau–Placzek ratio for concentration fluctuations becomes

$$R_{LP}(c) = \frac{T_f' \sum_{j,K=1}^{n-1} (\partial \varepsilon / \partial c_j)_{T,P,\{c'\}} (\partial \varepsilon / \partial c_K)_{T,P,\{c'\}} (\partial \tilde{\mu} / \partial c_K)_{T,P,\{c'\}}^{-1}}{T[\rho(\partial \varepsilon / \partial \rho)]_{T,\{c\}}^2 (v_{L,\infty}^2)^{-1}} \tag{40}$$

The above relation is somewhat cumbersome to use, and in order to understand the glassy state more clearly, it becomes necessary to specialize to binary systems. We shall examine the form of $(\partial \tilde{\mu} / \partial c)_{T,P,\{c'\}}$ more closely and use a model based on the behavior of liquid mixtures near the critical point of second-order phase transitions to formulate some relations for $(\partial \mu / \partial c)_{T,P}$. For a binary liquid mixture near any critical point, the order parameter, in this case the concentration c, describing the system will exhibit very large amplitude fluctuations which relax back to their equilibrium value ever more slowly as the critical point is approached. If the order parameter is now coupled to the optical dielectric constant ε, the large amplitude fluctuations in c will produce large fluctuations in ε, resulting in rather intense light scattering. It has been shown in Eqs. (20) and (21) that the concentration fluctuations take the form

$$\langle |\Delta c_k|^2 \rangle = (k_B T / \rho V')(\partial \mu / \partial c)_{P,T}^{-1} \tag{41}$$

where $(\partial\mu/\partial c)_{P,T}$ is the concentration derivative of the chemical potential of the two constituents. For a mixture at the critical composition, the Ornstein–Zernike–Debye (OZD) theory (Fisher, 1964; Chu et al., 1969) gives this in the form

$$(\partial\mu/\partial c)_{P,T} = A(T - T_c) \tag{42}$$

where T_c is the critical temperature and A is a constant related to the osmotic pressure. For the scattered intensity due to concentration fluctuations, the result is

$$I(\langle|\Delta c|^2\rangle) \propto \left(\frac{\partial\varepsilon}{\partial c}\right)^2_{P,T} \frac{(T/T_c)}{(T/T_c) - 1} \tag{43}$$

This holds only for mixtures at the critical composition. It may be extended to other concentrations by introducing the spinodal boundary concept from the theory of metastability (Langer, 1969, 1971; Binder and Stoll, 1973; Gaunt and Baker, 1970; Benedek, 1969; Litster, 1972). Consider the mixture at some (noncritical) composition at a temperature $T > T_c$. As the temperature is lowered the magnitude of the concentration fluctuations, i.e., $\langle|\Delta c_k|^2\rangle$, grows, apparently becoming divergent at some temperature T_s lower than T_c, that is, the behavior is approximately described by $\langle|\Delta c_k|^2\rangle \propto (T/T_s)[(T/T_s) - 1]^{-1}$. This is analogous to critical behavior with the spinodal temperature T_s, defined by the equation $[\partial\mu(T_s)/\partial c] = 0$, playing the role of the critical temperature. The curve of T_s versus c defined by this relationship locates the spinodal boundary of the binary system.

The proposed extension of Eq. (42) is

$$(\partial\mu/\partial c)_{P,T} = A(T - T_s) \tag{44}$$

The linear dependence on $T - T_s$ implied by Eq. (44) is a consequence of the mean-field character of the OZD theory and has been generalized; the nonclassical result may be expressed as

$$(\partial\mu/\partial c)_{P,T} = A(T - T_s)^\gamma \tag{45}$$

Here γ is a critical exponent. Substituting Eq. (45) into Eq. (42) the Landau–Placzek ratio for concentration fluctuations in a binary liquid is proportional to:

$$R_{LP}(c) \propto \frac{(\partial\varepsilon/\partial c)^2_{P,T}}{[\rho(\partial\varepsilon/\partial\rho)]^2_{T,c}} \frac{1}{(T - T_s)^\gamma} \tag{46}$$

Up to this point only equilibrium fluids were considered; to have Eq. (46) valid for use in glassy systems, further modifications are necessary to take into account their nonequilibrium nature. Again, the "fictive temperature" concept must be invoked but this time the fictive temperature characterizes the composition fluctuations of wave vector $\mathbf{k}$ that are frozen into the glass. Consequently, the temperature T in Eq. (46) now must become T_f', the fictive temperature characteristic of concentration fluctuations. The equation describing the Landau–Placzek ratio for concentration fluctuations in glasses becomes:

$$R_{LP}(c) \propto \frac{(\partial\varepsilon/\partial c)^2_{P,T}(T_f'/T)}{[\rho(\partial\varepsilon/\partial\rho)]^2_{T,c}(T_f' - T_s)^\gamma} \qquad (47)$$

Since $[\rho(\partial\varepsilon/\partial\rho)]_{T,c}$ is essentially constant, the two parameters $(\partial\varepsilon/\partial c)_{P,T}$ and $T_f'[T_f' - T_s]^{-\gamma}$ in Eq. (47) determine the magnitude of the Landau–Placzek ratio. As T_f' approaches T_s, $[T_f' - T_s]^{-\gamma}$ becomes very large. Ultimately, the Landau–Placzek ratio will diverge (with the exponent γ) unless $(\partial\varepsilon/\partial c)$ modifies this behavior by approaching zero.

At this point it is necessary to digress and consider the relationship of the "fictive temperatures" T_f' with T_f (or T_g). Because the relaxation time associated with concentration fluctuations is generally substantially longer than the structural relaxation time τ_s, one expects that T_f' is larger than T_f. For the kth Fourier component of the concentration fluctuations the appropriate relaxation time is $(Dk^2)^{-1}$, where D is the mass diffusion coefficient. For values of k appropriate in light scattering $(Dk^2)^{-1} \gg \tau_s$ so that concentration fluctuations are "arrested" at a relatively high temperature and $T_f' > T_f$ is a reasonable assumption. Usually T_f is taken to be the temperature at which the glass viscosity is $10^{13.5}$ P (Tool, 1946) or the glass transition temperature T_g. It has been estimated (Lai *et al.*, 1975) that for a glass the average structural relaxation time $\langle\tau_s\rangle$ is several orders of magnitude smaller than the mass diffusion relaxation time τ_D, specifically $\tau_D/\tau_s \approx 10^6$. Then as a first-order approximation, T_f' may be taken to be the temperature at which the glass viscosity is about 10^7–10^8 P.

It may be instructive to return to the expression for n-component systems (in terms of concentration fluctuations only) to see whether it is possible to rewrite these equations such that they can be applied to a three-component system. Introducing statistically independent concentration variables will allow us to express the concentration fluctuation part of the Landau–Placzek ratio as a sum of squares. For a three-component system it was shown (Lekkerkerker and Laidlaw, 1972) that the fluctuations in the variable $c_{1,red}$ are statistically independent of the fluctuations in the concentration variable c_2. $c_{1,red}$ is again defined as

$$c_{1,red} = c_1 - (\partial c_1/\partial c_2)_{P,T,\tilde{\mu}_1}c_2 \qquad (48)$$

Using this new variable and following Lekkerkerker's treatment (Lekkerker and Laidlaw, 1972) the concentration fluctuation part of the Landau–

Placzek ratio for a three-component glass may be written in the following way:

$$R_{\text{LP}}(c) = \frac{T_f'}{T} \frac{v_{L,\infty}^2}{[\rho(\partial\varepsilon/\partial\rho)]_{T,\{c\}}^2} \{(\partial\varepsilon/\partial c_1)_{P,T,c_2}^2 (\partial\tilde{\mu}_1/\partial c_1)_{P,T,c_2}^{-1}$$

$$+ [(\partial\varepsilon/\partial c_2)_{P,T,c_1}^2 + (\partial\varepsilon/\partial c_1)_{P,T,c_2}^2 (\partial c_1/\partial c_2)_{P,T,\tilde{\mu}_1}^2$$

$$+ 2(\partial\varepsilon/\partial c_2)_{P,T,c_1}(\partial\varepsilon/\partial c_1)_{P,T,c_2}(\partial c_1/\partial c_2)_{P,T,\tilde{\mu}_1}]$$

$$\times [(\partial\tilde{\mu}_2/\partial c_2)_{P,T,c_1} + (\partial\tilde{\mu}_2/\partial c_1)_{P,T,c_2}(\partial c_1/\partial c_2)_{P,T,\tilde{\mu}_1}]^{-1}\} \quad (49)$$

If c_2 is small, the last term in the square brackets becomes large, with the result that only fluctuations of the 1-component will make a substantial contribution to scattering from concentration fluctuations.

Also, if very dilute ternary solutions are considered, the ratios c_1/c_3 and c_2/c_3 are small, where c_3 is the solvent, and it can be shown that $(\partial c_1/\partial c_2)_{P,T,\tilde{\mu}_1}$ becomes very small. This allows rewriting Eq. (49) as a sum of two binary-like terms:

$$R_{\text{LP}}(c) = \frac{T_f'}{T} \frac{v_{L,\infty}^2}{[\rho(\partial\varepsilon/\partial\rho)]_{T,\{c\}}^2} [(\partial\varepsilon/\partial c_1)_{P,T,c_2}^2 (\partial\tilde{\mu}_1/\partial c_1)_{P,T,c_2}^{-1}$$

$$+ (\partial\varepsilon/\partial c_2)_{P,T,c_1}^2 (\partial\tilde{\mu}/\partial c_2)_{P,T,c_1}^{-1}] \quad (50)$$

It becomes difficult to analyze the Landau–Placzek ratio for a glass of more than three components. The final expression of the Landau–Placzek ratio of a binary glass is:

$$R_{\text{LP}} = \frac{T_f}{T} \left[\frac{K_{T,0}(T_f) - (\rho v_{L,\infty}^2)^{-1}}{(\rho v_{L,\infty}^2)^{-1}} \right] + \frac{T_f'}{TN''} \frac{v_{L,\infty}^2 (\partial\varepsilon/\partial c)_{P,T}^2 (\partial\tilde{\mu}/\partial c)_{P,T}^{-1}}{[\rho(\partial\varepsilon/\partial\rho)]_{T,c}^2} \quad (51)$$

The factor $(\partial\varepsilon/\partial c)$ is of interest. The Lorentz–Lorenz equation for a binary mixture may be written in the form (Cohen *et al.*, 1971; Dubois and Berge, 1971)

$$\frac{\varepsilon - 1}{\varepsilon + 2} = \frac{4\pi\rho}{3} \left[c\frac{\alpha_1}{m_1} + (1 - c)\frac{\alpha_2}{m_2} \right] = \frac{4\pi\rho}{3} \left[\left(\frac{\alpha_1}{m_1} - \frac{\alpha_2}{m_2} \right) c + \frac{\alpha_2}{m_2} \right] \quad (52)$$

where $\rho = \{(c/\rho_1) + [(1 - c)/\rho_2]\}^{-1}$ is the density of the mixture, m_i the mass of the molecules, and α_i the molecular polarizability (subscript 1 refers to the solute, subscript 2 refers to the solvent). After some manipulation,

$$\left(\frac{\partial\varepsilon}{\partial c} \right)_{T,P} = \frac{4\pi}{3} \left(\frac{\varepsilon + 2}{3} \right)^2 \rho \left[\frac{\alpha_1}{m_1} - \frac{\alpha_2}{m_2} - \frac{3}{4\pi}\frac{\varepsilon - 1}{\varepsilon + 2}\left(\frac{1}{\rho_1} - \frac{1}{\rho_2} \right) \right] \quad (53)$$

The validity of the above relation could be checked by simply plotting the measured values of the refractive indices ($\varepsilon = n^2$) as a function of concentration of the solute, and then the resulting slope of the curve gives a value for Eq. (53). Equation (52) shows that the concentration dependence of the dielectric constant may be eliminated for a binary system in which the effective polarizabilities of the two components are equal (i.e., $\alpha_1/m_1 = \alpha_2/m_2$). Hence the light scattering spectrum for a binary mixture will reduce to one with only density fluctuations.

Equation (51) shows that the concentration derivative of $\tilde{\mu}$ can be obtained from the Landau–Placzek ratio, and from $(\partial\tilde{\mu}/\partial c)_{T,P}$ and the Gibbs–Duhem relation, the activity coefficient of solute and solvent can be evaluated (Miller, 1967). At least for the the binary system the density fluctuation part of the Landau–Placzek ratio can be evaluated completely, while for the concentration fluctuation part $[\rho(\partial\varepsilon/\partial\rho)]^2$, $v_{L,\infty}^2$ can be measured from the Brillouin intensities and shifts, respectively, and $(\partial\varepsilon/\partial c)$, the composition dependence of the refractive index, is either obtained from Eq. (53) or from measured values of refractive index versus concentration. Therefore, in principle we have a method of obtaining the concentration derivative of the chemical potential with the added possibility of obtaining activity coefficients.

D.   The Scattered Spectrum of a Dense Medium at Various Viscosity Limits

In the above development expressions are given for the total integrated intensity of liquids (low viscosity) and glasses (high viscosity). From Eq. (34) it is seen that for a simple liquid the scattered spectrum is simply proportional to the isothermal compressibility. The spectrum for the case of a nonrelaxing low-viscosity liquid then is made up of the central Rayleigh line due to non-propagating entropy fluctuations, and the Brillouin doublets due to propogating adiabatic pressure fluctuations. For this case the total intensity is taken as

$$I_T \propto (K_{T,0} - K_{s,0}) + K_{s,0} \tag{54}$$

and using

$$K_{T,0} - K_{s,0} = \alpha_0{}^2 T/\rho c_{p,0}$$

the scattered spectrum is given schematically in Fig. 1a. Here the scattered light is observed at the same polarization as the incident light (polarized scattering).

As the viscosity of the liquid increases (i.e., $1-10^5$ P) some relaxation will appear (sonic dispersion) and the spectrum will gain a third component, the Mountain line (Mountain, 1966; Montrose et al., 1968). A schematic of this type of spectrum observed when structural relaxation occurs is shown in Fig. 1b. The central (Rayleigh) component is still dominated by the thermal

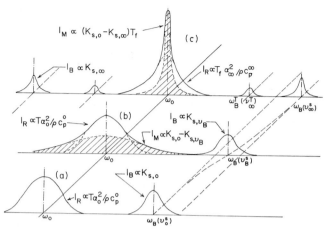

Fig. 1. Schematic representation of the scattered spectrum for a fluid at three viscosity limits. (a) Liquid (no relaxation) $\eta \simeq 10^{-2}$ P; (b) liquid (with relaxation) $\eta \simeq 1\text{--}10^4$ P, shaded portion represents the Mountain line; (c) glass $\eta \simeq 10^{13}$ P, shaded portion represents the Mountain line. Central linewidth is exaggerated for the purpose of illustration.

diffusion mode, as in the nonrelaxing case; however, an additional broad Rayleigh component will appear (shaded area in Fig. 1b). The width of this line is on the order of τ^{-1} (Mountain, 1966), where τ is the relaxation time for thermal relaxation. The shift of the Brillouin doublet will increase from the low-frequency limit value it had in the previous case to some new value $\omega_B(v_B{}^s)$. From Eqs. (35, 36) the total integrated intensity is obtained:

$$I_T \propto (T\alpha_0{}^2/\rho c_{p,0}) + (K_{s,0} - K_{s,v_B}) + K_{s,v_B} \tag{55}$$

where the adiabatic compressibilities are taken at the new frequency $(v_B{}^s)$. Again the sketch in Fig. 1b is for the polarized spectrum only; however, the depolarized part (incident polarization 90° from the scattered polarization) can give two shifted doublets that are Brillouin lines due to the presence of transverse waves, which now can propagate in the fluid as it becomes more viscous.

If the fluid has attained its glass transition temperature, its molecular configuration is "frozen in," and the relaxation times for any structural rearrangements are on the order of $10^8\text{--}10^{10}$ sec. From Eq. (38) the expression for the total intensity of a simple glass becomes

$$I_T \propto (T_f\alpha_0{}^2/\rho c_{p,0}) + T_f(K_{s,0} - K_{s,\infty}) + TK_{s,\infty} \tag{56}$$

The Brillouin doublets will be displaced by an amount proportional to the high-frequency limit value of the sound velocity $\omega(v_\infty{}^s)$ and their intensities will be related by the high-frequency limit value of the adiabatic compressibility $K_{s,\infty} = (\rho v_{L,\infty}^2)^{-1}$. The Mountain line will be much narrower than the line due to thermal diffusivity in the Rayleigh peak, but most of the intensity contribution comes from the Mountain line (Laberge *et al.*, 1973). A schematic of this case is given in Fig. 1c. We have not included any contributions from concentration (i.e., binary fluids) in the above schematics of scattered spectra of fluids at different viscosities for the sake of simplicity and clarity. As mentioned previously it is not easy to separate the Rayleigh line of a binary fluid into its components of many parameters, such as diffusion, heat flow, etc. However, for many binary solutions $\chi \gg D$ (discussed in the previous section) and the diffusion part appears as a sharp narrow peak superimposed over a much broader peak due to heat conduction (thermal diffusivity). Assuming reasonable values for the convolution of these two Lorentzians, one may frequently assign the measured peak solely to diffusion. In a later section of this chapter we shall treat the case of binary glasses (in a metastable state) to show how to separate diffusion effects from the remaining parts of the Rayleigh line and to obtain diffusion coefficients.

Table I summarizes the physical properties that can be studied by laser light scattering in simple and binary glass systems.

TABLE I

PHYSICAL PROPERTIES STUDIED BY LASER LIGHT SCATTERING FROM SINGLE-COMPONENT
AND BINARY GLASSES

	Single component glass	Binary glass
Total intensity	Isothermal compressibility $K_{T,0}(T_f)$	Activity coefficients[a] $(\partial\mu/\partial c)_{P,T}$
Landau–Placzek ratio	Isothermal compressibility $K_{T,0}(T_f)$ Scattering attenuation (α_s)	Activity coefficients[a] $(\partial\mu/\partial c)_{P,T}$ Scattering attenuation (α_s)
Width of Rayleigh line[b]	Thermal diffusivity (χ) $\chi = (\lambda/\rho c_P)$	Isothermic diffusion constant D[c]
Brillouin intensity	Pockels' coefficient (P_{ij})	Pockels' coefficient (P_{ij})
Shift of Brillouin line	Hypersonic velocity ($v_{L,\infty}$, $v_{T,\infty}$) and elastic constants C_{ij}	Hypersonic velocity ($v_{i,\infty}$, $v_{T,\infty}$) and elastic constants C_{ij}
Width of Brillouin line	Sound attenuation (phonon lifetimes)	Sound attenuation (phonon lifetimes)

[a] Independent measurements of ultrasonic compressibilities at T_f required.
[b] Only possible for sample above T_f'.
[c] $\chi \gg D$

III. Apparatus and Procedure

A. Introduction

This brief treatment of experimental techniques for measuring light scattering spectra of glass systems is restricted to two basic systems. One, a dispersive method to obtain Rayleigh and Brillouin spectra using some type of high-resolution optical spectrometer (Fabry–Perot) as an optical filter, and the other a system employed mostly for obtaining correlation functions (and thus probing the Rayleigh linewidths)—a type of digital correlation spectrometer. The Fabry–Perot spectrometer used with laser excitation and photon counting covers the entire temperature range of a glass system from liquid helium temperatures to well above the melting point of most glasses. This device resolves the Brillouin lines (measure their linewidths and shifts) and gives the Rayleigh intensity for most of the glasses of interest. Unfortunately, a measurement of scattered Rayleigh intensity only gives information about the magnitude of density and concentration fluctuations. Resolving the spectrum of the scattered Rayleigh light makes the dynamics of these fluctuations accessible. For a glass system at glass transition or below, the Rayleigh line is much too narrow to be resolved by any dispersive spectroscopic technique; however, an autocorrelation spectrometer often allows the study of the various kinds of thermodynamic fluctuation processes in glass at this temperature interval.

B. Dispersive Spectroscopy

The basic spectrometer system used with dispersive spectroscopy measurements of an optical spectrum is schematically shown in Fig. 2. The Fabry–Perot interferometer can be spherical or plane parallel, pressure swept or piezoelectrically scanned; the final choice depends upon the temperature region and the experimenters' funding. In glass systems, where the Rayleigh intensities are large and Brillouin intensities small, the instrumental contrast (discrimination) of the spectrometer can be more important than maximum transmission. Almost all glasses fluoresce, which leads to a large background in the spectrum, that can obscure some of the features of the Brillouin lines. Use of lasers that operate in the deep red region (i.e., He–Ne at 6328 Å, krypton at $\simeq 6500$ Å) and use of a narrow band filter (~ 10 Å) before the spectrometer will considerably reduce problems from fluorescence. High-resolution Fabry–Perot spectrometer systems have been described in great detail (Chiao and Stoicheff, 1964; Durand and Pine, 1968; Cummins, 1971; Schroeder *et al.*, 1973/1974).

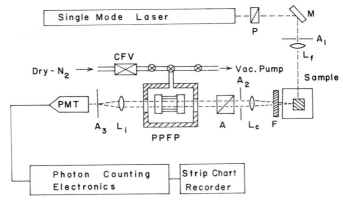

Fig. 2. Schematic diagram of the scattering geometry and detection system for Rayleigh–Brillouin spectroscopy. Nomenclature is as follows: PPFP (plane parallel Fabry–Perot); P, polarizer, Glan–laser prism; M, mirror; A_1 and A_2, apertures; A_3, pinhole; L_f, L_c, and L_i, lens—focusing, collecting, and imaging; F, narrowband filter; A, analyzer—Glan–Thompson prism; CFV, constant flow valve; PMT, photomultiplier tube.

C. Light Scattering Correlation (Autocorrelation Spectroscopy)

A schematic of a digital correlation spectrometer used by Lai *et al.* (1975) to study structural relaxation phenomena in glasses is shown in Fig. 3. Here the beam of an Argon–ion laser is focused into the sample volume. The light scattered at 90° is detected with a photomultiplier whose output pulses are processed by the digital correlation spectrometer. For details see Jakeman and Pike (1969), Jakeman *et al.* (1970); Nossal *et al.* (1971); and Lai (1973). Thus performing light scattering experiments by measuring the temporal autocorrelation function enables one to acquire data on thermodynamic

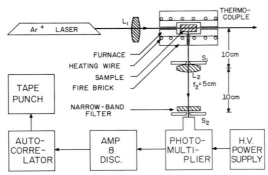

Fig. 3. Schematic of scattering geometry and detection system of a correlation spectrometer. (After Lai *et al.*, 1975.)

parameters and transport coefficients of the medium. Structural relaxation parameters are obtained from the correlation function describing light scattering by density fluctuations. If in addition to density fluctuations the system has concentration fluctuations and/or fluctuations in the anisotropy of the polarizability due to molecular reorientation, then it is possible to measure the thermal conductivity, the mass diffusion coefficient and reorientational aspects of molecular dynamics. Lai *et al.* (1975) have estimated that for a glass system the average structural relaxation time $\langle \tau_s \rangle$ is several orders of magnitude different from both the thermal diffusion relaxation time τ_χ and the mass diffusion relaxation time τ_D (i.e., $\tau_D / \langle \tau_s \rangle \approx 10^6$, $\tau_\chi / \langle \tau_s \rangle \approx 10^{-2}$). Hence, it is possible to measure and identify every contribution to the Rayleigh line for a binary glass system.

IV.   Results—Discussion

A.   Rayleigh and Brillouin Scattering from Simple Glasses at 300 K

The Rayleigh–Brillouin spectrum at room temperature for a single-component glass such as SiO_2 (Homosil) will be treated in full detail with respect to the different polarization modes that are possible in light scattering. The validity of Eq. (39) for $R_{LP}(\rho)$ will also be established and B_2O_3 and GeO_2 will be discussed. Typical traces of the SiO_2 spectrum for a 90° scattering angle are displayed in Fig. 4 for V and H incident light polarization. Here H lies in the scattering plane and V is normal to this plane. In the first two traces of Fig. 4 intensities from $VV + VH$ and $HH + HV$ contributions are shown; in traces in Fig. 4c an analyzer was used in conjunction with the polarizer to remove any degeneracy brought about by any overlap of intensities attributable to the different polarization modes. The $VV + VH$ spectrum exhibits the longitudinal Brillouin doublets and the central Rayleigh component similar to a liquid (except that the linewidths for the glass are extremely narrow), but in addition a transverse doublet (shear phonons) is also observed for the solid. The $HH + HV$ spectrum again exhibits two shifted sets of doublets and the central Rayleigh components. The doublet previously identified as the transverse component has the same height as in the $VV + HV$ case, whereas the doublet at the longitudinal position is greatly diminished in height. By using the analyzer we are able to identify what part of the spectrum belongs to each scattered polarization component. The middle trace in Fig. 4c shows that the transverse Brillouins are purely VH or HV; they also show that the HV and VH spectra are identical, in full agreement with the work of Mueller (1938). Moreover, there exists no $VH(HV)$ contribution to the longitudinal Brillouin components. The top and bottom traces in Fig. 4c confirm that the longitudinal

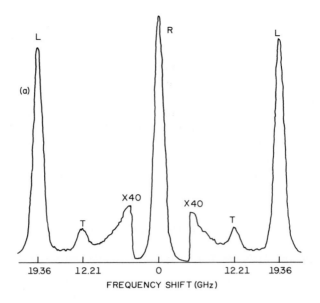

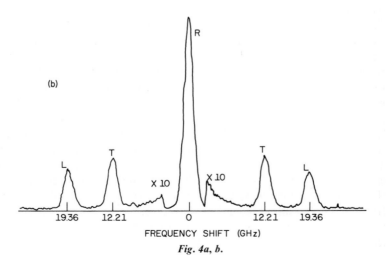

Fig. 4a, b.

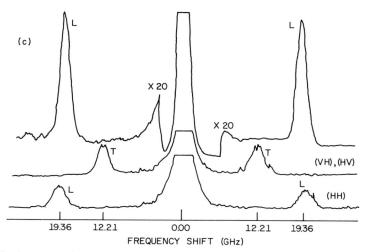

Fig. 4. Spectrum of scattered light from fused SiO_2 with different incident and scattered polarizations. (a) Observed Rayleigh–Brillouin scattering $VV + VH$. (b) Observed Rayleigh–Brillouin scattering $HH + HV$. (c) (*top*) Brillouin scattering (VV), (*middle*) Brillouin scattering (VH) or (HV) and (*bottom*) Brillouin scattering (HH). R, L, and T mark the Rayleigh line, longitudinal, and transverse Brillouin lines, respectively. (After Schroeder, 1974).

Brillouin components consist of pure VV and HH polarization only and the transverse Brillouins have no VV or HH part. The central or Rayleigh component exhibited both a polarized and depolarized component with the depolarized amplitude being about 5% of the polarized amplitude.

The above example has attributed Brillouin intensities of a simple glass to their respective modes, resulting in a set of "selection rules" for Brillouin scattering in glasses. This aspect of light scattering has been treated theoretically by a number of authors (Mueller, 1938; Born and Huang, 1954; Gammon, 1969; Schroeder, 1974) with which our empirical observations agree fully. In all of the above traces, the apparent linewidths of both Rayleigh and Brillouin lines are caused by a combination of the instrumental width of the laser line and Fabry–Perot spectrometer. To resolve the narrow linewidths of the glass one would need a single frequency laser together with a confocal spherical Fabry–Perot interferometer for the Brillouin lines, whereas the Rayleigh line would require a correlation spectrometer.

The light scattering data has been obtained for the three simple glasses (Schroeder, 1974); namely, SiO_2, B_2O_3, and GeO_2. For the first two, sufficient compressibility data was available to use Eq. (39) to calculate their Landau–Placzek ratios. A summary of this comparison of the Landau–Placzek ratios between measured and calculated values is shown in Table II. The depolarization ratio (I_{HV}/I_{VV}) for SiO_2 was measured to be about 0.05, whereas

TABLE II

COMPARISON OF LANDAU–PLACZEK RATIOS FOR SINGLE-COMPONENT GLASSES
AND NECESSARY PARAMETERS TO CALCULATE THE LANDAU–PLACZEK RATIO
FROM EQ. (39) FOR THESE GLASSES

Sample	T_f (K)	$K_{T,0}(T_f)$ (cm^2/dyn)	C_{11} (10^{10} dyn/cm^2)	R_{LP} (calc)	R_{LP} (meas)
SiO$_2$	1473[a]	6.80 × 10^{-12}[d]	77.857[f]	21.5	21.9[f]
B$_2$O$_3$	553[b]	39.0 × 10^{-12}[e]	20.71[f]	14.3	13.3[f]
GeO$_2$	828[c]	—	51.54[f]	—	24.3[f]

[a] Brückner (1964), Saka and MacKenzie (1971).
[b] Leidecker *et al.* (1971).
[c] Kurkjian and Douglas (1960).
[d] Laberge *et al.* (1973), Schroeder *et al.* (1973).
[e] Macedo and Litovitz (1965), Bucaro and Dardy (1974b).
[f] Schroeder (1974).

B$_2$O$_3$ gave a value of 0.30. This rather large value for B$_2$O$_3$ means that its measured Landau–Placzek ratio must be corrected according to a scheme proposed by Bucaro and Dardy (1974b) before any comparison with a calculated value will have validity. The small depolarization ratio of SiO$_2$ makes a correction for anisotropy scattering unnecessary. That SiO$_2$ should show so little anisotropic scattering compared to B$_2$O$_3$ can qualitatively be reconciled by considering the different structures now widely accepted for SiO$_2$ and B$_2$O$_3$, i.e., (SiO$_4$) tetrahedra versus (BO$_3$)$_N$ ribbons (Wemple *et al.*, 1973). Thus, optically the SiO$_4$ tetrahedra can be considered very nearly isotropic; however, the BO$_3$ ribbons (planar) are rather anisotropic (Bucaro and Dardy, 1974b). In Eq. (39) the factor $[\rho(\partial\varepsilon/\partial\rho)]^2_{T_f}/[\rho(\partial\varepsilon/\partial\rho)]_T^2$ is also included; nevertheless for B$_2$O$_3$ or SiO$_2$ it does not seem to vary by more than a few percent (Bucaro and Dardy, 1974a,b), so it was neglected entirely in these calculations.

The good agreement between measured and calculated values of the Landau–Placzek ratio for the two simple glasses does confirm the validity of the model used for light scattering from density fluctuations. Thus, the density fluctuations in the solid glass phase (in a nonequilibrium state) are governed by the equilibrium compressibility at the fictive temperature of the glass melt. No ultrasonic data for the GeO$_2$ melt was available, making the Landau–Placzek ratio comparison not feasible for this simple glass; nevertheless, the measured value is given. The depolarization ratio (I_{VH}/I_{VV}) for GeO$_2$ was measured to be about 0.31 (Schroeder, 1974) and again $\frac{4}{3}I_{HV}$ was subtracted from I_{VV} to give a measured value of the Landau–Placzek ratio that reflects only density fluctuations. This large anisotropy in GeO$_2$ may be reconciled by considering that GeO_2 consists of both sixfold and fourfold coordinated compounds (Wemple, 1973) and the sixfold coordinated GeO_2 (hexagonal) must be optically very anisotropic.

B. Rayleigh and Brillouin Scattering from Binary Glasses at 300 K

1. STATIC PROPERTIES—FROZEN-IN CONCENTRATION FLUCTUATIONS

Light scattering from some binary alkali–silicate glasses will be considered in this section. In Figs. 5, 6, and 7 the Landau–Placzek ratios are plotted as functions of concentrations for the binary alkali–silicates. For the two binary systems consisting of $K_2O\cdot SiO_2$ and $Na_2O\cdot SiO_2$, respectively, it is apparent from Fig. 5 and Fig. 6 that a major portion of the scattering takes place in a specific concentration region for each glass family. The $K_2O\cdot SiO_2$ system exhibits intense Rayleigh scattering from a few mole % K_2O to about 20 mole % K_2O; in the $Na_2O\cdot SiO_2$ system this region extends almost from pure silica to about 25 mole % Na_2O. Besides the sharply peaked contribution there seems to be a broad background spread out over the entire concentration range for both of the above samples; this is denoted by the broken line. For the $Li_2O\cdot SiO_2$ and $TiO_2\cdot SiO_2$ systems in Fig. 7 no such pronounced

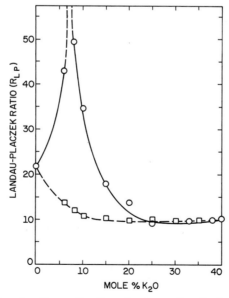

Fig. 5. Landau–Placzek ratio as a function of concentration. Dashed lines show contribution from density fluctuations computed from Eq. (39). $\bigcirc = R_{LP}$ (total); $\square = R_{LP}$ ($\langle \Delta \rho^2 \rangle$) (After Schroeder *et al.*, 1973)

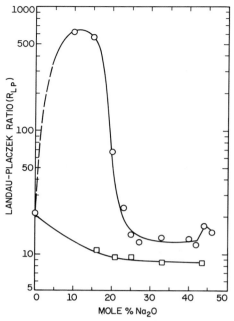

Fig. 6. Landau–Placzek ratio as a function of concentration. Dashed lines show contribution from density fluctuations computed from Eq. (39). $\bigcirc = R_{LP}$ (total); $\square = R_{LP}$ ($\langle \Delta \rho^2 \rangle$). (After Schroeder, 1974.)

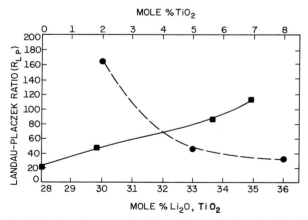

Fig. 7. Landau–Placzek ratio as a function of concentration for the $Li_2O \cdot SiO_2$ and $TiO_2 \cdot SiO_2$ systems. $\blacksquare = (X)TiO_2(100-X)SiO_2$; $\bullet = (Y)Li_2O(100-Y)SiO_2$, (After Schroeder, 1974.)

effects could be detected since in both systems samples of adequate optical quality could be obtained only for limited concentration ranges. The $Li_2O \cdot SiO_2$ measurements range from about 29 to 37 mole % Li_2O with phase separation (no longer a homogeneous glass) causing opalescence at the lower concentration edge and crystallization making the sample opaque beyond the 37 mole % Li_2O.

The values of the Landau–Placzek ratio ran from a low of 9.24 for $25K_2O \cdot 75SiO_2$ to a high value of 618.41 for $10Na_2O \cdot 90SiO_2$. In order to be able to study only the part due to concentration fluctuations the broad background, attributable to density fluctuations, was calculated with the aid of Eq. (39). High-temperature ultrasonic data (Bockris *et al.*, 1955; Bockris and Kojonen, 1960; Bloom and Bockris, 1957; Laberge *et al.*, 1973) and high-temperature density data from Shartsis *et al.* (1952) and Shermer (1956) was used to calculate the equilibrium compressibility at the fictive temperature for every glass system considered. Fictive temperatures were obtained from the viscosity data of Poole (1949) and Schnaus *et al.* (1976) and the elastic constants C_{11} are from the hypersonic Brillouin data of Schroeder (1974). Superimposed on this broad background due to density fluctuations is a rather pronounced peak in $Na_2O \cdot SiO_2$ and $K_2O \cdot SiO_2$ which is attributed to concentration fluctuations. Figures 5 and 6 suggest that density fluctuations are responsible for almost all of the light scattered for alkali concentrations greater than 25 mole % in both the $Na_2O \cdot SiO_2$ and $K_2O \cdot SiO_2$ glasses. The scattering attributed to concentration fluctuations reaches maximum values at 10 and 7 mole % alkali oxide for $Na_2O \cdot SiO_2$ and $K_2O \cdot SiO_2$, respectively. For concentrations less than these maximum values, the intensity decreases rapidly until the pure SiO_2 value is reached. For the $Li_2O \cdot SiO_2$ and $TiO_2 \cdot SiO_2$ glasses given in Fig. 7, the concentration region covered was not large enough to make a comparable analysis, but qualitatively the trend is the same as for the two binary glasses.

To account quantitatively for the observed excess scattering a model was developed in a previous section of this chapter which will now be applied to the binary alkali–silicate glasses. The model for concentration–fluctuation scattering is based on a spinodal temperature (boundary), which will be reviewed briefly.

Gibbs (1906) considered the stability of a phase to infinitesimal fluctuations and established a stability criterion for the existence of metastable equilibrium. Gibbs considered two categories of infinitesimal changes to which a metastable phase must be immune. The first is a microscopic part of a new and more stable phase. If the surface tension is positive, then the phase is always metastable to the former fluctuation, and only a finite fluctuation may bring about instability to the original phase. This finite fluctuation is called a nucleus. The second is infinitesimal composition or density fluctuations. A necessary condition for stability of a phase to microscopic composition or density fluctuations is that the chemical potential of each component increases as the density of that component increases (Cahn and Charles, 1965). Certain systems exhibit a temperature, which is a function of composition, below which this criterion is violated and this limit to metastability has been denoted by the van der Waal school (Prigogine and Defay, 1954; Skripov, 1974) as the spinodal curve. Hence, the spinodal is merely the stability boundary of a metastable phase relative to continuous changes in state, and it may be approached without disturbing the macro-scopic homogeneity of the substance (Skripov, 1974). In Fig. 8 is shown a hypothetical temperature–composition curve for a binary system. The spinodal line is defined by the locus of points given by $(\partial^2 G/\partial c^2)_{P,T} = 0$, and it separates the unstable and metastable parts of a two phase region. Hilliard (1970) has shown that if the initial composition lies outside the spinodal, $(\partial^2 G/\partial c^2)_{P,T} > 0$ and an infinitesimal fluctuation increases the free energy, making the system metastable. Inside the spinodal $(\partial^2 G/\partial c^2)_{P,T} < 0$, and any small fluctuation decreases the free energy of the system, making it unstable. From Eqs. (20) and (21)

$$I(\mathbf{k}, \omega) \propto \langle|\Delta c_k|^2\rangle = (k_B T_f'/N'')(\partial\mu/\partial c)_{P,T_f}^{-1}$$

so that when the composition of a binary glass system is near the spinodal boundary, where $(\partial\mu/\partial c)_{P,T}$ is small, the scattering should become large.

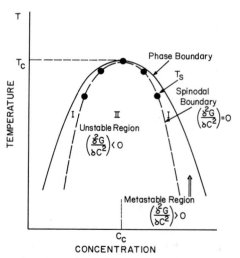

Fig. 8. Schematic phase diagram for a binary glass system. Solid lines represent the phase boundary (coexistence curve) and dashed lines show the spinodal curve. I denotes the metastable region and II the unstable region.

Neilsen (1969), Tomozawa *et al.* (1969, 1970), Andreev *et al.* (1970), and Hammel (1965) have studied phase separation in $Na_2O \cdot SiO_2$ glasses by small-angle x-ray scattering, electron microscopy, and in the case Andreev, x-ray scattering together with visible light scattering and explained the results in terms of spinodal decomposition.

Haller *et al.* (1974) have calculated the spinodal boundary for the $Na_2O \cdot SiO_2$ system. The fictive temperature that characterizes the composition fluctuations of the same system has been determined from available viscosity–temperature measurements of Schnaus *et al.* (1976). These results are plotted in Fig. 9 and we have based an explanation for the excess scattering in the $Na_2O \cdot SiO_2$ system on Eq. (47) and Fig. 9. It is seen that the fictive temperature T_f' and the spinodal temperature T_s almost coincide in the concentration range from about 8 to 20 mole % Na_2O, which indicates that the factor $[T_f' - T_s]^{-1} T_f'$ becomes very large in this concentration range. Consequently, the scattered light should be very large in this region and, indeed, this is borne out by the measured results for the $Na_2O \cdot SiO_2$ system.

For the $K_2O \cdot SiO_2$ system the existence of an immiscibility dome at the low alkali concentration has been confirmed by Schroeder *et al.* (1975) and

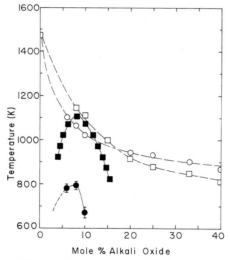

Fig. 9. Comparison of the spinodal temperatures (T_s) and the fictive temperatures (T_f') for concentration fluctuations for the $Na_2O \cdot SiO_2$ and $K_2O \cdot SiO_2$ glass systems. ● = denotes T_s for $K_2O \cdot SiO_2$ and ■ = denotes T_s for $Na_2O \cdot SiO_2$. ○ = denotes T_f' for $K_2O \cdot SiO_2$ and □ = denotes T_f' for $Na_2O \cdot SiO_2$. The spinodal temperatures for $Na_2O \cdot SiO_2$ were obtained from Haller *et al.* (1974) and for $K_2O \cdot SiO_2$ from Schroeder *et al.* (1975). The fictive temperatures (T_f') for $K_2O \cdot SiO_2$ are from Schnaus *et al.* while the T_f' for $Na_2O \cdot SiO_2$ are from Schnaus *et al.* (1976).

with available viscosity–temperature data (Schnaus *et al.*) one may apply the same kind of reasoning as used in the $Na_2O \cdot SiO_2$ system to explain the light scattering behavior of this particular system. In Fig. 9 we have plotted the T_s and T_f' temperatures as a function of concentration for the $K_2O \cdot SiO_2$ system. The minimum observed at 25 mole % K_2O is plausible and as the K_2O content is reduced the scattered intensity increases, since the fictive temperature T_f' approaches the spinodal temperature T_s, causing an increase in the scattering from concentration fluctuations as predicted by Eq. (47).

For the $Li_2O \cdot SiO_2$ system the spinodal temperatures as a function of concentration are available from Charles (1967) and Haller *et al.* (1974); T_f' can be estimated from viscosity data of Schnaus *et al.* If T_f' and T_s are considered as functions of concentration, the light scattering for the $Li_2O \cdot SiO_2$ system can be explained from the miscibility gap for concentrations greater than 30 mole % Li_2O.

For the binary system $TiO_2 \cdot SiO_2$ the scattering data appear to indicate the existence of an immiscibile boundary. However, the density fluctuation part of the scattering cannot be calculated exactly, because the low-frequency compressibility is not known. The concentration range studied was also too limited to make any definite conclusions.

2.   KINETICS OF CONCENTRATION FLUCTUATIONS IN GLASS SYSTEMS

The intensity of the scattered light is determined by the time average of a fluctuating quantity, so measurement of the intensity alone cannot provide any information about the dynamics of these fluctuations. To study the dynamics of a system the spectral distribution of the scattered light must be measured. The direct measurement of the Rayleigh linewidth is not feasible in a glass at high viscosities, but there is a method based on annealing studies that allows the determination of relaxation times (i.e., information that previously could only be obtained in an equilibrium liquid from line-width measurements) and the spinodal temperatures of this system. This discussion follows the work of Schroeder *et al.* (1975) where glasses in nonequilibrium states are heat treated to reach equilibrium states at the particular heat-treatment temperatures.

In a viscous liquid, the diffusion coefficient can be quite small (less than 10^{-10} cm²/sec) so that for typical values of $\mathbf{k}$ ($\sim 10^5$ cm^{-1} or so) the line-widths are too narrow to be resolved by dispersive spectroscopic techniques. Consequently, only intensity, i.e., thermodynamic, information can be extracted from the data.

If the viscosity of the liquid is increased to about 10^{12} P by rapid under-cooling so that it becomes a glass, the time scale of molecular motions becomes so long that the material behaves as an amorphous solid for normal observation times. At some temperature—designated by T_f'—the variations or fluctuations in composition are arrested, "permanently" set in the material, so that further cooling produces no change in their magnitude. (T_f' depends upon the rate of cooling: for slow cooling T_f' will be less than for sudden quenching.) The intensity I measured at a temperature $T < T_f'$ is given (at least approximately) by

$$I \propto \langle |\Delta c_k|^2 \rangle_{T_{f'}} = k_B T_f'(N'')^{-1}(\partial\mu/\partial c)_{P,T_{f'}}^{-1}$$

where $\langle \ \rangle_{T_{f'}}$ denotes an average over the equilibrium ensemble characteristic of the temperature T_f'.

If a sample is held to equilibrium at a temperature T_1, the fluctuations in composition are given by Eq. (32), with $T = T_1$. At some time $t = 0$ the sample (in equilibrium at T_1) is placed in contact with a thermal bath at a temperature $T_2 < T_1$. In time the state of the system, initially characterized by T_1, will evolve to a state characterized by the temperature T_2. Therefore, by measuring the light scattering intensity I at various times $t > 0$, the mean square composition fluctuations from $\langle |\Delta c_k|^2 \rangle_{T_1}$ at $t = 0$ to $\langle |\Delta c_k|^2 \rangle_{T_2}$ as $t \to \infty$ can be studied.

These "instantaneous" measurements are performed by interrupting the heat treatment at various times and suddenly quenching the sample to a low temperature. If the sample is sufficiently viscous, this sudden quench to some low temperature at which molecular movements are slow leaves the structure essentially unchanged and the instantaneous configuration of the material is "frozen in." Thus the composition fluctuation dynamics of highly viscous liquids (or molten glasses) can be studied on time scales as long as hundreds of hours, without placing impossibly severe resolution criteria on the optical scattering spectrum analyzer. Schroeder *et al.* (1975) have carried out such experiments for various temperatures (values of T_2) in several alkali–silicate glass melts.

To analyze experiments of this type some theoretical framework that describes the behavior of the composition fluctuations during the relaxation to equilibrium of a system from an initial nonequilibrium configuration must be developed. For a system in equilibrium at temperature T the mean square composition fluctuation is:

$$\langle |\Delta c_k|^2 \rangle_T = \frac{1}{Q_T} \int dX \, f_T(X) \, \Delta c_k(X) \, \Delta c_k(X) \tag{57}$$

where X denotes all the phase space position and momentum variables, $\Delta c_k(X)$ gives the fluctuation in the kth component of the concentration as a function of these variables, $f_T(X)$ is the equilibrium distribution function and $Q_T = \int dX \, f_T(X)$. The state of a nonequilibrium system is characterized by a distribution function $f(X, t)$ that depends explicitly on time. The experimental conditions described above correspond to the requirements that

$$f(X, t) = f_1(X) \qquad t \leq 0 \tag{58a}$$

and

$$f(X, t) = f_2(X) \qquad t \to \infty \tag{58b}$$

where $f_1(X)$ and $f_2(X)$ are the equilibrium distribution functions corresponding to temperatures T_1 and T_2, respectively.

If even in its nonequilibrium states the light scattering intensity is proportional to the instantaneous mean square composition fluctuations; that is, the intensity observed at time t is

$$I(t) \propto \langle |\Delta c_k|^2 \rangle_t \tag{59}$$

where $\langle \ \rangle_t$ denotes the average with respect to the instantaneous nonequilibrium ensemble, then

$$\langle |\Delta c_k|^2 \rangle_t = \int [dX \, f(X, t) \, \Delta c_k(X) \, \Delta c_k(X) \Big/ \int dX \, f(X, t)] \tag{60}$$

The approach of composition fluctuations to their equilibrium value involves the dynamics of the quantity.

$$\langle A_k \rangle_t = \langle |\Delta c_k|^2 \rangle_t - \langle |\Delta c_k|^2 \rangle_{T_2} \tag{61}$$

For departures from equilibrium sufficiently small that $f(X, t)$ differs from $f_2(X)$ only through terms linear in the deviations, e.g., in the A_k, (Zwanzig, 1965; Mori, 1964):

$$\langle A_k \rangle_t = (\langle A_k \rangle_1 / \langle |A_k|^2 \rangle_2) \langle A_k(0) A_k(t) \rangle_2 \tag{62}$$

where $A_k(t) = \Delta c_k^2(t) - \langle |\Delta c_k|^2 \rangle_2$, and $\langle A_k(0) A_k(t) \rangle$ is its time autocorrelation function in the equilibrium ensemble at temperature T_2. Treating Δc_k as a Gaussian random variable (Davenport and Root, 1958; VanKampen, 1965) gives

$$\langle A_k(0) A_k(t) \rangle_2 = 2 \langle \Delta c_k(0) \, \Delta c_k(t) \rangle^2 \tag{63}$$

and if one assumes a simple diffusion model

$$\langle \Delta c_k(0) \, \Delta c_k(t) \rangle = \langle \Delta c_k^2 \rangle \exp(-Dk^2 t) \tag{64}$$

Using Eqs. (63) and (64) in Eq. (62) gives

$$\langle A_k \rangle_t = [\langle |\Delta c_k|^2 \rangle_1 - \langle |\Delta c_k|^2 \rangle_2] \exp(-2Dk^2 t) \tag{65}$$

This equation gives the relaxation of the nonequilibrium system in terms of equilibrium fluctuations with the relaxation time

$$\tau = (2Dk^2)^{-1} \tag{66}$$

In terms of the scattered light, Eqs. (58) and (59) show that $\langle A_k \rangle_t \propto I(t) - I(\infty)$ and $\langle |\Delta c_k|^2 \rangle_1 - \langle |\Delta c_k|^2 \rangle_2 \propto I(0) - I(\infty)$. Consequently, using Eq. (65), the time variation of the intensity is

$$I(t) = I(\infty) + [I(0) - I(\infty)]e^{-t/\tau} \qquad (67)$$

This equation can be compared readily with experimental results. In Fig. 10 typical scans versus time are shown for $K_2O \cdot SiO_2$ glasses.

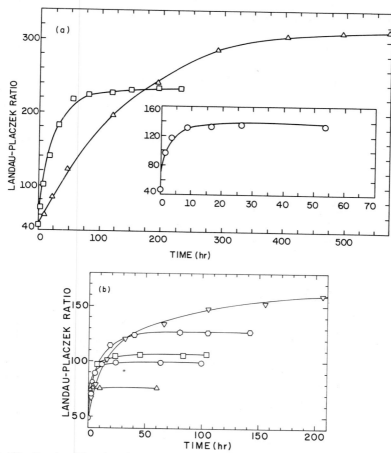

Fig. 10. Landau–Placzek ratios as a function of times and temperatures for the $K_2O \cdot SiO_2$ system. (a) $6K_2O \cdot 94SiO_2$ glass—the symbols for the different heat treatment temperatures are: $\bigcirc$ for $T = 637.8°C$, $\square$ for $T = 604.8°C$, and $\triangle$ for $T = 579.0°C$. (After Schroeder, *et al.* 1975.) (b) $8K_2O \cdot 92SiO_2$ glass—the symbols for the different heat treatment temperatures are: $\triangle$ for $T = 627.0°C$, $\bigcirc$ for $T = 613.5°C$, $\square$ for $T = 603.5°C$, $\bigcirc$ for $T = 589.7°C$, and $\triangledown$ for $T = 573.3°C$.

Only contributions from local concentration fluctuations are shown, and any contributions from density fluctuations have been calculated with the aid of Eq. (39) and subtracted from the total measured values. The contribution from density fluctuations is found to be independent of both time and heat-treatment temperature, since they are frozen in at a much lower temperature than the composition fluctuations; namely, at the annealing or glass transition temperature corresponding to a viscosity of $10^{13.5}$ P.

In fitting the data to Eq. (67), the linearity embodied in Eqs. (62)–(66) has been tacitly assumed. In Fig. 11 the quality of these fits is illustrated. The assumption of linearity is supported by the work of Mohr and Macedo (1975) in which similar heat treatments were performed. These authors observed the approach to equilibrium at a temperature T of a borosilicate glass, using several initial temperatures T_1. Values of $T_1 > T$ and $T_1 < T$ were examined in their experiments. They found that the relaxation times were independent of T_1 and depended only on the equilibrium temperature for values of $|T_1 - T|$ comparable to those used in this study. Moreover, the intensity changes $|I(\infty) - I(0)|$ were in some instances larger than those

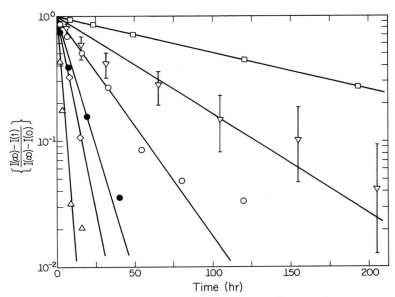

Fig. 11. The relaxation function $[I(\infty) - I(t)]/[I(\infty) - I(0)]$ versus time for some select heat treatment temperatures in several alkali silicate glasses. The symbols denote the following: for $6K_2O \cdot 94SiO_2$, $\square$ at $T = 579.0°C$, $\bigcirc$ at $T = 604.8°C$, $\triangle$ at $T = 637.8°C$; for $8K_2O \cdot 92SiO_2$, $\bullet$ at $T = 589.7°C$, $\triangledown$ at $T = 573.3°C$; and for $10K_2O \cdot 90SiO_2$, $\diamondsuit$ at $T = 573.4°C$. (After Schroeder *et al.*, 1975.)

reported here. Thus, the results can be interpreted in terms of linear relaxation theory as given by Eq. 67.

From Eqs. (45) and (41), we obtain an expression of the form

$$I(\infty) \propto k_B T/(T - T_s)^\gamma \qquad (68)$$

which describes the dependence of the equilibrium intensity on the heat treatment temperature T. Here $T_s(c)$ is the spinodal temperature for the concentration c, and γ is the exponent characterizing the divergence of $(\partial \mu / \partial c)$. Classically, $\gamma = 1$, whereas in modern critical point theories $\gamma = 1.25$. From the experimental variation of $I(\infty)$ with T, values of $T_s(c)$ and γ can be obtained. In Fig. 12 are shown typical curves of $[T/I(\infty)]^{1/\gamma}$ versus T for $8K_2O \cdot 92SiO_2$ glass samples. From Eq. (68) the temperature axis intercept is T_s. The plots show that the data are described equally well by either $\gamma = 1$ or 1.25. The inability to specify γ from the data arises because of the limited range of temperatures over which measurements could be made. For the data in Fig. 12 the reduced temperature $(T - T_s)/T_c$ varied by a factor of only about 2.5 (Schroeder et al., 1975), with $T_c = 800$ K. For the $8K_2O \cdot 92SiO_2$ sample, T_s values of 801.5 K and 788.0 K were obtained for the $\gamma = 1$ and $\gamma = 1.25$ assumptions, respectively.

It is possible that the linear extrapolation breaks down near the spinodal boundary, which however, would not lead to substantial deviations in the

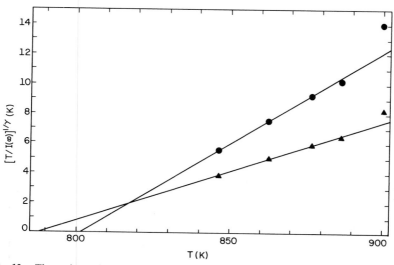

Fig. 12. The reciprocal equilibrium intensity and temperature $[T/I(\infty)]^{1/\gamma}$ as a function of temperature for the $8K_2O \cdot 92SiO_2$ system; ● denotes $\gamma = 1$ was used and ▲ signifies that $\gamma = 1.25$ was employed. (After Schroeder et al., 1975.)

determination of T_s. For the three compositions studied, the spinodal temperatures and probable uncertainties are (Schroeder *et al.*, 1975)

$$C_1 = 6K_2O \cdot 94SiO_2 \qquad T_s = 783 \pm 15 \text{ K}$$
$$C_2 = 8K_2O \cdot 92SiO_2 \qquad T_s = 793 \pm 15 \text{ K}$$
$$C_3 = 10K_2O \cdot 90SiO_2 \qquad T_s = 673 \pm 25 \text{ K}$$

Since the glass transformation temperatures for these three compositions are approximately 890, 820, and 781 K, respectively, the actual phase separation implied by these spinodal temperatures is not practically observable because of kinetic limitations. Nevertheless, it has been shown that light scattering combined with annealing studies made it possible to detect the spinodal boundary for the $K_2O \cdot SiO_2$ glass system. Locating the spinodal temperatures or showing that an immiscibility dome does exist for $K_2O \cdot SiO_2$ was almost impossible by convential methods (Haller *et al.*, 1974) of clearing temperatures or maximum opalescence temperatures. Charles (1967) estimated the critical point for this system to be about 350°C at a K_2O mole fraction of 0.045 and as a result the experimental demonstration of phase separation in $K_2O \cdot SiO_2$ would not be expected. However, light scattering, which is sensitive to changes in the local dielectric constant caused by changes in the magnitude of the concentration fluctuations, which in turn are determined by $(\partial\mu/\partial c)_{P,T}$, can still provide information about an immiscible system, even if this immiscible system has a spinodal boundary with a critical temperature less than the glass transition temperature.

From Fig. 11 the relaxation times of the $K_2O \cdot SiO_2$ glasses increase dramatically as the spinodal temperatures are approached. Such behavior is expected from the Cahn–Cook theory (Cahn, 1961; Cook, 1970) of spinodal decomposition, where the relaxation times (in the long wavelength limit) are given by

$$1/\tau \propto M(\partial\mu/\partial c)_{P,T}k^2 \qquad (69)$$

where M is the diffusional mobility, and $M \propto T/\eta$, where η is the shear viscosity. Since $(\partial\mu/\partial c)_{P,T} \propto (T - T_s)^\gamma$:

$$(\tau T/\eta) \propto (T - T_s)^{-\gamma} \qquad (70)$$

$(\tau T/\eta)$ should diverge in the same fashion as the scattered intensity $I(\infty)$.

Alternatively Kawasaki (1970) suggested a mode–mode coupling approach to critical point kinetics. In his description the relaxation times are governed by the correlation length ξ in addition to the viscosity. Specifically, for $\xi k \ll 1$

$$(\tau T/\eta) = (6\pi/k_B)(\xi/k^2) \propto (T - T_s)^{-\nu} \qquad (71)$$

where we have made use of the relations $\xi \propto (T - T_s)^{-\nu}$ and $2\nu = \gamma$ and have written T_s the spinodal temperature, in place of the critical temperature T_c. Kawasaki considered only critical composition systems; Eq. (71) is proposed as a reasonable generalization to other concentrations not far from critical.

Since the Cahn–Cook and Kawasaki approaches imply markedly different temperature dependences for the quantity $(\tau T/\eta)$, it should be possible to test their applicability to the $K_2O \cdot SiO_2$ binary system from the temperature variation of the product $(\tau T/\eta)\,[T/I(\infty)]$. In the Cahn–Cook theory this quantity is independent of the temperature, whereas Kawasaki predicts proportionality to $(T - T_s)^{\gamma/2}$. A comparison is provided in Fig. 13. Shear viscosity values were obtained from Schnaus et al. From the data in the figure it is not possible to distinguish between the two approaches. This result is not too surprising in view of the relatively large reduced temperatures in this study, i.e., $0.066 \lesssim (T - T_s)/T_c \lesssim 0.13$ for the 8% K_2O sample with T_c taken $\simeq 800$ K.

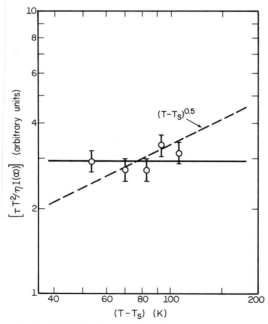

Fig. 13. The product $(\tau T/\eta)\,[T/I(\infty)]$ plotted as a function of $(T - T_s)$ for the $8K_2O \cdot 92SiO_2$ sample. The error bars on the points indicate our estimate of the most probable errors in the data. (After Schroeder et al., 1975.)

C. Light Scattering from Pseudobinary, Ternary, and Other Multicomponent Glass Systems

In multicomponent glasses the application of Eqs. (39) and (40) to quantitative results is quite difficult. High-temperature ultrasonic data are usually not available for the specific systems, so calculations of the density fluctuation contributions are not possible. It is also difficult to determine $(\partial \varepsilon / \partial c_j)(\partial \varepsilon / \partial c_K)$ and $(\partial \tilde{\mu}_j / \partial c_K)$ for multicomponent systems. Nevertheless, much qualitative information can be gained from light scattering measurements of multicomponent systems. A few selected systems and applications to the selection of glasses for optical fiber fabrication are discussed below.

The experimental Landau–Placzek ratios for the mixed alkali systems $(X) \cdot Na_2O \cdot (25-X)K_2O \cdot 75SiO_2$ and $(X) \cdot Li_2O \cdot (33-X)K_2O \cdot 67SiO_2$ are given in Table III. Treating these glasses as pseudobinaries one may investigate the effect that varying the Na_2O and Li_2O to K_2O concentration ratio has on the light scattering. The most striking feature is the occurrence of the maximum near a concentration ratio $r \simeq 0.5$ (r is defined as the ratio of the number of moles of Na_2O, or Li_2O to the number of moles of K_2O). The fictive temperature exhibits no unusual behavior as a function of r, varying smoothly with a shallow minimum near $r = 2$ for the $9K_2O \cdot 16Na_2O \cdot 75SiO_2$ composition. To account for the peak from density fluctuations would therefore require a compressibility more than twice that of either of the 25% single alkali glasses. This is unreasonable. Thus, one must assume that the peak is associated in some fashion with local fluctuations in concentration. In ternary systems there are at least two contributions to the composition fluctuations $\langle |\Delta c_1|^2 \rangle$ and $\langle |\Delta c_2|^2 \rangle$, where c_1 and c_2 are the

TABLE III

MEASURED LANDAU–PLACZEK RATIOS AS A FUNCTION OF ALKALI OXIDE CONCENTRATION FOR MIXED ALKALI OXIDE GLASS SYSTEMS[a]

$A \cdot Na_2O \cdot (25-A)K_2O \cdot 75SiO_2$ (mole %)	R_{LP} (total)	$B \cdot Li_2O \cdot (33-B)K_2O \cdot 67SiO_2$ (mole %)	R_{LP} (total)
0/25/75	9.2	0/33/67	9.8
0.5/24.5/75	18.5	8/25/67	23.2
2.5/22.5/75	12.7	16.5/16.5/67	19.6
6.25/18.75/75	14.7	25/8/67	32.0
9/16/75	22.9	33/0/67	47.8
12.5/12.5/75	18.4		
18.25/6.25/75	16.7		
20/5/75	14.7		
25/0/75	14.8		

[a] Results taken from Schroeder (1974).

concentrations of K_2O and Na_2O, respectively [see Eq. (50)]. Alternatively, the material may be characterized by an overall alkali concentration $c = c_1 + c_2$ and the previously defined ratio $r = c_2/c_1$. Thus, the scattering is analyzed in terms of the contributions

$$I_1 = (\partial\varepsilon/\partial c)^2_{P,T,r}\langle|\Delta c|^2\rangle \tag{72}$$

and

$$I_2 = (\partial\varepsilon/\partial r)^2_{P,T,c}\langle|\Delta r|^2\rangle \tag{73}$$

This breakup is convenient because I_2 is expected to be negligible, since $(\partial\varepsilon/\partial r)^2_{P,T,c}$ should be vanishingly small. The substitution of Na_2O for K_2O in the scattering volume produces a negligible change in ε, since the specific refractivities of these two constituents are nearly identical. The behavior of I_1 is now examined. The factor $(\partial\varepsilon/\partial C)_{P,T,r}$ can be calculated using the Gladstone–Dale relationship (Gladstone and Dale, 1864; Pinnow, 1970) and the Huggins–Sun formula (Huggins and Sun, 1943) which describe the variation of refractive index as a function of c. It is found (Schroeder, 1974) that $(\partial\varepsilon/\partial c)_{P,T,r}$ varies monotonically as a function of r with no hint of a maximum near $r = 0.5$, where the maximum in the scattering exists. Thus this maximum in R_{LP} probably represents anomalous behavior of $\langle|\Delta c|^2\rangle$ as a function of r. Let us explore what implications can be drawn from this observation.

The existence of an immiscibility dome for the mixed alkali–silicate system $Li_2O\cdot Na_2O\cdot SiO_2$ has been documented by Moriya et al. (1967). Evidence for the existence of a similar dome for the $K_2O\cdot SiO_2$ system has been found (Schroeder et al., 1975), therefore, it is not unreasonable to assume that $Na_2O\cdot K_2O\cdot SiO_2$ (and $Li_2O\cdot K_2O\cdot SiO_2$) will also exhibit such an immiscibility dome. Unfortunately, exact values for the spinodal boundaries are not available for these systems so that this part of the discussion will be limited to a qualitative treatise.

If it is assumed that the spinodal boundary is characterized (for a given ratio Na_2O to K_2O) by a spinodal temperature $T_s(c)$ and recall that $\langle|\Delta c|^2\rangle \propto T_f'/(T_f' - T_s)$; then for $c = 25$ mole %, the scattering results suggest that for $r \approx 0.5$, the quantity $(T_f' - T_s)$ is a minimum. Noting that T_f' changes relatively little with r, this appears to mean that the value of $T_s(c)$ for the $16K_2O\cdot 9Na_2O\cdot 75SiO_2$ glass ($r \approx 0.5$) is larger than that for either the $25K_2O\cdot 75SiO_2$ ($r = 0$) or the $25Na_2O\cdot 75SiO_2$ ($r = \infty$) glasses.

For the $Li_2O\cdot K_2O\cdot SiO_2$ system again an immiscibility dome has been postulated (Moriya et al., 1967) and a few of the miscibility (i.e., coexistence) temperatures T_m have been empirically determined by assuming a monotonous variation of T_m with concentration in this system of constant silica content. Moriya assumed that the $K_2O\cdot SiO_2$ glass system has an immiscibility dome. No data on the fictive temperatures of this system are available at this point, hence only the simplest inferences may be drawn from the above picture. The scattered intensity increases rapidly from the binary $K_2O\cdot SiO_2$ value with the maximum value being reached at the $Li_2O\cdot SiO_2$ glass composition. Again the low spinodal temperatures of the $K_2O\cdot SiO_2$ system substantially reduce the immiscibility temperatures or the spinodal boundary of the mixed alkali phase producing the low scattered intensity.

Next we shall examine the ternary glasses consisting of $MgO\cdot Na_2O\cdot SiO_2$, $Al_2O_3\cdot K_2O\cdot SiO_2$ and $Al_2O_3\cdot Na_2O\cdot SiO_2$, and in Table IV both the experimental Landau–Placzek ratios and the attenuation coefficient due to molecular light scattering processes in the material as a function of the concentration are found (Schroeder, 1974). The optical scattering loss is determined by the Brillouin spectroscopic method of Rich and Pinnow (1972) where the measured Landau–Placzek ratio is employed in the following way:

$$\alpha_s = \alpha_B(R_{LP} + 1) \tag{74}$$

and

$$\alpha_B = \frac{8\pi^3}{3} \frac{k_B T}{\lambda^4} (n^4 P_{12})^2 \frac{1}{\rho v_{L,\infty}^2} \tag{75}$$

where ρ is the density, λ the wavelength of the laser, P_{12} the longitudinal Pockels' coefficient, $v_{L,\infty}$ the longitudinal hypersonic velocity, n the refractive index, k_B Boltzmann's constant, and T the temperature. The Pockels coefficient and the hypersonic velocity can be determined from the measured Brillouin intensity ratio and from the measured Brillouin splitting, respectively. Thus, the measured Landau–Placzek ratio, and the Brillouin intensity and splitting allows the calculation of the total scattering attenuation coefficient, which may simply be expressed in decibels per kilometer by the relation, Loss (dB/km) = $-4.34 \times 10^5 \alpha_s$.

We will consider the glasses in Table IV from the standpoint of binary glasses to which MgO or Al_2O_3 have been added as a minor constituent or dopant. First, it is necessary to consider the influence of this minor constituent on the density fluctuations. It has been determined from high-temperature velocity measurements (Laberge, 1973) that for the binary alkali–silicates the contribution to density fluctuations $\langle |\Delta\rho|^2 \rangle$ decreases with decreasing molecular weight of the alkali oxide because of the combination of lowering the fictive temperature T_f and the low-frequency compressibility. It was also shown (Laberge, 1973) that the addition of CaO or Al_2O_3 to alkali–silicates lowers the magnitude of the density fluctuations. But, of course, it is possible that the decreased scattering arising from this lowering of the density fluctuations may be more than offset by an increase

TABLE IV

Landau–Placzek Ratios and Scattering Attenuation
Coefficients for Some Ternary Glasses[a]

Glass Composition (mole %)	R_{LP}	Attenuation (at 6328 Å, dB/km)
$Al_2O_3 \cdot K_2O \cdot SiO_2$		
0/25/75	9.2	2.2
5/25/75	12.8	2.5
10/25/65	19.7	3.4
15/25/60	26.9	3.6
0/33/67	9.8	2.0
5/33/62	12.4	1.9
10/33/57	11.9	1.7
15/33/52	17.5	2.1
0/40/60	10.2	1.8
10/40/50	11.9	1.5
$Al_2O_3 \cdot Na_2O \cdot SiO_2$		
0/25/75	14.8	2.5
5/25/70	18.8	2.6
10/25/65	17.8	2.5
0/33/67	13.7	1.7
5/33/62	13.0	1.5
10/33/57	37.7	4.0
15/33/52	16.9	1.6
0/40/60	13.8	1.5
5/40/55	14.4	1.3
10/40/50	15.8	1.3
15/40/45	15.6	1.2
20/40/40	21.1	1.3
$MgO \cdot Na_2O \cdot SiO_2$		
0/20/80	67.6	11.0
10/20/70	21.3	3.4
20/20/60	24.2	2.7
30/20/50	37.7	3.2
0/37.5/62.5	13.0	1.5
20/37.5/42.5	18.7	1.8

[a] Results taken from Schroeder (1974).

in scattering from the concentration fluctuations. As an example, Hammel (1967) has shown that the addition of CaO to $Na_2O \cdot SiO_2$ raises the immiscibility dome for this system with the consequence that the magnitude of the concentration fluctuations for the $CaO \cdot Na_2O \cdot SiO_2$ system is quite large.

This effect is clearly exhibited if the Landau–Placzek ratio of NBS standard glass 710 containing 11.6 wt % CaO, 16.5 wt % alkali and the remainder SiO_2 is compared with a $9CaO \cdot 30K_2O \cdot 61SiO_2$ glass. The NBS 710 glass has a Landau–Placzek ratio of 45 (Schroeder *et al.*, 1973/1974) while the $9CaO \cdot 30K_2O \cdot 61SiO_2$ takes on a value of about 15 (Schroeder, 1974) which is in agreement with Hammel's findings.

If we now examine the addition of MgO to the $Na_2O \cdot SiO_2$ system, the effect is opposite to that of CaO. The scattered intensity that is observed is lower than in the binary system. This can be explained by assuming that MgO additions depress the $Na_2O \cdot SiO_2$ immiscibility dome thereby decreasing the magnitude of the concentration fluctuations $\langle |\Delta c|^2 \rangle$. As the concentration of MgO is increased beyond about 10%, the total scattering intensity increases; this can be attributed to an increase in the zero frequency compressibility and the fictive temperature. For the higher Na_2O concentrations in the $Na_2O \cdot SiO_2$ system, i.e., 40% Na_2O, devitrification is likely to occur. At these compositions the addition of MgO inhibits the crystallization by increasing the stability of the ternary glass; the MgO bonds apparently impede the rearrangement of the silicate structure which occurs when crystallization sets in. The viscosity of the melt is also increased due to the addition of MgO or CaO which is an indication that these ternary glasses have a higher glass transition region with the consequence that both fictive temperatures T_f and T_f' will be raised. All of the above observations seem to indicate that a reduction in the composition variations within the structure of the glass has taken place.

Next we shall consider the Al_2O_3 doped glasses, where it has been shown that the critical point of a sodium–borosilicate system may be substantially changed by the addition of small amounts of Al_2O_3 (Tomozawa and Obara, 1973; Simmons *et al.*, 1970) while the viscosity of the melt is only slightly affected. These findings shall be drawn upon to explain the results for the Al_2O_3 doped binary alkali–silicates (Gupta *et al.*, 1975). The Al_2O_3 addition to the $K_2O \cdot SiO_2$ and $Na_2O \cdot SiO_2$ systems will lower the spinodal temperature T_s since the critical point temperature of the entire system is lowered, but T_f and T_f' will be only slightly increased, since the viscosity does not change too very remarkably. Hence, in this domain $\langle |\Delta c|^2 \rangle$ will be kept low. However, beyond about 5 mole % Al_2O_3 the viscosity is increased and T_f and T_f' begin to increase with concentration. The increase of T_f now affects $\langle |\Delta\rho|^2 \rangle$ with the result that the gain obtained from a reduced $\langle |\Delta c|^2 \rangle$ is cancelled. One further effect that the addition of Al_2O_3 to $Na_2O \cdot SiO_2$ and $K_2O \cdot SiO_2$ produces is an increase in the low-frequency sound velocity which implies a decrease in $K_{T,0}(T_f)$; this has a tendency to offset the increase due to the T_f increase in the magnitude of the density fluctuations $\langle |\Delta\rho|^2 \rangle$.

With the addition of Al_2O_3 to $K_2O \cdot SiO_2$ we see that at 25 mole % K_2O the system is already quite far away from the immiscibility region; hence the added Al_2O_3 only increases the melt bulk viscosity of the samples with an increase of T_f thereby causing $\langle |\Delta\rho|^2 \rangle$ to increase uniformly. This is reflected in the behavior of the results given in Table IV. At $33K_2O$, Al_2O_3 doping gives at first a slight decrease in the scattered intensity up to around 5 mole % Al_2O_3 brought about by a faster decrease of $K_{T,0}(T_f)$ while T_f increases at a slower rate. But beyond 5 mole % Al_2O_3 viscosity effects take over and T_f increases much faster than the decrease in $K_{T,0}(T_f)$. Again this is rather nicely borne out by the measured results (Schroeder, 1974). The $Al_2O_3 \cdot K_2O \cdot SiO_2$ systems studied had a K_2O content in excess of 25 mole %. For these concentrations of K_2O the spinodal temperatures are quite far below the glass transition temperatures. This means that the concentration fluctuations are quite small and that Al_2O_3 doping will primarily affect $\langle |\Delta\rho|^2 \rangle$. The experimental data show that it does so in just the fashion anticipated above.

For the $Al_2O_3 \cdot Na_2O \cdot SiO_2$ system the situation is somewhat more interesting. The critical temperature is much higher (about 850°C) and the immiscibility dome extends beyond 20 mole % Na_2O. However, even in these systems the scattering is dominated by density rather than composition fluctuations. Certainly the scattering in the 33 and 40 mole % Na_2O families seems to be primarily controlled by density fluctuations; the slight decreases observed for small Al_2O_3 doping are attributable to a decrease in $K_{T,0}(T_f)$; at the higher Al_2O_3 concentrations, viscosity effects take over and the scattering increases.

Pinnow et al. (1975) have carried out a study on the $Al_2O_3 \cdot Na_2O \cdot SiO_2$ system, where the main emphasis was to consider this system for optical fiber applications. One of their conclusions was that some compositions can be made where the attenuation of the $Al_2O_3 \cdot Na_2O \cdot SiO_2$ glass is about one-fourth that of SiO_2, which agrees rather well with the findings given above. When Pinnow et al. (1975) compared the measured results of concentration fluctuation scattering to calculated results based on the assumption of random mixing of the components of molecules, the measured values were less. This is strong evidence for local ordering in $Al_2O_3 \cdot Na_2O \cdot SiO_2$ glass systems.

A number of authors, namely, Maurer (1973), Keck et al. (1972), and Keck and Tynes (1972) have concluded that intrinsic scattering is the major loss mechanism in bulk glasses and optical fibers since absorption (primarily due to transition elements, impurities and OH in the glass) can be controlled and eliminated. Hence, the importance of light scattering measurements on n-component glasses is realized since it will provide the Landau–Placzek ratio and acoustic velocity (both hypersonic shear and longitudinal velocities) and with this information the optical attenuation due to intrinsic scattering in the glass can be determined simply and reliably.

In the above treatment we have considered glasses only in the single-phase region (homogeneous glasses); a certain amount of light scattering study (usually intensity as a function of scattering angle) has been made on phase-separated glass samples. However, we shall not touch on this area and the interested reader is referred to the review by Hammel (1972) and the work of Becherer *et al.* (1971).

D. Low-Temperature Brillouin Scattering in Glasses

The Brillouin spectrum of the scattered light from thermal phonons allows the evaluation of the phase velocity of sound from the shift and the acoustic absorption from the linewidth. Thus, thermal Brillouin scattering is well suited to examine absorption in glasses in the microwave frequency regime, provided one is able to resolve the natural linewidth of the Brillouin line. Normal ultrasonic techniques have been applied to glasses up to about 1 GHz (Jones *et al.*, 1964). Beyond this frequency, these techniques are no longer applicable due to the relatively high attenuation. A number of experimentors have detected Brillouin spectra from glasses (Krishnan, 1950; Pesin and Fabelinskii, 1960, 1961; Flubacher *et al.*, 1960; Shapiro *et al.*, 1966; Sabirov *et al.*, 1968), primarily fused silica; confirming that high-frequency sound can propagate in glasses. However, the first work that applied high-resolution techniques and resolved the natural Brillouin linewidth of fused silica, allowing the measurement of the longitudinal hypersound attenuation, was that of Durand and Pine (1968). Subsequently, Pine (1969) carried out a temperature-dependent study of fused silica where the velocity and attenuation of longitudinal hypersonic waves were measured for temperatures between 80 and 600 K. It has been shown that stimulated Brillouin scattering from fused silica also yields the same information about hypersonic velocity and attenuation (Tannenwald, 1966; Walder and Tany, 1967; Heinicke *et al.*, 1971; Heinicke and Winterling, 1967). However, transient effects may make the analysis for the attenuation much more difficult (Heinicke and Winterling, 1967).

From the Brillouin data of fused silica, Pine (1969) has found a broad low-temperature peak (peak temperature ≈ 130 K) in the absorption, and two models are reviewed that could account for the attenuation of sound in SiO_2. First, the anharmonic or three phonon interaction of Bömmel and Dransfeld (1960); secondly, the structural relaxation mechanism of Anderson and Bömmel (1955) are compared to the available acoustic absorption data of fused silica. The anharmonic model gives the damping rate Γ for longitudinal waves in the form

$$\Gamma = (\gamma_0^2 T c_v / 8\rho v_L^2)[\omega^2\tau/(1 + (\omega\tau)^2)] \tag{76}$$

Here γ_0 is the averaged dimensionless Grüneisen constant which measures the anharmonicity, T the temperature, ρ the density, ω and v_L the frequency and velocity of a sound wave, and c_v and τ are the heat capacity and relaxation time, respectively, of a thermal phonon mode. The damping rate Γ is related to the Brillouin linewidth δv and the acoustic attenuation constant α by $\Gamma = \alpha v_L = \pi \, \delta v$.

The structural relaxation mechanism advanced by Anderson and Bömmel (1955) to explain ultrasonic absorption in fused silica has the damping rate Γ given by

$$\Gamma = \frac{G\omega(\omega/\Omega)}{1 + (\omega/\Omega)^2} \tag{77}$$

Here G is a temperature-independent constant representing the strength of the relaxation and Ω is the transition rate or frequency of a structural relaxation. The particular structural transition postulated by Anderson and Bömmel is a reordering of Si–O–Si bond angle. Here a part of the oxygen atoms can perform a transverse motion between bonding Silicon atoms for which two potential minima exist (double potential well). The double potential well has a central barrier of activation energy E which can only be overcome by a thermally activated process, then the transition rate for jumping between minima is the product of the zero point vibrational frequency ω_1 in one well and the Boltzmann probability of jumping the barrier; namely,

$$\Omega = \omega_1 \exp(-E/k_B T) \tag{78}$$

Or if $\omega_1 \ll k_B T/\hbar$, then the Eyring form $\Omega = (k_B T)/\hbar \, \exp(-E/k_B T)$ is applicable.

In testing both models, Pine (1969) finds that the anharmonic model appears to fit the data better (with fewer parameters), whereas the acoustic losses are not uniquely explained by a structural relaxation mechanism.

The work of Durand and Pine (1968), Pine (1969) has shown that the acoustic properties of glasses in the 1–100 GHz frequency range may suitably be studied by thermal Brillouin scattering. The next region of interest is that of very low temperatures (around liquid helium temperature). Impetus to explore the acoustic properties of glasses around liquid helium temperatures is given by the observed anomalies in the specific heat and thermal conductivity of glasses. It is well known that for crystalline solids the low-temperature lattice specific heat agrees to within a few percent of that calculated from the Debye model using measured sound velocities. In contrast the low-temperature situation for noncrystalline solids (i.e., glasses) is markedly different, and the Debye law is not obeyed (Zeller and Pohl, 1971; Stephens, 1973; Pohl *et al.*, 1974). At low temperatures the specific heat of the crystalline solid approaches a T^3 law, whereas specific heat data of a glass can best be described by a polynomial expression (Pohl *et al.*, 1974),

$$c_v = C_1 T + C_3 T^3 \tag{79}$$

and C_3 is up to twice as large as predicted by the Debye model. According to Pohl *et al.* (1974) two possible explanations for the specific heat behavior of glass may be postulated. Firstly, Debye phonons do exist in the glasses but also some other excitations exist which mask the specific heat of the Debye phonons. The specific heat of glasses at low temperatures now takes the form

$$c_v = C_1 T + (C_3' + C_{\text{Debye}})T^3 \tag{80}$$

where both $C_1 T$ and $C_3' T^3$ are attributable to the disorder. Second, Debye phonons might not be good normal modes of the glass and the specific heat is determined by some unknown excitations. A number of low-temperature ultrasonic attenuation measurements on glasses in the frequency range of 1 GHz (Hunklinger *et al.*, 1972, 1973; Golding *et al.*, 1973; Arnold *et al.*, 1974) have aroused the suspicion that thermal Debye phonons might be too short lived to be considered as normal modes in the glass, thereby invalidating the separation of C_3 into the form given by Eq. (80). Several theoretical approaches exist to explain these observed anomalies of specific heat and thermal conductivity in amorphous materials; e.g., a macroscopic model of Fulde and Wagner (1971) and the microscopic theory of Phillips (1972) and Anderson *et al.* (1972). Both the Phillips (1972) and Anderson *et al.* (1972) models were highly successful in explaining the results from ultrasonic measurements. They suggest phonon scattering by two-level tunneling states characteristic of glasses. These two-level systems are perhaps structural defects which perform quantum mechanical tunneling between two equilibrium configurations. It should be mentioned that although the tunneling state model is successful in ultrasonic attenuation, it cannot be used without further change with Brillouin data (Jackson *et al.*, 1976).

The first evidence for the existence of thermal phonons was provided by Love (1973). Here Brillouin scattering was used to search for thermally excited Debye phonons with frequencies of about 20 GHz. From the light scattering measurements of fused silica and a borosilicate glass in the temperature range of 300–1.7 K, Love was able to detect the longitudinal Brillouin lines for both glasses down to 1.7 K, and for the borosilicate glass a trace of the transverse Brillouin peak was seen at the lowest temperature. From the widths of the Brillouin lines the phonon lifetimes are estimated to be at least ten vibrational periods at all temperatures. Thus, Love concludes that Debye phonons do exist as good normal modes in glasses as low as 1.7 K and the experimental specific heat may indeed be written in the form of Eq. (80). Similarly Jackson *et al.* (1976) have used the technique of high-resolution thermal Brillouin scattering to measure phonon lifetimes in fused silica at temperatures down to 2.4 K. They conclude from their measurements that longitudinal phonons ($v \simeq 33$ GHz) are well-defined excitations whose lifetimes increase as the temperature is lowered.

Pelous and Vacher (1975) have carried out a Brillouin scattering investigation from the attenuation of longitudinal hypersound in fused quartz from 77 to 300 K. Again this work overlaps that of Pine (1969) and good agreement is found between the results of both authors. The hypersonic attenuation goes through a maximum at 120 ± 10 and 115 ± 10 K for frequencies of 33.7 and 24.7 GHz, respectively. Pelous and Vacher (1975) also conclude that the low-temperature behavior of the hypersonic attenuation is governed by a broad peak and this confirms the necessity of applying a complex distribution of relaxation times for fitting the experimental results. The same authors, Vacher and Pelous (1976), also studied thermal phonons in fused silica and borosilicate glass by a Brillouin scattering technique in the temperature range of 300–4.2 K. Again the conclusion is reached that phonons in glasses are well defined excitations.

E.   High-Temperature Rayleigh and Brillouin Scattering in Glasses

Two experimental methods are available to carry out light scattering measurements from glasses at elevated temperatures. First, temporal autocorrelation spectroscopy probes the central or Rayleigh line and second, dispersive spectroscopy (i.e., by means of a Fabry–Perot interferometer) examines both the Rayleigh and Brillouin lines but usually only the Brillouin linewidths are fully resolved. The first method gives a correlation function from which thermodynamic parameters and transport coefficients of the medium are obtained. For light scattered by density fluctuations, the correlation function gives structural relaxation parameters such as the average relaxation time $\langle \tau \rangle$ and the distribution of relaxation times $G_m(t)$ (Demoulin *et al.*, 1974). If concentration fluctuations are involved in the scattering process, it becomes possible to separate out both the thermal diffusion relaxation time and the mass diffusion relaxation time. In essence it is possible to probe the Rayleigh line and obtain information concerning the spectral distribution of this line, as described by Eq. (32).

The dispersive spectroscopic technique can only resolve the Brillouin linewidth. Even at very high temperatures where the glass may have viscosities comparable to a liquid, the fine details of the Rayleigh line are obscured by the immense black-body radiation and the Rayleigh linewidth resolution is not possible. Nevertheless, an important quantity, the Landau–Placzek ratio, can be obtained from the scattered intensities and by studying this ratio as a function of temperature the validity of Eq. (39) may be examined. The Brillouin linewidths as a function of temperature give again (as in the low-temperature case) acoustic absorption data. From the Landau–Placzek ratio and Eq. (39) it is possible to calculate the isothermal compressibility and study its temperature dependence above the glass transition temperature.

Bucaro and Dardy (1974a–c) have performed light scattering measurements from SiO_2 and B_2O_3 at high temperatures. SiO_2 was studied through the glass transition temperature up to about 2000 K while B_2O_3 was taken up to about 800 K. In Fig. 14 their typical VV and VH spectral traces (Bucaro and Dardy, 1974b) for B_2O_3 at about 700 K are displayed. In the VV traces the Brillouin doublets caused by the phonon modes and the more intense central line (Rayleigh line) are clearly seen. The VV Rayleigh line is primarily attributable to nonpropagating density fluctuations, whereas the central line in the VH trace is due to anisotropy scattering. The smaller doublets in VH are the shear or transverse Brillouins. The Landau–Placzek ratio as a function of temperature for both SiO_2 and B_2O_3 from the work of

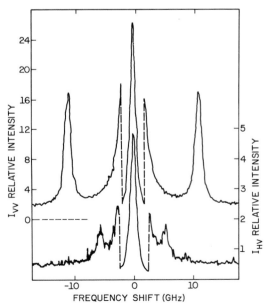

Fig. 14. VV and VH spectra of light scattered from B_2O_3 at 700 K (*scattering* angle of 90°), recorded with an instrumental width of 1 GHz. (After Bucaro and Dardy, 1974b.)

Bucaro and Dardy (1974a,b) are shown in Fig. 15. For the SiO_2 glass, this ratio decreases rapidly from its room temperature value of 21.9 to about 8 at 1400 K, and this is reasonable and may largely be attributed to the decrease of the (T_f/T) factor in Eq. (39) as the system rises in temperature. A minimum is reached around T_f and then the Landau–Placzek ratio remains constant up to 2000 K. Above the glass transition temperature Eq. (39) simply becomes

$$R_{LP}(\rho) = K_{T,0}(T)(\rho v_{L,\infty}^2) - 1 \tag{81}$$

and a measure of the Landau–Placzek ratio and the Brillouin shift allows the calculation of the static isothermal compressibility of the melt. Bucaro and Dardy (1974a) calculate $K_{T,0}(T)$ from measured high-temperature data for SiO_2 with the aid of Eq. (81) and find that the static isothermal compressibility is relatively temperature independent above T_f. The $K_{T,0}(T)$ value for SiO_2 (8.5×10^{-12} cm^2/dyn) and its temperature independence seem to counter the conclusion of Bloom and Bockris (1957) that in the alkali–silicate glasses above T_f, $K_{T,0}(T)$ must undergo a large decrease below 20 mole % alkali oxide. This conclusion was based on the discrete ion theory developed by Bockris and coworkers (1954, 1955, 1960) for the liquid silicate lattice, and its validity has been assailed by the above results. The expected lessening of the compressibility in transiting from low alkali oxide content to pure silica has not been realized. Again no significant temperature dependence has been detected for the hypersonic moduli of SiO_2 above T_f (Bucaro and Dardy, 1974a) and this points to the absence of any significant structural changes in this temperature region.

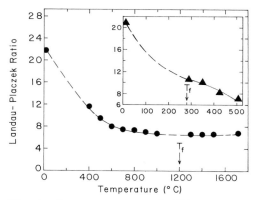

Fig. 15. Landau–Placzek ratio versus temperature for SiO_2 and B_2O_3 glass. Here both plots contain the contribution from anisotropic scattering. ● = denotes SiO_2 and ▲ = denotes B_2O_3. (After Bucaro and Dardy, 1974a, b.)

For B_2O_3 Bucaro and Dardy (1974b) determine the isothermal compressibility in the melt. Above T_f a large increase in the isothermal compressibility is found reaching a maximum around 770 K. Due to the large anisotropy scattering found in this glass, the Rayleigh intensity has to be corrected (Kielich, 1967) before the Landau–Placzek ratio reflects only contributions from pure density fluctuations. By carrying out light-scattering studies of SiO_2 and B_2O_3 both in the melt near T_f and in the solid phase, Bucaro and Dardy find that the part of the density fluctuations connected with the relaxational part of the isothermal compressibility is frozen in at the fictive temperature and the scattering intensity from density fluctuations is described by Eq. (39). Thus, the magnitude of the density fluctuations in the solid depend simply on the equilibrium compressibility at the fictive temperature of the melt.

The high-temperature correlation spectroscopy is exemplified by the studies done on B_2O_3 glass (Bucaro et al., 1975); (Macedo and Montrose, 1974) and a mixed alkali–silicate glass (Lai et al., 1975). In each case the authors obtained a normalized correlogram proportional to $|G_m(t)|^2$, which may be fit by a function characterizing the structural relaxation. $|G_m(t)|$ is usually taken as

$$|G_m(t)| = \exp[-(t/\tau_\beta)^\beta] \tag{82}$$

with β and τ_β being adjustable parameters. With this equation the average relaxation time is then given by

$$\langle \tau \rangle \equiv \int_0^\infty dt |G_m(t)| = (\tau_\beta/\beta)\Gamma(1/\beta) \tag{83}$$

For B_2O_3 at 310°C, Bucaro et al. (1975) have found typical values of τ_β and β to be 12.2 and 0.65 sec, respectively.

F. The Krishnan Effect in Glasses

Experimental studies of light scattering in isotropic systems have yielded two empirical relationships among four possible intensity measurements [i.e., $I_R(VV)$, $I_R(VH)$, $I_R(HV)$, and $I_R(HH)$]. Specifically, it has been found that for 90° scattering, the ratio

$$\rho_h \equiv \frac{I_R(VH)}{I_R(HH)} = 1 \tag{84}$$

and the VH and HV intensities are equal, i.e.,

$$I_R(VH) = I_R(HV) \tag{85}$$

The second of these, the Rayleigh reciprocity relation, follows directly as long as the material being considered is not optically active. No such firm theoretical foundation exists for the equality presented in Eq. (84).

In fact, Eq. (84) is substantially less general than Eq. (85). In a series of investigations published during the 1930s R. S. Krishnan (1934, 1935, 1936a,b, 1937, 1938) measured a number of materials for which

$$\rho_h \equiv \frac{I_R(VH)}{I_R(HH)} < 1 \tag{86}$$

The observation of this polarization anomaly has been termed the "Krishnan effect" and the quantity ρ_h is called the "Krishnan ratio".

In Table V are shown values of ρ_h as a function of concentration for both $K_2O \cdot SiO_2$ and $Na_2O \cdot SiO_2$ systems and some NBS standard glasses. These data show that $\rho_h \approx 1$ for pure SiO_2 and $\rho_h < 1$ as alkali is added. The minima near 8 mole % in the $K_2O \cdot SiO_2$ system where $\rho_h \approx 0.35$ and near 10 mole % in the $Na_2O \cdot SiO_2$ system where $\rho_h \approx 0.07$ are quite distinct. The minima result chiefly from a rather pronounced maxima in the curves of $I_R(HH)$ versus concentration as opposed to any anomalous behavior in the $I_R(VH)$ variation (Schroeder, 1974). Huang and Wang (1974, 1975) have observed the Krishnan effect in polybutadiene and polypropylene glycol, as these liquids are cooled below the glass transition temperature. These authors also observed that the decrease in ρ_h is mainly associated with an increase of the HH intensity; whereas the VH intensity remains fairly constant throughout the temperature range studied.

TABLE V

DEPOLARIZATION RATIO (KRISHNAN RATIO) ρ_h AS A FUNCTION OF CONCENTRATION FOR $K_2O \cdot SiO_2$ AND $Na_2O \cdot SiO_2$ GLASSES AND SOME NBS STANDARD OPTICAL GLASSES[a]

Type of glass (mole %)	Krishnan ratio (ρ_h)	Type of glass (mole %)	Krishnan ratio (ρ_h)
SiO_2 (Homosil)	0.98	$27Na_2O \cdot 73SiO_2$	0.88
$6K_2O \cdot 94SiO_2$	0.62	$33Na_2O \cdot 67SiO_2$	0.81
$8K_2O \cdot 92SiO_2$	0.35	$40Na_2O \cdot 60SiO_2$	0.93
$10K_2O \cdot 90SiO_2$	0.51	NBS #710	0.99
$15K_2O \cdot 85SiO_2$	0.68	NBS #711	0.97
$25K_2O \cdot 75SiO_2$	0.77	NBS #713	1.01
$33K_2O \cdot 67SiO_2$	0.68	NBS #716	0.46
$10Na_2O \cdot 90SiO_2$	0.07	NBS #717	1.00
$15Na_2O \cdot 85SiO_2$	0.81		

[a] Results taken from Schroeder (1974).

A number of attempts have been made at a theoretical explanation of the Krishnan effect. Gans (1936, 1937) developed a theory of scattering by molecular clusters, and he finds that the Krishnan effect demands the existence of optically anisotropic clusters of strong aspheric shape. Moreover, the Gans theory predicts $\rho_h > 1$, making it unacceptable as an explanation of the Krishnan effect. Mueller (1938) has proposed a theory of strain scattering of light in which it is assumed that the frozen-in nonpropagating strains differ only from their propagating counterparts in the manner in which they participate in equipartition. A longitudinal mode has an associated energy U_L (i.e., $k_B T_f$) while a transverse mode has energy U_T. The result for the Krishnan ratio becomes

$$\rho_h = \frac{U_T}{U_L} \frac{c_{11}}{2c_{44}} \tag{87}$$

Several features of the above are unappealing. Chief among these is the need for assigning less than $k_B T_f$ of the energy to the transverse modes since to a large extent it is the freezing in of these modes near T_f that is responsible for the glass transition. There also seems to be no grounds on which to decide in which glasses one should observe $\rho_h = 1$ as opposed to $\rho_h < 1$. Goldstein (1959) also proposed a model based on local anisotropies; however, the effect calculated is much too small to account for the Krishnan effect.

None of the above treatments seem to give a satisfactory explanation for the Krishnan effect. From the observations for $K_2O \cdot SiO_2$ and $Na_2O \cdot SiO_2$ systems, it is seen that ρ_h goes to a minimum when the scattering from $\langle |\Delta c_k|^2 \rangle$ is a maximum. Specifically, this happens because $I_R(HH)$ is a maximum at this concentration—the concentration dependence of $I_R(HV)$ being relatively unimportant. The correspondence in the location (with respect to concentration) of these maxima with the critical composition suggests that the $I_R(HH)$ intensity reflects some aspect of the critical phase-separation process. Moreover, the magnitude of the deviation from unity of ρ_h for a given composition is dependent on how close in temperature the spinodal boundary (which plays the role of the critical temperature for compositions other than the critical composition) is approached. This behavior is exemplified for the $K_2O \cdot SiO_2$ system in Fig. 16, where ρ_h is presented as a function of $(T - T_s)$ with T being the temperature at which each particular sample was heat treated and T_s is the familiar spinodal temperature at the specific concentration of every sample. These results very strongly suggest some coupling of HH to the composition fluctuations $(\langle |\Delta c_k|^2 \rangle)$ for the glasses involved. The work of Rank and Douglas (1948) also reinforces the idea that critical phenomenon are an integral feature of the explanation of the Krishnan effect. Their samples (DF-3 and BSC-2 in their nomenclature) are similar in composition to NBS standard glasses

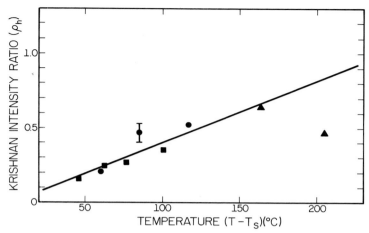

Fig. 16. The Krishnan ratio (ρ_h) versus ($T - T_s$) for $K_2O \cdot SiO_2$ glasses. ● = denotes the $6K_2O \cdot 94SiO_2$ glass, ■ = denotes the $8K_2O \cdot 92SiO_2$ glass, and ▲ = denotes $10K_2O \cdot 90SiO_2$ glass. (After Schroeder, 1974.)

#711 and #716, respectively (Schroeder *et al.*, 1973/1974). Rank and Douglas observed a Krishnan effect in these glasses. NBS #716 is a Sodium–aluminoborosilicate glass of a composition for which phase separation is not unlikely; indeed, $\rho_h < 1$ for this sample.

Hence, it seems as if somehow the isotropic composition fluctuations $\langle|\Delta c_k|^2\rangle$ are coupled with longitudinal "frozen-in" fluctuations of the dielectric susceptibility. The important assumption is the fact that these latter fluctuations, which are described by a tensor dielectric constant $\Delta\varepsilon_{xy}$, can be affected by the composition fluctuations which are described by a scalar. A similar instance of such a coupling has been noted and explored by Kawasaki (1966) in his discussion of the behavior of the shear viscosity of a system near its critical point. We suggest that some similar mode–mode coupling approach may in fact provide the key to the Krishnan effect. Swift (1968) and Ferrell (1970) have applied mode–mode coupling to binary fluid mixtures. An expression similar to that obtained by Goldstein $\rho_h = (1 + Q)^{-1}$ can be expected to result, in which Q rather than resulting from the coupling of density variations to the dielectric susceptibility (as Goldstein discusses) describes the coupling of the susceptibility to composition fluctuations. Near a critical point (or an immiscibility dome characterized by a spinodal boundary) Q is expected to diverge as $(T - T_s)^{-\gamma}$ reflecting the divergence of $\langle|\Delta c_k|^2\rangle$. Huang and Wang (1974) propose a model based on the gradual frozen-in fluctuations of the anisotropy and this should account for the increased *HH* scattering. Their model is formulated to account for the Krishnan effect in amorphous polymers and organic glasses, and they state that the effect is associated with the orientational motion and not the frozen-in density fluctuations.

G. Glass Material Constants and Transport Coefficients Determined by Light Scattering

In a glass the elastic strain produced by a small stress can be described by two elastic constants, C_{11} and C_{44}. Using the Cauchy relation that $2C_{44} = C_{11} - C_{12}$ allows one to determine C_{12}. For pure longitudinal waves

$$C_{11} = \rho v_{L,\infty}^2 \tag{88}$$

and for pure transverse waves

$$C_{44} = \rho v_{T,\infty}^2 \tag{89}$$

where ρ is the density, and $v_{L,\infty}$ and $v_{T,\infty}$ the longitudinal and transverse (shear) velocities, respectively. As shown in a previous section these velocities can be calculated from the Brillouin shifts with the aid of the Brillouin equation given by (Brillouin, 1922)

$$v_{L,\infty \atop T,\infty} = (c_0/2v_0)[\Delta v_{L,T}/(\sin \theta/2)](1/n) \tag{90}$$

Here $\Delta v_{L,T}$ is the appropriate Brillouin splitting (either longitudinal or shear) in GHz, v_0 the incident laser frequency, c_0 velocity of light in vacuum, n the refractive index at the laser frequency, and θ the scattering angle.

The coupling between light (photons) and sound (phonons) is phenomenologically described by the Pockels' tensor (Nye, 1957) which gives the intensity and polarization properties for Brillouin scattering. For a glass only two independent Pockels' coefficients exist and they are P_{12} and P_{44} and $2P_{44} = P_{11} - P_{12}$. The scattering intensity for the various Brillouin components become (Born and Huang, 1954)

$$I_{VV} = I_0\varepsilon_0^4(\omega_0/c_0)^4(Vk_BT/32\pi^2)(P_{12}^2/C_{11}) \tag{91}$$

$$I_{HH} = I_0\varepsilon_0^4(\omega_0/c_0)^4(Vk_BT/32\pi^2)(P_{44}^2/C_{11}) \tag{92}$$

and

$$I_{VH} = I_{HV} = I_0\varepsilon_0^4(\omega_0/c_0)^4(Vk_BT/32\pi^2)(P_{44}^2/2C_{44}) \tag{93}$$

The notation used here is the same as throughout the chapter. The VV and HH components are due to scattering from longitudinal phonons and consequently depend on normal strains, but the VH and HV components originate exclusively from shearing strains which act in planes normal to the scattering plane. Thus, the choice of incident light polarization to scattered light polarization selects specific acoustic modes. Consequently, two sets of material constants determine the Brillouin spectrum: the elastic constant C_{ij} selects the magnitude of the Brillouin shifts; while the Pockels' coefficient P_{ij} scales the Brillouin intensities.

In practice one does not measure absolute Brillouin intensities to obtain the Pockels' coefficients, but a relative intensity measurement is made (Schroeder *et al.*, 1973) (i.e., for SiO_2 the longitudinal Brillouin intensity ratio with respect to toluene can be taken and once P_{12} of SiO_2 is determined, SiO_2 becomes the standard for all the rest of the glass sample measurements of the P_{ij}). Typical values of the P_{ij} and C_{ij} for select glass samples are given in Table VI.

Once the elastic constants for a glass have been determined from Brillouin scattering data a number of related constants may be calculated. Namely, the adiabatic bulk modulus given by $\beta_s = C_{11} - \frac{4}{3}C_{44}$ and the adiabatic compressibility $K_s = (\beta_s)^{-1}$, Young's modulus E is given by

$$E = C_{44}(3C_{11} - 4C_{44})/2(C_{11} - C_{44}) \tag{94}$$

and Poisson's ratio σ. Poisson's ratio may be calculated by two independent methods; one using the elastic constants

$$\sigma_1 = (C_{11} - 2C_{44})/2(C_{11} - C_{44}) \tag{95}$$

and the other makes use of the intensity ratio $\rho_B = I_B(VH)/I_B(HH)$

$$\sigma_2 = (1 - \rho_B)/(1 - 2\rho_B) \tag{96}$$

In Table VII calculated values of β_s, E, σ_1, and σ_2 are listed for some typical glasses.

TABLE VI

POCKELS COEFFICIENTS (P_{12}, P_{44}) AND ELASTIC CONSTANTS (C_{11}, C_{44}) FOR SOME SELECTED GLASS SAMPLES AT $T = 296$ K AND 6328 Å[a]

Sample (mole %)	P_{12}	$\|P_{44}\|$	C_{11} (10^{10} dyn/cm^2)	C_{44} (10^{10} dyn/cm^2)
SiO_2	0.270	0.0718	77.857	30.966
$25K_2O \cdot 75SiO_2$	0.229	0.028	56.87	18.23
$33K_2O \cdot 67SiO_2$	0.188	0.038	93.89	32.88
GeO_2	0.254	0.065	51.54	18.28
B_2O_3	0.298	0.024	20.71	5.91
$25Na_2O \cdot 75SiO_2$	0.214	0.040	69.95	24.47
$10MgO \cdot 20Na_2O \cdot 70SiO_2$	0.222	0.043	78.76	27.52
$7TiO_2 \cdot 93SiO_2$	0.246	0.073	72.07	29.81

[a] Results taken from Schroeder (1974).

TABLE VII

BULK MODULUS β_s, YOUNG'S MODULUS E, AND POISSON'S RATIO σ FOR
SOME SELECTED GLASSES[a,b]

Sample (mole %)	β_s (10^{10} dyn/cm^2)	E (10^{10} dyn/cm^2)	σ_1	σ_2
SiO$_2$	36.57	72.45	0.170	0.158
25K$_2$O · 75SiO$_2$	32.57	46.08	0.264	0.266
25Na$_2$O · 75SiO$_2$	37.33	60.24	0.231	0.235
B$_2$O$_3$	12.83	15.37	0.300	—
33Li$_2$O · 67SiO$_2$	50.05	80.92	0.231	—

[a] Results taken from Schroeder (1974).
[b] β_s, E, and σ_1 are calculated from the elastic constants C_{12}, C_{44} while σ_2 is calculated from Brillouin intensities.

In a previous section of this chapter it has been shown that by means of Rayleigh scattering combined with annealing studies, the relaxation times of a glass as a function of temperature may be obtained. From the relaxation times τ the diffusion coefficient for a glass at a specific temperature can be determined. In Table VIII, some typical values of D_{exp} for some glass samples are shown. Also included are diffusion constants calculated from the Stokes–Einstein equation:

$$D_{SE} = k_B T / 6\pi r \eta \qquad (97)$$

where r is the particle radius (taken to be 3 Å), η is the shear viscosity, and T the temperature. From Table VIII, the Stokes–Einstein formula clearly fails rather dramatically; the measured values are 3–4 orders of magnitude greater than those predicted by Eq. (97). There is no obvious interpretation to this except to comment that the microscopic friction coefficients appropriate to the viscous flow and diffusional processes are apparently quite different.

On the other hand, one may interpret the results in terms of a so-called "jump diffusion" model, as described by Egelstaff (1967). Here the molecules are assumed to perform a random walk with rms step size l^*; the diffusion coefficient is given by

$$D_j = l^* / 6\tau_j \qquad (98)$$

where τ_j is the average time between steps. As a first approximation we might set τ_j equal to the shear relaxation time τ_s:

$$\tau_j \simeq \tau_s = \eta_s / C_{44} \qquad (99)$$

With η_s being the shear viscosity and C_{44} the shear elastic constant. Equation (95) allows the calculation of the step size l^*. Values for l^* are given in

TABLE VIII

DIFFUSION COEFFICIENTS FROM LIGHT SCATTERING FOR SOME SELECTED GLASSES

Temperature (°C)	Viscosity (P)	D_{exp} $(10^{-16} cm^2/sec)$	D_{SE} $(10^{-16} cm^2/sec)$	l^* (Å)
$6K_2O \cdot 94SiO_2{}^a$				
579.0	1.75×10^{13}	0.52	1.19×10^{-4}	14.2
604.8	4.5×10^{12}	2.5	4.77×10^{-4}	15.9
637.8	6.0×10^{11}	31	3.71×10^{-3}	19.0
$8K_2O \cdot 92SiO_2{}^a$				
573.3	8.5×10^{12}	1.2	2.43×10^{-4}	15.7
589.7	3.0×10^{12}	5.1	7.02×10^{-4}	18.9
603.5	1.3×10^{12}	15	1.53×10^{-3}	21.3
613.5	7.5×10^{11}	23	2.89×10^{-3}	20.0
627.0	3.6×10^{11}	75	6.11×10^{-3}	25.5
$10K_2O \cdot 90SiO_2{}^a$				
573.4	1.8×10^{12}	9	1.15×10^{-3}	20.3
$19.5Na_2O \cdot 80.5SiO_2{}^b$				
555	1.74×10^{10}	81.0	2.32×10^{-1}	—
565	9.34×10^{9}	133	4.38×10^{-1}	—
575	4.9×10^{9}	247	8.45×10^{-1}	—
$13Na_2O \cdot 11CaO \cdot 76SiO_2{}^b$				
615	3.90×10^{10}	55.2	1.12×10^{-1}	—
645	5.52×10^{9}	250	8.20×10^{-1}	—
674	9.55×10^{8}	761	4.86	—

[a] Schroeder et al. (1975).
[b] Hammel (1972).

Table VIII. These are typically on the order of 20 Å which is a factor of 6–20 larger than what would be expected. This may mean that the simple assumption of Eq. (99) rather seriously overestimates the jump time. Hammel (1965, 1967, 1972) (Hammel et al., 1968; Hammel and Ohlberg, 1965) in his diffusion work on glasses also found similar results and his conclusions are in agreement with ours. This is not unreasonable in the sense that a single jump cannot really be expected to effect a relaxation of local shear stresses. However, an alternate interpretation is also possible: Eq. (98) is derived under the assumption that a single jump of a particle is sufficient to cause complete loss of memory of the particles' initial phase–space configuration. This assumption can be relaxed somewhat if the factors in Eq. (98) are correspondingly reinterpreted: τ_j is taken to be the average time required for the particle to "forget" its original configuration and l^* is the corresponding rms correlation distance. This would imply that perhaps 6 ∼ 20 individual steps are involved in this loss of correlation. In fact, the actual behavior lies somewhat intermediate to these two interpretations. It is also evident from the work of Hammel and coworkers (1965, 1968, 1972) that the diffusion behavior of glasses cannot simply be ascribed to models based on viscous forces.

V. Summary and Conclusion

The fundamental task of this chapter was to show how Rayleigh–Brillouin scattering may be employed in the study of glass. We began by treating the scattering problem for a liquid and then showed what changes are necessary to make the liquid theory applicable to glasses. The Rayleigh–Brillouin measurements for a number of families of single component, binary and pseudobinary alkali–silicate glasses in the concentration interval from pure silica to highest alkali concentrations for which glasses could be made have been presented. These measurements indicate that in the vicinity of an immiscibility dome a glass system exhibits excess scattering, which appears to be associated with the disappearance of the concentration derivative of the chemical potential.

From the light scattering results for single-component glasses (i.e., SiO_2, B_2O_3, and GeO_2) at room temperature one may conclude that the scattered intensity is properly described by a model involving the fictive temperature, lattice temperature, the hypersonic sound speed, and the relaxational part of the isothermal compressibility. This confirms earlier assumptions of "frozen-in" density fluctuations and the concept of a fictive temperature characterizing the magnitude of these fluctuations. The high-temperature light scattering undeniably reinforces the validity of the above model. Hence, the intensity discrepancy that has existed between earlier measured and calculated light scattering results for single-component glasses has been resolved.

For the binary glass systems a model based on the Landau–Ginzburg theory of critical fluctuations to account for the excess scattering found in the region of immiscibility domes was presented. This model, after modification to include the spinodal temperature and fictive temperature for concentration fluctuations, accounts for the magnitude and variation of the Rayleigh scattering (quasielastic scattering) as a function of the concentration in the region of the immiscibility dome. This reinforces the concept of using the spinodal boundary in place of the critical temperature for concentrations other than the critical concentration. Thus, it has been shown that density and composition fluctuations primarily are responsible for the light scattering observed from glasses and the contributions from composition fluctuations cannot simply be evaluated by a random mixing model, but rather the position of the immiscibility dome as a function of concentration must be considered. The composition fluctuations diverge inside a two-phase region and take on large magnitudes near the spinodal boundary in the single-phase region, then proceed to fall off very rapidly for concentrations far from the critical. The density fluctuations on the other hand are not strongly sensitive to the composition.

A central linewidth investigation, which is useful in revealing the exact nature of the fluctuations responsible for the observed scattering, was shown to be possible by means of a correlation spectrometer if certain criteria involving the viscosity of the glass system to be studied is fulfilled. Often the extreme narrowness of the Rayleigh line negates the correlation spectrometer approach. However, the nonequilibrium nature of the glass allows one to devise an experiment giving equivalent information; namely, relaxation times for the particular system. This same experimental technique allows the determination of the spinodal temperatures for certain glass systems. Hence, light scattering coupled to an annealing study provides a technique that complements electron microscopy for locating the spinodal boundary of a glass (limited to the single-phase region).

Finally, Brillouin scattering from glasses at low temperatures has been dealt with and a number of models to account for the observed behavior were reviewed. We also considered the depolarized quasielastic scattering. For concentrations near the critical composition, the Krishnan effect was quite pronounced and we looked at the Krishnan ratio as a function of both temperature and concentration. To explain the results we proposed a type of mode–mode coupling, where the isotropic concentration fluctuations are coupled with longitudinal "frozen in" fluctuations of the dielectric susceptibility.

ACKNOWLEDGMENTS

The author is much indebted to Dr. C. J. Montrose of the Catholic University of America, Dr. J. A. Bucaro of the Naval Research Laboratory and Dr. J. W. Haus of the National Bureau of Standards for their many valuable discussions and suggestions.

References

Anderson, O. L., and Bömmel, H. E. (1955). *J. Am. Ceram. Soc.* **38**, 125.
Anderson, P. W., Halperin, B. I., and Varma, C. M. (1972). *Phil. Mag.* **27**, 1.
Andreev, N. S., Aver'yanov, V. E., and Voishvillo, N. A. (1960). *In* "The Structure of Glass," Vol. 2, pp. 205–208. Consultants Bureau, New York.
Andreev, N. S., Boiko, G. G., and Bokov, N. A. (1970). *J. Non-cryst. Solids* **5**, 41.
Are'fev, I. M., Kopylovskii, B. D., Mash, D. S., and Fabelinskii, I. L. (1967). *JETP Lett.* **5**, 535.
Arnold, W., Hunklinger, S., Stein, S., and Dransfeld, K. (1974). *J. Non-cryst. Solids* **14**, 192.

Becherer, G., Göcke, W., and Herms, G. (1971). *Naturforsch.* **26a**, 1177.

Benedek, G. B. (1966). *In* "Statistical Physics, Phase Transition and Superfluidity," Vol. 2, pp. 1–98. Brandeis Univ. Summer Inst. in Theoret. Phys.

Benedek, G. B. (1969). *In* "Polarisation, Matiére et Rayonement" (French Phys. Soc., ed.), pp. 49–84. Press Univ. de France, Paris.

Benedek, G. B., and Fritsch, K. (1966). *Phys. Rev.* **149**(2), 647.

Berge', P., Calmettes, P., Dubois, M., and Laj, C. (1970). *Phys. Rev. Lett.* **24**, 89.

Binder, K., and Stoll, E. (1973). *Phys. Rev. Lett.* **31**, 47.

Bloom, H., and Bockris, J. O'M. (1957). *J. Phys. Chem.* **61**, 515.

Bockris, J. O'M., and Lowe, D. C. (1954). *Proc. Roy. Soc. (London)* **226A**, 423.

Bockris, J. O'M., and Kojonen, E. (1960). *J. Am. Chem. Soc.* **82**, 4493.

Bockris, J. O'M., MacKenzie, J. D., and Kitchner, J. A. (1955). *Trans. Faraday Soc.* **51**, 1734.

Bömmel, H. E., and Dransfeld, K. (1960). *Phys. Rev.* **117**, 1245.

Born, M., and Huang, K. (1954). *In* "Dynamical Theory of Crystal Lattices," pp. 373–381. Oxford Univ. Press, London and New York.

Brillouin, L. (1922). *Ann. Phys. (Paris)* **17**, 88.

Brückner, R. (1964). *Glastech. Ber.* **37**, 413.

Bucaro, J. A., and Dardy, H. D. (1974a). *J. Appl. Phys.* **45**, 5324.

Bucaro, J. A., and Dardy, H. D. (1974b). *J. Appl. Phys.* **45**, 2121.

Bucaro, J. A., and Dardy, H. D. (1974c). *J. Chem. Phys.* **60**, 2559.

Bucaro, J. A., Dardy, H. D., and Corsaro, R. D. (1975). *J. Appl. Phys.* **46**, 741.

Cahn, J. W. (1961). *Acta Met.* **9**, 795.

Cahn, J. W., and Charles R. J. (1965). *Phys. Chem. Glasses* **6**(5), 181.

Charles, R. J. (1967). *J. Am. Ceram. Soc.* **50**(12). 631.

Chiao, R. Y., and Stoicheff, B. P. (1964). *J. Opt. Soc. Am.* **54**, 1286.

Chu, B. (1967a). *Phys. Rev. Lett.* **18**, 200.

Chu, B. (1967b). *J. Chem. Phys.* **47**, 3816.

Chu, B. (1974). "Laser Light Scattering," Chapter III. Academic Press, New York.

Chu, B., Schoenes, F. J., and Fisher, M. E. (1969). *Phys. Rev.* **185**, 219.

Cohen, C., Sutherland, W. W. H., and Deutch, J. M. (1971). *Phys. Chem. Liq.* **2**, 213.

Cook, H. E. (1970). *Acta Met.* **18**, 297.

Cummins, H. Z. (1969). *Proc. Int. School Phys. "Enrico Fermi"* (R. J. Glauber, ed.), Course XLII, Quantum Optics, p. 247. Academic Press, New York.

Cummins, H. Z. (1971). *Proc. Int. Conf. Light Scattering Solids, 2nd* (M. Balkanski, ed.), p. 3. Flammarion Sci., Paris.

Cummins, H. Z., and Swinney, H. L. (1970). *In* "Progress in Optics" (E. Wolf, ed.), Vol. VIII. North–Holland Publ., Amsterdam.

Davenport, W. B. Jr., and Root, W. L. (1958). "An Introduction to the Theory of Random Signals and Noise," Chapter 8, pp. 145–170. Academic Press, New York.

Debye, P., and Bueche, A. M. (1949). *J. Appl. Phys.* **20**, 518.

DeGroot, S. R., and Mazur, P. (1962). "Non-Equilibrium Thermodynamics," p. 25. North–Holland Publ., Amsterdam.

Demoulin, C., Montrose, C. J., and Ostrowsky, N. (1974). *Phys. Rev.* **A 9**, 1740.

Dubois, M., and Berge, P. (1971). *Phys. Rev. Lett.* **26**, 121.

Durand, G. E., and Pine, A. S. (1968). *IEEE J. Quantum Electron.* **QE–4**(9), 523.

Egelstaff, P. A. (1967). "An Introduction to the Liquid State," p. 130. Academic Press, New York.

Einstein, A. (1910). *Ann. Phy. (Leipzig)* **38**, 1275.

Fabelinskii, I. L. (1968). "Molecular Scattering of Light," p. 375. Plenum Press, New York.

Ferrell, R. A. (1970). *Phys. Rev. Lett.* **24**, 1169.

Fisher, M. E. (1964). *J. Math. Phys.* **5**, 944.

Flubacher, P., Leadbetter, A. J., Morrison, J. A., and Stoicheff, B. P. (1960). *Int. J. Phys. Chem. Solids.* **12**, 53.

Fulde, P., and Wagner, H. (1971). *Phys. Rev. Lett.* **27**, 1280.

Gammon, R. W. (1969). *In* "Light Scattering Spectra of Solids" (G. B. Wright, ed.), Paper G-2, p. 579. Springer–Verlag, New York.

Gans, R. (1936). *Physics* **37**, 19.

Gans, R. (1937). *Physics* **38**, 625.

Gaunt, D. S., and Baker, G. A. Jr. (1970). *Phys. Rev. Bull.* **1**, 1184.

Gibbs, J. W. (1906). "The Scientific Papers of J. Willard Gibbs," Vol. I, Thermodynamics, p. 354. Longmans, Green, London.

Gladstone, J. H., and Dale, T. P. (1864). *Phil. Trans. R. Soc.* (*London*) **153**, 337.

Golding, B., Graebner, J. E., Halperin, B. I., and Schutz, R. J. (1973). *Phys. Rev. Lett.* **30**, 223.

Goldstein, M. (1959). *J. Appl. Phys.* **30**, 493.

Gross, E. (1930a). *Nature* (*London*) **126**, 201, 400, 603.

Gross, E. (1930b). *Z. Phys.* **63**, 685.

Gross, E. (1932). *Nature* (*London*) **129**, 722.

Gupta, P. K., Schroeder, J., Aggarwal, I. D., and Macedo, P. B. (1975). *Proc. Symp. Opt. Acoust. Micro-Electron.* (J. Fox, ed.), Vol. XXIII, pp. 175–183. Polytechnic Press, New York.

Haller, W., Blackburn, D. H., and Simmons, J. H. (1974). *J. Am. Ceram. Soc.* **57**, 120.

Hammel, J. J. (1965). *Proc. Int. Congr. Glass, Brussels, 26 June 1965* Paper No. 36. Gordon and Breach, New York.

Hammel, J. J. (1967). *J. Chem. Phys.* **46**, 2234.

Hammel, J. J. (1972). *In* "Light Scattering From Glass" (T. L. Hench and D. B. Dove, eds.), Chapter 28, pp. 963ff. Dekker, New York.

Hammel, J. J., and Ohlberg, S. M. (1965). *J. Appl. Phys.* **36**, 1442.

Hammel, J. J., Mickey, J., and Golob, H. R. (1968). *J. Colloid Interface Sci.* **27**, 329.

Haus, J. W. Jr. (1974). *J. Chem. Phys.* **60**, 2638.

Heinicke, W., and Winterling, G. (1967). *Appl. Phys. Lett.* **11**, 231.

Heinicke, W., Winterling, G., and Dransfeld, K. (1971). *Proc. Int. Conf. Light Scattering Solids, 2nd* (M. Balkanski, ed.), p. 463. Flammarion Press, Paris.

Herzfeld, K. F., and Litovitz, T. A. (1959). "Absorption and Dispersion of Ultrasonic Waves," pp. 439–446. Academic Press, New York.

Hilliard, J. E. (1970). "Phase Transformations" (M. E. Aaronson, ed.), p. 497. Am. Soc. for Metals, Cleveland, Ohio.

Huang, Y. Y., and Wang, C. H. (1974). *J. Chem. Phys.* **61**, 1868.

Huang, Y. Y., and Wang, C. H. (1975). *J. Chem. Phys.* **62**, 120.

Huggins, M. L., and Sun, K. H. (1943). *J. Am. Ceram. Soc.* **24**, 4.

Hunklinger, S., Arnold, W., Stein, S., Nava, R., and Dransfeld, K. (1972). *Phys. Lett.* **42A**, 253.

Hunklinger, S., Arnold, W., and Stein, S. (1973). *Phys. Lett.* **45A**, 311.

Jackson, H. E., Walton, D., and Rand, S. (1976). *In Proc. Int. Conf. Light Scattering Solids, 3rd* (M. Balkanski, R. C. C. Leite, and S. P. S. Porto, eds.), p. 683. Flammarion Sci. France.

Jakeman, E., and Pike, E. R. (1969). *J. Phys. A.: Gen. Phys.* **2**, 411.

Jakeman, E., Oliver, C. J., and Pike, E. R. (1970). *J. Phys. A.: Gen. Phys.* **3**, 145.

Jones, C. K., Klemens, P. G., and Rayne, J. A. (1964). *Phys. Lett.* **8**, 31.

Jordan, P. C., and Jordan, J. R. (1966). *J. Chem. Phys.* **45**, 2492.

Kawasaki, K. (1966). *Phys. Rev.* **150**, 291.

Kawasaki, K. (1970). *Ann. Phys.* **61**, 1.

Keck, D. B., and Tynes, A. R. (1972). *Appl. Opt.* **11**, 1502.

Keck, D. B., Schultz, P. C., and Zimar, F. (1972). *Appl. Phys. Lett.* **21**, 215.

Kielich, S. (1967). *J. Chem. Phys.* **46**, 4090.

Kolyadin, A. I. (1956). *Opt. Spektrosc.* **1**, 907.

Kolyadin, A. I. (1960). *In* "The Structure of Glass," Vol. 2, pp. 202–204. Consultants Bureau, New York.

Komarov, L. I., and Fisher, I. Z. (1963). *JETP* **16**(5), 1358.

Krishnan, R. S. (1934). *Proc. Indian Acad. Sci.* **1**, 211, 717, 782, 915.

Krishnan, R. S. (1935). *Proc. Indian Acad. Sci.* **2**, 221.

Krishnan, R. S. (1936a). *Proc. Indian. Acad. Sci.* **3**, 211.

Krishnan, R. S. (1936b). *Proc. Indian Acad. Sci.* **3**, 126, 556.

Krishnan, R. S. (1937). *Proc. Indian Acad. Sci.* **5**, 94, 305, 407, 499, 551, 557.

Krishnan, R. S. (1938). *Proc. Indian Acad. Sci.* **7**, 21.

Krishnan, R. S. (1950). *Nature (London)* **165**, 933.

Krishnan, R. S., and Rao, P. V. (1944). *Proc. Indian Acad. Sci.* **20A**, 109.

Kurkjian, C. R., and Douglas, R. W. (1960). *Phys. Chem. Glasses* **1**, 19.

Laberge, N. L. (1973). Ph.D. Dissertation, The Catholic Univ. of Am., Washington, D.C., unpublished.

Laberge, N. L., Vasilescu, V. V., Montrose, C. J., and Macedo, P. B. (1973). *J. Am. Ceram. Soc.* **56**(10), 506.

Lai, C. C. (1973). Light Intensity Correlation Spectroscopy and Its Application to Study of Critical Phenomena and Biological Problems. D. Sc. thesis, M. I. T., Cambridge, Massachusetts.

Lai, C. C., Macedo, P. B., and Montrose, C. J. (1975). *J. Am. Ceram. Soc.* **58** (3–4), 120.

Landau, L. D., and Lifschitz, E. M. (1969). "Statistical Physics," p. 345. Addison-Wesley, Reading, Massachusetts.

Langer, J. S. (1969). *Ann. Phys. (N.Y.)* **54**, 258.

Langer, J. S. (1971). *Ann. Phys. (N.Y.)* **65**, 53.

Leidecker, H. W., Simmons, J. H., Litovitz, T. A., and Macedo, P. B. (1971). *J. Chem. Phys.* **55**(5), 2028.

Lekkerkerker, H. N. W., and Laidlaw, W. G. (1972). *Phys. Chem. Liq.* **3**, 1332.

Lekkerkerker, H. N. W., and Laidlaw, W. G. (1973). *Phys. Rev. A* **7** (4), 1332.

Leontovich, M. (1931). *Z. Phys.* **72**, 247.

Levin, D. I. (1955). *In* "Rayleigh Scattering in Glasses and the Structure of Glass" (*Proc. Conf. Structure Glass, Leningrad*), p. 198. Izd. Akad. Nauk. SSSR, Acad. Sci. USSR Press, Moscow-Leningrad.

Litster, J. D. (1972). "Dynamical Aspects of Critical Phenomena" (J. I. Budnick and M. P. Kawatra, eds.), pp. 152–164. Gordon and Breach, New York.

Love, W. F. (1973). *Phys. Rev. Lett.* **31**(13), 822.

Macedo, P. B., and Litovitz, T. A. (1965). *Phys. Chem. Glasses* **6**(3), 69.

Macedo, P. B., and Montrose, C. J. (1974). *In Proc. Int. Congr. Glass, 10th, Jpn.* pp. 68–72.

Mandelstam, L. I. (1926). *J. Russ. Phys. Chem. Soc.* **58**, 381.

Maurer, R. D. (1956). *J. Chem. Phys.* **25**, 1206.

Maurer, R. D. (1973). *Proc. IEEE* **61**(4), 452.

McIntyre, D., and Sengers, J. V. (1968). *In* "Physics of Simple Liquids (H. N. V. Temperley, ed.), Chapter II, p. 447. North–Holland Publ., Amsterdam.

Miller, G. A. (1967). *J. Phys. Chem.* **71**, 2305.

Mohr, R. K., and Macedo, P. B. (1975). In "Optical Properties of Highly Transparent Solids" (S. S. Mitra and B. Bendow, eds.), p. 279. Plenum Press, New York.

Montrose, C. J., Soloveyev, V. A. and Litovitz, T. A. (1968). *J. Acoust. Soc. Am.* **43**(1), 117.

Mori, H. (1964). *Progr. Theor. Phys.* **33**, 423.

Moriya, Y., Warrington, D. H., and Douglas, R. W. (1967). *Phys. Chem. Glasses* **8**(1), 14.

Mountain, R. D. (1966). *J. Res. Nat. Bur. Std.* **70A**, 207.
Mountain, R. D., and Deutch, J. M. (1969). *J. Chem. Phys.* **50**(3), 1103.
Mueller, H. (1938). *Proc. Roy. Soc. (London)* **166A**, 425.
Münster, A. (1974). "Statistical Thermodynamics," Vol. 2, pp. 612–615. Academic Press, New York.
Neilsen, G. F. (1969). *Phys. Chem. Glasses* **10**, 54.
Nossal, R., Chen, S. H., and Lai, C. C. (1971). *Opt. Commun.* **4**, 35.
Nye, J. F. (1957). "Physical Properties of Crystals." Oxford Univ. Press, London and New York.
Pecora, R. (1964). *J. Chem. Phys.* **40**, 1604.
Pelous, J., and Vacher, R. (1975). *Solid State Commun.* **15**, 279.
Pesin, M. S., and Fabelinskii, I. L. (1960). *Sov. Phys.-Dokl.* **4**, 1264.
Pesin, M. S., and Fabelinskii, I. L. (1961). *Sov. Phys.-Dokl.* **5**, 1290.
Phillips, W. A. (1972). *J. Low Temp. Phys.* **7**, 351.
Pine, A. S. (1969). *Phys. Rev.* **185**(3), 1187.
Pinnow, D. A. (1970). *IEEE. J. Quant. Electron.* **QE–6**(4), 233.
Pinnow, D. A., Candau, J. S., LaMacchia, J. T., and Litovitz, T. A. (1968). *J. Acoust. Soc. Am.* **43**(1), 131.
Pinnow, D. A., VanUitert, L. G., Rich, T. C., Ostermayer, F. W., and Grodkiewicz, W. H. (1975). *Mater. Res. Bull.* **10**, 133.
Pohl, R. O., Love, W. F., and Stephens, R. B. (1974). *Proc. Int. Conf. Amorphous Liq. Semicond.* (J. Stuke and W. Brenig, eds.), p. 1121. Taylor and Francis Ltd., London.
Poole, J. P. (1949). *J. Am. Ceram. Soc.* **32**, 230.
Prigogine, I., and Defay, R. (1954). "Chemical Thermodynamics." p. 246. Wiley, New York.
Prod'homme, L. (1957). *Acad. Sci. Paris* **245**, 300.
Raman, C. V. (1927). *J. Opt. Soc. Am.* **15**, 185.
Raman, C. V., and Rao, B. V. R. (1937). *Nature (London)* **139**, 58.
Raman, C. V., and Rao, B. V. R. (1938). *Nature (London)* **141**, 242.
Ramm, W. (1934). *Phys. Z.* **35**, 111, 756.
Rank, D. H., and Douglas, A. E. (1948). *J. Opt. Soc. Am.* **38**, 966.
Rayleigh, Lord (J. W. Strutt) (1919). *Proc. Roy. Soc. (London)* **A95**, 476.
Rich, T. C., and Pinnow, D. A. (1972). *Appl. Phys. Lett.* **20**, 264.
Ritland, H. N. (1959). *J. Am. Ceram. Soc.* **37**, 370.
Sabirov, L. M., Starunov, V. S., and Fabelinskii, I. L. (1968). *JETP Lett.* **8**, 246.
Sakka, S., and MacKenzie, J. D. (1971). *J. Non-cryst. Solids* **6**, 145.
Schnaus, U. E., Schroeder, J., and Haus, J. W. (1976). *Phys. Lett.* **57A**, 92.
Schnaus, U. E., Haus, J. W., and Schroeder, J. To be published.
Schroeder, J. Mohr, R. K., Macedo, P. B., and Montrose, C. J. (1973). *J. Am. Ceram. Soc.* **56**, 510.
Schroeder, J., Mohr, R. K., Montrose, C. J., and Macedo, P. B. (1973/74). *J. Non-Cryst. Solids* **13**, 313.
Schroeder, J. (1974). Rayleigh and Brillouin Scattering in Amorphous Solids: Silicate Glasses. Ph.D. Dissertation, The Catholic Univ. of Am., Washington, D.C., unpublished.
Schroeder, J. Montrose, C. J., and Macedo, P. B. (1975). *J. Chem. Phys.* **63**(7), 2907.
Shapiro, S. M., Gammon, R. W., and Cummins, H. Z. (1966). *Appl. Phys. Lett.* **9**(4), 157.
Shartsis, L., Spiner, S., and Capps, W. (1952). *J. Am. Ceram. Soc.* **35**, 155.
Shermer, H. F. (1956). *J. Res. Nat. Bur. Std.* **57**, 97.
Simmons, H. H., Napolitano, A., and Macedo, P. B. (1970). *J. Chem. Phys.* **53**(3), 1165.
Skripov, V. P. (1974). "Metastable Liquids," Chapters 9 and 10. Wiley, New York.
Stephens, R. B. (1973). *Phys. Rev. B* **8**, 2896.
Stevens, J. R., Bowell, J. C., and Hunt, J. L. (1972). *J. Appl. Phys.* **43**, 4354.

Swift, J. (1968). *Phys. Rev.* **173**, 257.

Tannenwald, P. E. (1966). "Physics of Quantum Electronics" (P. L. Kelley, ed.), p. 223 McGrw-Hill, New York.

Tomozawa, M., Herman, H., and MacCrone, R. K. (1969). *Proc. Conf. Mech. Phase Trans.* Inst. of Metals, London.

Tomozawa, M., and Obara, R. A. (1973). *J. Am. Ceram. Soc.* **56**, 378.

Tomozawa, M., MacCrone, R. K., and Herman, H. (1970). *J. Am. Ceram. Soc.* **53**, 62.

Tool, A. Q. (1946). *J. Am. Ceram. Soc.* **29**, 240.

Vacher, R., and Pelous, J. (1976). *Proc. Int. Conf. Phonon Scattering Solids, 2nd* (L. J. Challis *et al.*, eds.). Plenum Press, New York.

Van Hove, L. (1954). *Phys. Rev.* **95**, 249.

Van Kampen, N. G. (1965). *In* "Fluctuation Phenomenon in Solids" (R. E. Burgess, ed.), pp. 139–177. Academic Press, New York.

Van Kampen, N. G. (1969). *Proc. Int. School Phys. "Enrico Fermi"* (R. J. Glauber, ed.), Course XLII, Quantum Optics. Academic Press, New York.

Velichkina, T. S. (1953). *Izv. Akad. Nauk SSSR Ser. Fiz.* **17**, 546.

Velichkina, T. S. (1958). *Fiz. Inst. Akad. Nauk. SSSR* **9**, 59.

Venkateswaran, C. S. (1942). *Proc. Indian Acad. Sci.* **A15**, 362.

Voishvillo, N. A. (1962). *Opt. Spectrosc. (USSR)* **12**, 225.

Von Smoluchowski, M. (1908). *Ann. Phys. (Leipzig)* **25**, 205.

Walder, J., and Tany, C. L. (1967). *Phys. Rev. Lett.* **19**, 263.

Wemple, S. H. (1973). *Solid State Commun.* **12**, 701.

Wemple, S. H., Pinnow, D. A., Rich, T. C., Jaeger, R. E., and VanUitert, L. G. (1973). *J. Appl. Phys.* **44**, 5432.

White, J. A., Osmundson, J. S., and Ahn, B. H. (1966). *Phys. Rev. Lett.* **16**, 639.

Zeller, R. C., and Pohl, R. O. (1971). *Phys. Rev. B* **4**, 2029.

Zwanzig, R. (1965). *Ann. Rev. Phys. Chem.* **16**, 67.

Resonance Effects in Glasses

P. CRAIG TAYLOR

US Naval Research Laboratory
Washington, D.C.

I. Introduction

Unlike the singular success of diffraction techniques in determining the structure of crystalline solids, no one comprehensive technique exists for unraveling the structure of glasses. Instead, the structural characteristics of glasses must be inferred indirectly by applying plausible physical models to the results of several different experimental techniques such as x-ray, electron and neutron diffraction, viscosity, infrared aborption, electron spin resonance (ESR), nuclear magnetic resonance (NMR), and gamma resonance (Mössbauer) spectroscopy. Despite considerable research using many different experimental techniques, details concerning the basic structural properties of glasses remain almost as elusive today as they were before publication of the first x-ray diffraction results on SiO_2 almost 50 years ago (Zachariesen, 1932; Warren *et al.*, 1936). Diffraction measurements can usually determine the number of bonding nearest neighbors of the constituent atoms in a glass, but there is still considerable disagreement over the extent of local order beyond first nearest neighbors. In SiO_2 Zachariesen (1932) and Warren *et al.* (1936) originally suggested that the largest definable fundamental structural unit of glassy SiO_2 was an SiO_4 tetrahedron and that these tetrahedra were linked together in a random fashion to form the bulk glass (random network model). More recently Konnert and Karle (1972) have proposed a tridymite-type structural arrangement for glassy SiO_2. In B_2O_3 both the BO_3 planar triangle (Warren *et al.*, 1936) and the boroxyl ring (Warren, 1972) have been suggested as the largest definable structural entities. Similar differences in interpretation exist for any number of other glasses.

In any description of the structure of glasses randomness is certainly of fundamental importance, but the absence of long-range structural periodicity does not preclude the existence of local structural order as evidenced by some well-defined local structural unit. The basic question which must be answered in order to determine the structure of any glass is: Over what range can local structural order be defined and how are these ordered regions linked together to form the bulk material? Mössbauer, NMR, and ESR spectroscopy are three probes of local structural order which can aid in answering this question.

It is the purpose of this chapter to outline the usefulness of these three resonance techniques, ESR, NMR, and Mössbauer spectroscopy, for elucidating details of the local structural order in glassy solids. In addition it will be shown that these resonance techniques are useful in probing the vibrational elementary excitations of glasses, the details of which are quite different from those of crystalline solids. Finally the usefulness of one of these resonance effects, in particular ESR, in studying (1) paramagnetic impurities, (2) defects induced by radiation, rapid quenching, grinding or other means, and (3) localized electronic states within the forbidden energy gap will be discussed.

One might well ask why these three particular resonance techniques are grouped together for discussion. The answer is not so much because they are all resonance or spectroscopic techniques, but because they are all local probes which are affected by the same local electric and magnetic interactions (electric field gradient, magnetic shielding and antishielding of bound and itinerant electrons, magnetic dipole–dipole effects, and so forth). As experimental techniques these three resonance effects probe more localized regions in much greater detail than x-ray and other scattering techniques.

There are many excellent treatments of NMR, ESR, and Mössbauer spectroscopy (see, e.g., Abragam, 1961; Poole, 1967; Goldanskii and Makarov, 1968, for discussions and reference lists of NMR, ESR, and Mössbauer spectroscopy, respectively). There are also several reviews of the application of these three techniques to the study of glasses. Several authors have discussed the application of NMR to the study of both oxide glasses (Bray, 1967, 1970; Müller-Warmuth, 1965a,b; Taylor et al., 1975, 1976a and chalcogenide glasses (Bishop, 1974; Taylor et al., 1976a. The uses of ESR techniques to study oxide glasses have been reviewed by Griscom (1973/74, 1976) and by Taylor et al. (1975). Kurkjian (1970) and Coey (1974) have described the application of Mössbauer techniques to the study of oxide glasses, and Taneja et al. (1973) have described Mössbauer studies of chalcogenide glasses. In addition, Ruby (1972) has also briefly reviewed Mössbauer studies of glasses.

Weeks (1974) and Wong and Angell (1971, 1976) have discussed all three resonance techniques as applied to vitreous materials. These previous reviews have concentrated on the use of these techniques to infer local structural order in oxide glasses and have by and large not discussed the application of the techniques to structural inferences in nonoxide glasses or to the determination of elementary vibrational excitations in glasses. Most of the cited reviews of NMR, ESR, and Mössbauer studies in glasses contain copious references to the literature including numerous citations of work on complicated, multicomponent glass systems, and the reader is referred to these sources for comprehensive descriptions of the materials which have been studied. When one considers the excellent reviews already available, there is little need for yet another compendium of experimental results, so the emphasis of the present chapter is on the simplest glasses, especially the oxides, (B_2O_3, SiO_2), and the chalcogenides (Se, As_2Se_3, $GeSe_2$), and a critical examination of what can be determined about the local structural order or the elementary vibrational properties of these glasses using NMR, ESR, or Mössbauer techniques. The discussion will not include organic glasses, amorphous metals and alloys, or amorphous ferromagnetic or antiferromagnetic materials. No attempt will be made to present an exhaustive list of experimental results, but rather this chapter will concentrate on a limited number of illustrative examples and will develop them in enough detail to reveal the essential physics and chemistry of the problem and to give the reader an appreciation for both the power and the limitations of the techniques. Comparisons with crystalline materials will be made where appropriate. An extensive, although by no means exhaustive, reference list is included to provide the reader with quick access to most of the recent pertinent work. The usefulness of these three resonance techniques will be illustrated by citing simple materials where the techniques have proven most effective without the undue complications imposed by more complex materials. Since the kinds of information which can be determined do not in general change from material to material, there is no loss in generality in employing this approach. In fact, it is usually the case that the more simple the system the more information which can be extracted.

The extraction of meaningful information from NMR, ESR, or Mössbauer spectra of glasses ultimately depends on the adoption of a physically and chemically reasonable model through which the experimental data may be interpreted. This dependence on reasonable physical models results from our current lack of a comprehensive theoretical understanding of the structure of glasses. This limitation is not unique to these three experimental techniques but applies to the interpretation of any experimental measurements on glasses, including x-ray diffraction data. In the examples to be presented in this chapter, care will be taken to interpret the results using models which are simple enough to be tractable and yet sophisticated enough to reflect the details of the local structural order with some accuracy.

II. Analysis of Magnetic Resonance and Mössbauer Spectra

Even for single crystals, the analysis of NMR, ESR, and Mössbauer spectra can be complicated when several interactions must be considered. In magnetic resonance (NMR and ESR), and Mössbauer spectroscopy, many interactions depend in rather complicated fashions on the relative orientations of the crystalline axes with respect to the applied magnetic field. In Mössbauer spectroscopy one must be concerned with the additional complication of the directions of the crystalline axes with respect to the emitted γ quanta. When the sample is polycrystalline, the conditions for resonance must be averaged over all equally probable orientations of the crystalline axes to obtain the predicted absorption. Such an average is called a powder pattern in magnetic resonance and will be discussed in Section II,C. In glasses one must also take account of the variations in local bonding configurations which often require an additional average over an ensemble of randomly distorted paramagnetic or nuclear sites. A discussion of these ensembles is also presented in Section II,C. A more detailed presentation of magnetic resonance Hamiltonians, resonance conditions, and powder patterns is available in Taylor *et al.* (1975).

A. Basic Hamiltonians

For the majority of applications of NMR, ESR, and Mössbauer spectroscopy to glasses there is a great similarity between terms appearing in the three respective Hamiltonians. A basic theoretical understanding of these interactions is assumed, and in this section we will only review briefly these three Hamiltonians with an emphasis on their similarities. In the case of magnetic resonance (NMR and ESR), there is a formal similarity between the two respective Hamiltonians. In matrix notation, the following expression contains the three terms which are usually most important in the ESR effective spin Hamiltonian (electronic Zeeman term, fine structure term, and hyperfine term)

$$\mathscr{H} = \beta \mathbf{S} \cdot \mathbf{g} \cdot \mathbf{H} + \mathbf{S} \cdot \mathbf{D} \cdot \mathbf{S} + \mathbf{S} \cdot \mathbf{A} \cdot \mathbf{I} \tag{1}$$

where g, D, and A are the gyromagnetic, fine structure, and hyperfine structure tensors, respectively, S the electron spin operator, β the Bohr magneton, and **H** the applied magnetic field.

In similar notation the NMR Hamiltonian can be written as the sum of three terms (nuclear Zeeman term, magnetic shift term, and quadrupolar term)

$$\mathscr{H} = -\gamma h \mathbf{I} \cdot \mathbf{H} + \gamma h \mathbf{I} \cdot \boldsymbol{\sigma} \cdot \mathbf{H} + \mathbf{I} \cdot \mathbf{Q} \cdot \mathbf{I} \tag{2}$$

where $\boldsymbol{\sigma}$ is a magnetic shift (chemical shift, paramagnetic shift or Knight shift) tensor, Q is the quadrupolar tensor, γ is the nuclear gyromagnetic ratio, and **I** is the nuclear spin operator. A comparison of Eqs. (1) and (2) illustrates the formal similarity of several of the NMR and ESR interactions. In particular, the electronic Zeeman term is identical in form with the sum of the nuclear Zeeman and magnetic shift terms, and the fine structure term is formally identical with the quadrupolar term. The similarity is not complete, however, since there is no term in the NMR Hamiltonian equivalent to the ESR hyperfine term.

Although there is a formal similarity between corresponding terms in Eqs. (1) and (2), the physical origin of the terms can be somewhat different. For example, the isotropic contributions (those which do not depend on the orientation of the spin with respect to the applied magnetic field) to both the electronic and nuclear Zeeman terms result from the same type of magnetic dipole interaction of the form $-\mu \cdot \mathbf{H}$. On the other hand, anisotropies in the g tensor of Eq. (1) result from an admixture of excited states into the ground state wave function of the unpaired electron via the spin orbit interaction while the anisotropies introduced by the magnetic shift of Eq. (2) result from shifts in the local magnetic field at the nuclear site due to the presence of bonding, isolated paramagnetic, or conduction band electrons. The hyperfine term is a dipole–dipole interaction between electronic and nuclear spins and depends often only on the s and p character of the wave function of the unpaired electron.

The absorption which one observes in magnetic resonance is between two energy levels both of which derive from the ground state of the nucleus or the electron, and these energies typically fall in the radio frequency and microwave ranges, respectively. Mössbauer spectroscopy, on the other hand, involves recoil-free emission and resonant absorption between nuclear ground and excited states and occurs in the gamma ray region of the electromagnetic spectrum. The energy source for the gamma absorption is the emission of fluorescence from nuclear levels in a source material. A spectrum is obtained in magnetic resonance by sweeping either the frequency or the applied magnetic field. In principle the magnetic field could also be swept in Mössbauer spectroscopy in some cases, but usually one obtains a spectrum by driving the source at a suitable velocity (~ 1 cm/sec) with respect to the absorber, thereby changing the frequency of the emitted gamma quanta.

In Mössbauer spectroscopy the Hamiltonian contains terms similar to those which are present in NMR with the important distinction that the Mössbauer effect measures differences between an excited state and the ground state of the nucleus, while NMR involves only properties of the ground state. A simple illustration of this statement is presented in Fig. 1 where the six allowed Mössbauer transitions for ^{57}Fe in the presence of a magnetic field are shown by solid arrows and the single NMR transition by a dashed arrow. The levels g and e in Fig. 1 represent the ground and excited states, respectively. The Zeeman interaction splits these levels into four excited-state levels and two ground-state levels, and the quadrupolar interaction shifts the four excited state levels as indicated on the right-hand side of the diagram.

In Mössbauer terminology the Zeeman interaction is often referred to as a magnetic hyperfine interaction and the quadrupolar interaction is often termed an electric hyperfine interaction. In addition, the chemical shift is often called an isomer shift in Mössbauer spectroscopy.

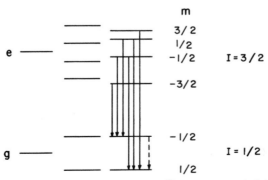

Fig. 1. Ground and excited state energy levels for ^{57}Fe in a magnetic field. Allowed Mössbauer transitions are indicated by the solid arrows. The allowed NMR transition is indicated by a dashed arrow.

The probability of observing a Mössbauer transition is given in the harmonic approximation by

$$f' = \exp\{-4\pi\langle \mathbf{i} \cdot \mathbf{x}\rangle^2/\lambda^2\} \tag{3}$$

where $\mathbf{x}$ is the overall amplitude of the nuclear vibrations, $\mathbf{i}$ the normalized direction of the gamma quantum, and λ the gamma wavelength. In the Debye approximation f' is independent of temperature at low T and decreases exponentially with T at high temperatures.

B. Resonance Conditions

Two simple resonance conditions, one pertaining to ESR and one to NMR, are presented here to facilitate the illustration in the next section of the basic features of powder patterns. Powder patterns are in principle also important in Mössbauer studies of polycrystals or glasses, but the experimental results accumulated to date are not detailed enough to warrant this added analytical complication.

The orientation of the applied magnetic field H with respect to the principal axes of the tensor quantities of Eqs. (1) and (2) are specified by the Euler angles θ and ϕ. Nuclear spin quantum numbers are denoted by I and m and electronic spin quantum numbers by S and M. In this notation the following ESR resonance condition for the $M - 1 \rightarrow M$, $\Delta m = 0$ allowed transitions is obtained from Eq. (1) by ignoring the second term (fine structure) and evaluating the third term (hyperfine structure) to first order in perturbation theory

$$\nu = g\beta H/h + Am/gh \tag{4}$$

where

$$g = [g_1^2 \sin^2 \theta \sin^2 \phi + g_2^2 \sin^2 \theta \cos^2 \phi + g_3^2 \cos^2 \theta]^{1/2}$$

$$A = [A_1^2 g_1^2 \sin^2 \theta \sin^2 \phi + A_2^2 g_2^2 \sin^2 \theta \cos^2 \phi + A_3^2 g_3^2 \cos^2 \theta]^{1/2}$$

The quantities g_1, g_2, g_3 and A_1, A_2, A_3 are the three principal components of the g and hyperfine tensors, respectively. In a similar fashion, a simple NMR resonance condition is obtained from Eq. (2) by neglecting the second term (magnetic shift) and by evaluating the third term (quadrupolar term) to first order in perturbation theory.

$$\nu = (\gamma H/2\pi) - \tfrac{1}{2}(m - \tfrac{1}{2})\nu_Q(3\cos^2\theta - 1 - \eta \cos 2\phi \sin^2 \theta) \tag{5}$$

where

$$\nu_Q = 3e^2 qQ/2I(2I - 1)h$$

The quadrupole coupling constant $e^2 qQ/h$ contains the quantities Q, the quadrupole moment of the nucleus and q, the maximum gradient of the electric field at the nuclear site. The asymmetry parameter η is the difference in the field gradients in the other two principal directions normalized to the maximum component.

C. Random Orientations and Statistical Distortions of Sites

In general, the field at which resonance occurs in either NMR or ESR of glasses, whether it is obtained from perturbation theory as in the case of Eqs. (1) and (2) or from numerical diagonalization of the exact Hamiltonian, can be written formally as $H_m = H_m(\mu, \phi, S_i)$ where $\mu = \cos\theta$ and ϕ are the Euler angles which specify the orientation of the site with respect to the applied magnetic field, m denotes the appropriate resonance transition and S_i denotes the set of Hamiltonian parameters which are appropriate for that particular site. In a powdered polycrystalline sample all orientations are equally probable and the variables μ and ϕ vary randomly from site to site. In this case the observed resonance line shape reflects the average over this ensemble of randomly oriented sites which is called a powder pattern. Singularities (shoulders and divergences) generally occur in powder patterns at magnetic fields which correspond to the evaluation of the resonance condition [Eqs. (3) or (4)] along the three principal axes. The theoretical powder patterns obtained for the ESR case where only the Zeeman interaction is present are shown in Fig. 2. In Fig. 2a the g tensor is axially symmetric and singularities occur at fields $H_\perp = h\nu/g_\perp\beta$ and $H_\parallel = h\nu/g_\parallel\beta$ where ν is the microwave spectrometer frequency. The case of complete

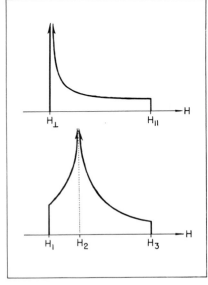

Fig. 2. ESR powder patterns for (a) an axially symmetric g tensor and (b) a completely asymmeteric g tensor (after Taylor *et al.*, 1975).

asymmetry is shown in Fig. 2b. A simple example of a powder pattern in a polycrystalline sample is presented in the next section. Further details concerning powder patterns in magnetic resonance and on how to extract the Hamiltonian parameters from them are available elsewhere (Taylor *et al.*, 1975).

In a disordered or glassy material an additional complication arises due to the basic randomness of the structure itself. In these materials, there often exists a continuously random variation in the local environments surrounding any particular nuclear or paramagnetic site. This variation of local environments can result in a continuous variation of the Hamiltonian parameters describing the magnetic resonance spectra of these sites. While the absorption in a powdered crystalline material can be accurately represented by a randomly oriented ensemble of otherwise identical sites, the absorption in a glassy or disordered solid must often be characterized by an additional ensemble of sites with differing local environments. The assumption which is often made in constructing the absorption envelope for glassy or disordered materials is that these two ensembles are statistically independent. That is, there exists an identical ensemble of sites with random variations in local environments for every particular orientation, and conversely, there exists an identical ensemble of randomly oriented sites for every particular local environment. Formally this absorption envelope can be written as (Jellison *et al.*, 1976; Peterson and Kurkjian, 1972),

$$S(H) = \sum_m I_m p_{H',m}(H_m) \tag{6}$$

where $p_{H',m}$ is the probability distribution (density) function for the Hamiltonian parameters and I_m is the transition probability for transition m.

The dependent variables of Eq. (6) may be correlated (that is, an increase in one Hamiltonian parameter from one site to another in the glass may occur with the concomitant increase or decrease of another Hamiltonian parameter), but they are all statistically independent.

Because of the statistical independence of the dependent variables the order of integration may be altered from that indicated in Eq. (6) for actual calculations. (See Jellison *et al.*, 1976). In particular, it has proven advantageous because of an economy in the numerical computational methods to separate explicitly the integrations over $d\mu \, d\phi$ and those over random variations in Hamiltonian parameters (Taylor and Bray, 1970, 1972a), in which case, if we neglect the convolution to account for isotropic dipolar broadening, Eq. (6) becomes (Jellison *et al.*, 1976; Taylor and Bray, 1970)

$$S(H) = \frac{1}{4\pi} \frac{1}{dH} \sum_m \prod_i \int_{S_i} p(S_i) \, dS_i \int_H^{H+dH} I_m(\Omega) \, d\Omega [H_m(S_i)] \tag{7}$$

where $d\Omega = d\mu \, d\phi$.

Computer simulation techniques are essential to the evaluation of magnetic resonance spectra observed in glasses and disordered materials because closed mathematical expressions are not generally possible to describe the more complicated averages over both orientation and local environment. Both a finite number of orientations and a finite number of sites possessing discrete Hamiltonian parameters are used to approximate the continuously random variation of local environments (Taylor and Bray 1968, 1970, 1972a; Lefebvre and Maruani 1965a,b; Kopp and Mackey, 1969; Mackey *et al.*, 1969; Maruani, 1964).

In favorable situations only some of the Hamiltonian parameters (and thus only some of the singular points of the powder spectrum) are sensitive to the existing variations in local environments. In the most favorable of situations the variations in Hamiltonian parameters can be calculated from an assumed model of the nuclear or paramagnetic site. One basic principle has proved to be extremely useful in evaluating the magnetic resonance spectra of glasses, namely, that the best starting point for fitting a glass spectrum is the spectrum observed in the corresponding crystalline compound, when one exists. Several illustrations of this general procedure will be given in the ESR and NMR sections which follow.

Because of the experimental techniques employed, both NMR and ESR spectra are commonly recorded as derivatives of the power absorbed as a function of the magnetic field while Mössbauer spectra are usually presented as absorption of gamma quanta as a function of the velocity of the source. In order to compare calculated NMR and ESR spectra with experiment, the absorption envelopes are usually differentiated and presented as derivative traces.

The sharp divergences and shoulders shown in the powder patterns of Fig. 2 are washed out in actual experimental traces due to random dipolar interactions or relaxation processes between spins or, in glassy materials, due also to statistical distributions of the Hamiltonian parameters. If the distributions of Hamiltonian parameters are not important, then the dipolar or relaxation effects can usually be taken outside of the integral sign in Eq. (6) or (7) and approximated by convoluting the powder pattern against a Gaussian or Lorentzian broadening function (Taylor and Bray 1970; Taylor *et al.*, 1975).

III. ESR in Glasses

There are at least three types of paramagnetic sites in glasses which can give rise to ESR responses. These types include radiation-induced defects (color centers), paramagnetic impurities, and optically induced paramagnetic states. Irradiation of both crystals and glasses with x rays, γ rays, neutrons, or electrons yields paramagnetic electrons which are located at defect sites. These defects are quite similar in both crystals and glasses and are often atomic vacancies which have trapped either an electron or a hole. Paramagnetic ions such as Fe^{3+} or Mn^{2+} in a glass will also yield an ESR response. In addition, a fundamentally different type of ESR center has recently been observed in semiconducting glasses after a few minutes exposure to a few milliwatts of infrared light at low temperature (Bishop et al., 1975, 1976a,b,). These optically induced centers do not occur in crystalline materials as the other two classes do, are less stable than the radiation induced centers, and do not appear to be directly associated with atomic vacancies. For these reasons, this class of paramagnetic site appears to be more "intrinsic" to the glassy state than either of the other two.

A. Radiation-Induced Defects

This section begins with a simple example of the kinds of differences one most often encounters between the ESR spectra of radiation-induced color centers in polycrystalline and glassy solids. This example involves the analysis of the $O_2{}^-$ molecular ion in polycrystalline sodium peroxide (Griscom, 1976 and in amorphous peroxyborates (Edwards et al., 1969). Figure 3 shows the powder pattern and derivative spectrum in the absence of dipolar or relaxation broadening (solid lines with arrows) calculated for $O_2{}^-$ in polycrystalline sodium peroxide. The spin Hamiltonian for this example contains only the Zeeman term. This figure also shows a comparison between the experimental trace (noisy curve) and the calculated derivative trace including isotropic Lorentzian broadening (dotted curve). The agreement between experiment and calculation is quite good.

In Fig. 4 the ESR spectrum attributed to $O_2{}^-$ in an amorphous peroxyborate (noisy curve) is compared to that observed in polycrystalline Na_2O_2 (dashed curve). At first glance the two experimental curves appear to be quite

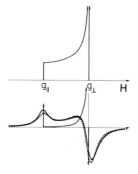

Fig. 3. ESR powder pattern for axially symmetric g tensor (*top*) and first derivative (*bottom*). The noisy curve is the experimental spectrum of the $O_2{}^-$ center in polycrystalline sodium peroxide. The dotted curve is a computer simulation assuming axial symmetry and a Lorentzian convolution function. (After Griscom, 1976.)

Fig. 4. Experimental ESR derivative spectrum of O_2^- in an amorphous peroxyborate (noisy trace). The dashed curve is the experimental spectrum of O_2^- in polycrystalline Na_2O_2 on the same scale. The dotted line is a computer simulation to the glass spectrum using a simple model described in the text and the distribution of $g_{||}$ values indicated in the inset. (After Griscom, 1976.)

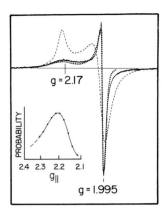

different, but it has been shown (Griscom, 1976) that these differences are due to a statistical distribution of $g_{||}$ values for the various O_2^- sites in the amorphous host and are thus *not* due to a characteristically different center in the two materials. The distribution of $g_{||}$ values is shown in the inset and the theoretical spectrum obtained using the distribution is indicated by the dotted line. The agreement between experimental and calculated traces in this case also is quite good.

A conclusion from this simple example is that ESR spectra in amorphous solids can sometimes be fit with the same accuracy as spectra in corresponding polycrystalline solids. The fit may not be unique and the distribution of the spin-Hamiltonian parameter $g_{||}$ may not make physical or chemical sense. To examine these two questions one must invoke a reasonable physical model for the O_2^- defect and examine the consequences of distorting the defect in the amorphous host. Using the g-value theory developed by Konzig and Cohen (1959), and a Gaussian distribution of energy splittings between the two $2p\pi_g$ orbitals, Griscom (1976) has calculated the skewed $g_{||}$ distribution function shown in the inset to Fig. 4. and a much narrower distributon function for $g_\perp$. The dotted curve of Fig. 4 represents the result of the model calculation. Furthermore, the halfwidth of the statistical distribution of energy splittings necessary to fit the spectrum is small. Thus not only can the O_2^- spectrum in amorphous peroxyborates be fit, but it can be fit within the rather stringent constraints of a simple chemical model which is based on knowledge of the O_2^- ion in crystalline hosts. In addition, within the constraints of this model the distributions of spin–Hamiltonian parameters (both $g_{||}$ and $g_\perp$) are unique.

Not all ESR or NMR spectra observed in glasses are as simple as the one in the above example, and the model calculations are in many cases less convincing. Nonetheless, it is in the spirit of this simple example that the more complex glass spectra should be viewed. Detailed structural information can be obtained from ESR and NMR spectra of glasses provided that appropriate model calculations can be developed. In developing these physical or chemical models the most important guideline yet devised is to examine first the behavior in a corresponding crystalline solid where all of the sites are identical.

1. SiO$_2$ AND SILICATE GLASSES

The enumeration and preliminary identification of the radiation-induced paramagnetic defects in SiO$_2$ and silicate glasses are primarily the results of the early work of Weeks (1956, 1967, 1974) and Schreurs (1967). In this section we discuss the three most prominent radiation-induced defects—an electron center designated as E' by analogy with a similar center occurring in crystalline quartz and two hole centers designated HCl and HC2 by Schreurs.

Perhaps the most studied radiation-induced defect in oxide glasses is the E' center. Early studies of crystalline α-quartz have identified the E' center in this material as an electron trapped in a nonbonding sp^3 hybird orbital on a silicon atom at the site of an oxygen atomic vacancy (Weeks, 1956; Silsbee, 1961; Feigl, 1970). This identification rests primarily on a detailed analysis of the hyperfine interaction of the unpaired electron with a ^{29}Si nucleus which is only 4.7% abundant. In SiO$_2$ glass the similarity of the ESR spectrum (in the absence of the hyperfine interaction) with that observed in poly-crystalline quartz for electrons localized on nonmagnetic ^{28}Si and ^{30}Si atoms has suggested the presence of the E' center in this glass (Weeks and Nelson, 1960; Weeks and Sonder, 1963). Figure 5 indicates the agreement between the observed ESR derivative spectrum in irradiated silica glass (solid curve) and the curve calculated using the parameters measured in crystalline α-quartz (dashed curve). This calculation, due to Friebele *et al.* (1974), does not include any statistical distribution of the spin-Hamiltonian parameters which accounts for the subtle differences between the dashed and solid curves.

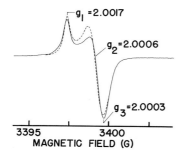

Fig. 5. ESR derivative spectrum of the E' center associated with nonmagnetic ^{28}Si and ^{30}Si nuclei in glassy SiO$_2$ (solid line). Dashed curve is a computer simulation using spin-Hamiltonian parameters appropriate for the E' center in crystalline α-quartz. (After Friebele *et al.*, 1974.)

A more positive identification of the E' center in SiO_2 glass has been provided by the observation (Griscom *et al.*, 1974) of a ^{29}Si hyperfine interaction similar to that which was definitive in the identification of the E' center in crystalline SiO_2. These authors used SiO_2 enriched to 95% in ^{29}Si in order to study the details of this hyperfine interaction.

Figure 6 shows the ^{29}Si hyperfine structure observed for naturally abundant ^{29}Si in SiO_2 glass. There are $2I + 1$ allowed hyperfine transitions for a nucleus of spin I which means that for ^{29}Si where $I = \frac{1}{2}$ one observes two lines. The resonance, which is off scale in the central portion of this figure, is due primarily to the E' center for those spins which are localized on non-magnetic silicon nuclei (^{28}Si and ^{30}Si).

The dotted curve of Fig. 6 represents a computer fit to the experimental data which employs the crystalline model for the E' center and essentially a Gaussian distribution in the bond angle between the unpaired spin and one of the Si–O bonds. Although the values of the atomic s and p contributions to the hyperfine coupling constant are not known precisely enough to calculate the average bond angle accurately, the deviations from this average angle can be accurately ascertained using the simple model. The results are shown in Fig. 7 where the distribution of angles which provided the best-fit curve of Fig. 6 is displayed.

The most important conclusions to be drawn from this simple, yet elegant, calculation are that the E' centers in crystalline α-quartz and SiO_2 glass are essentially identical and that very small distortions ($\sim 0.7°$) in the defect–Si–O bond angle account for the broadened resonance observed in the glass.

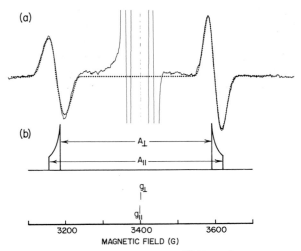

Fig. 6. ESR spectrum of irradiated SiO_2 showing ^{29}Si hyperfine structure attributed to the E' center (solid curve). The dashed curve is a computer simulation using a superposition of powder patterns of the type indicated at the bottom of the figure. (After Griscom *et al.*, 1974.)

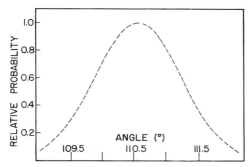

Fig. 7. Distribution of defect–Si–O bond angles for E' centers in irradiated SiO_2 as determined from the model described in the text. (After Griscom *et al.*, 1974.)

Once again we note that within the contraints of this simple and highly plausible physical model the distribution of spin-Hamiltonian parameters is uniquely determined subject only to experimental uncertainty.

In irradiated alkali silicate and other silicate glasses two paramagnetic hole centers usually dominate the ESR response. Both centers involve only the electronic Zeeman term of Eq. (3). The center HCl is characterized by a completely asymmetric (orthorhombic) g tensor and dominates the alkali silicate spectra for glasses of low alkali oxide content. The spectra of HCl observed by Schreurs (1967) in K, Na, Rb, and Cs silicate glasses are shown in Fig. 8. Note that in the potassium silicate spectrum, where the resolution is the best, the three features characteristic of the derivative of the orthorhombic powder pattern of Fig. 2b can be discerned. There is a bump in the derivative spectrum of Fig. 8 near $g = 2.02$ which corresponds to the low-field shoulder in the powder pattern of Fig. 2b, a dispersive feature near

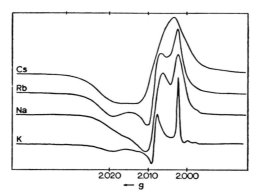

Fig. 8. Experimental derivative spectra of the ESR center HCl in irradiated alkali silicate glasses. The alkali ion is labeled on the left-hand side of each trace. (After Schreurs, 1967.)

$g = 2.01$ which corresponds to the divergence, and a sharp bump near $g = 2.00$ which corresponds to the high-field shoulder.

Detailed calculations by Griscom (1976), which have been reproduced in Fig. 9, indicate that in this case also a distribution of spin-Hamiltonian values is necessary to fit the experimental spectrum accurately. The distribution functions for g_3 and g_2 are shown in the bottom portion of this figure where it was assumed that the statistical fluctuations of g_3 and g_2 increase and decrease together (positive correlation between the two variables). A small response due to the Si E' center just described is also indicated in Fig. 9 as well as two weak features attributed to a small number of HCl centers which undergo an interaction with ^{29}Si (5% abundant) nuclei. The spectrum of Fig. 9 was taken at 9 GHz. Similar agreement was obtained using the same distribution functions at 35 GHz (Griscom, 1976).

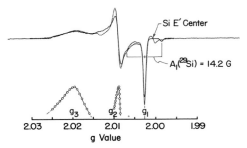

Fig. 9. Experimental ESR derivative spectrum of HCl in an irradiated potassium silicate glass (solid line at top). The dotted line is a computer simulation of the spectrum using the g value distributions indicated at the bottom. (After Griscom, 1976.)

Sidorov and Tyul'kin (1967) and Griscom (1976) have investigated the behavior of HCl in silicate glasses enriched in ^{29}Si. From these experiments it can be concluded that the hyperfine interaction is with a single Si nucleus although detailed fits to the spectra are presently incomplete. Because the hyperfine interaction is with a single Si nucleus, the center cannot be a hole on a bridging oxygen atom (oxygen bonded to two Si atoms on adjacent tetrahedra). Schreurs (1967) suggested that the center HCl was a hole localized on two nonbridging oxygen atoms on a Si–O tetrahedron. More recent calculations (Griscom, 1976) suggest that the hole is on a single nonbridging oxygen atom. In this model it is the g_3 component which is related to a statistical distribution of energy splittings between the two oxygen nonbonding $2p_z$ and $2p_y$ orbitals, and it is suggested that the large halfwidth of this distribution is due to the presence of a nearby alkali ion. This identification of the hole center HCl is capable of correlating more experimental evidence than the model originally suggested by Schreurs, but the analysis is not as complete as in the case of the E' center because the analogous center to HCl in crystalline materials has not been studied in detail.

A second hole center, called HC2, also occurs in alkali silicate glasses and is dominant in glasses with high alkali content. The ESR response of this center is shown in Fig. 10 for potassium and sodium silicate glasses. Schreurs (1967) has suggested that this center is due to a hole trapped on three non-bridging oxygen atoms on a SiO_4 tetrahedron. The lack of any great asymmetry or resolved structure in the ESR curves of Fig. 10 makes it difficult at present to test this hypothesis. Certainly the detailed nature of the ESR center in silicate glasses known as HC2 is presently unknown.

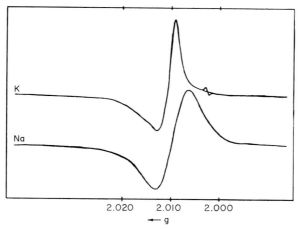

Fig. 10. Experimental derivative spectra of the ESR center HC2 in potassium and sodium silicate glasses. (After Schreurs, 1967.)

2. B_2O_3 and Borate Glasses

Some of the radiation damage centers occurring in borate glasses have analogs in the irradiated silicate glasses but there is also strong evidence for the presence of additional centers which are not present in the silicate glasses. In this section we describe (1) several hole centers which appear in the alkali borate system, (2) the analog of the silicon E' electron center which is localized on a boron atom, (3) electron centers associated with alkali ions, and (4) halogen-associated electron and hole centers in glasses doped with halogen atoms.

The first systematic investigations of radiation induced paramagnetic defects in borate glasses were performed by Karapetyan and Yudin (1963) and by Lee and Bray (1963, 1964). These authors succeeded in separating the effects of two different hole centers which are characteristic of low- and high-alkali content glasses. Many more recent investigations by a number of authors (see Griscom, 1973/74 for a review of these results) have refined the spin-Hamiltonian parameters and their statistical distributions and have provided at least tentative identifications for most of these hole centers.

As usual we start the discussion of the borate glass hole centers with a brief description of the center occurring in related alkali borate crystalline compounds. The X-band ESR spectra observed in γ-irradiated 1:2 and 1:4 lithium borate compounds are shown in Fig. 11. The thin solid lines represent computer simulations of the spectra assuming a single site in both compounds while heavy traces represent the experimental results. There are obvious discrepancies between the experimental and computer-simulated traces under the simplifying assumption of a single site; more detailed experiments indicate that there are at least two primary sites in both compounds. (Griscom et al., 1968; Taylor and Griscom, 1971). Nonetheless, the major features of the spectra observed in the 1:2 and 1:4 lithium borate compounds are accurately reproduced assuming a single primary site in each case.

The primary site in the 1:2 compound has been determined from the details of the ^{11}B hyperfine interaction to be a hole on an oxygen atom weakly interacting with two identical 4-coordinated boron atoms (Taylor and Griscom, 1971). This spectrum has also been observed in the strontium diborate compound where all of the borons are 4-coordinated. Unlike the 1:2 compound, the structural units present in the 1:3, 1:4, and 1:5 alkali

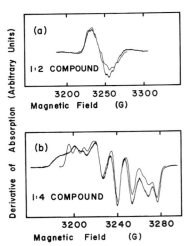

Fig. 11. Experimental ESR derivative spectra observed in the irradiated 1:2 and 1:4 lithium borate polycrystalline compounds (heavy traces). The thin lines are computer simulations which assume a hyperfine interaction with two equivalent boron atoms (curve a) and with one boron atom (curve b) as described in the text. (After Taylor and Bray, 1972a.)

borate compounds contain one 4-coordinated boron per structural unit which is linked exclusively to 3-coordinated borons. The structural units in the 1:3 and 1:5 alkali–borate compounds are very similar while the 1:4 compound consists of a mixture of 1:3 and 1:5 structural units. (Krogh-Moe, 1962, 1965). The primary ESR site in the 1:4 compound (see Fig. 11b) is the same as that in the 1:3 compound (Griscom *et al.*, 1968), and has been identified as a hole on an oxygen atom weakly interacting with a single boron nucleus. This center can thus be associated with the 1:3 type structural units.

As in the irradiated silicate glasses, the major features of the ESR spectra observed in the alkali borate glasses can be accurately reproduced assuming a simple superposition of the 1:2 and 1:3 or 1:4 type sites of Fig. 11, but detailed fits of course require one to consider the statistical distributions of spin-Hamiltonian parameters caused by the disorder in the glassy phase. The agreement obtained in the lithium borate system ignoring distributions of spin-Hamiltonian parameters is shown in Fig. 12. The computer simulations for the 1:5, 1:4, and 1:3 glass spectra are in reasonable agreement with experiment, especially when one considers the accuracy with which the crystalline compound spectra of Fig. 11 were fit. In the 1:2 glass composition additional discrepancies near 3225 and 3275 G are apparent. These discrepancies reflect the appearance near the 1:2 composition of substantial numbers of paramagnetic hole–centers characteristic of the 1:1 glass. Compare the 1:1 glass spectrum at the top of Fig. 12 with the 1:2 glass spectrum and computer simulation. Recent experiments on irradiated crystalline calcium and strontium metaborate (1:1 compounds) indicate the presence of a four line spectrum similar to that of the 1:1 glass (Taylor and Bray, 1972a).

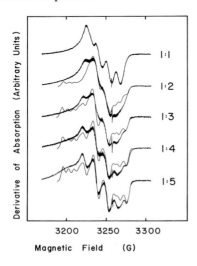

Fig. 12. Experimental ESR derivative spectra observed in lithium borate glasses whose lithium oxide–boron oxide ratios are as indicated (heavy lines). Thin solid lines represent computer-simulated spectra obtained by assuming a weighted average (as described in the text) of the computer simulations for the 1:4 and 1:2 compounds of Fig. 11. (After Taylor and Bray, 1972a.)

It is probable that the four line spectrum is characteristic of all chainlike metaborate compounds and is indicative of the 1:1 type structural unit. One may infer from these results that the discrepancies in fitting the glass spectra of high alkali oxide content probably result from the increasing presence of an ESR site similar to that which is found in the chainlike metaborate compounds.

One interesting point can be made concerning the computer simulations of the data of Figs. 11 and 12. Within the limited accuracy of the "discrete site" approximation to the glass spectra, the results support the general predictions of Krogh-Moe (1962, 1965) that the structural units present in the glass are characteristic of those present in the crystalline compounds of similar compositions. The relative number of 1:3 type sites characteristic of the 1:5, 1:4, and 1:3 compounds decreases continuously with increasing alkali content over the range shown in Fig. 12 while the 1:2 type sites increase over this range. A more detailed and definitive substantiation of the Krogh-Moe hypothesis is presented in the NMR section below.

Now that the predominance of the 1:3 type spectrum (sometimes called the "five line plus a shoulder" spectrum or the boron oxygen hole center BOHC in the literature) in glasses of low alkali oxide content has been illustrated, we turn to a more sophisticated analysis of this characteristic hole center. Through the use of isotopic substitution, Lee and Bray (1963, 1964) were the first to determine that the observed hyperfine structure of the 1:3 hole center is due to a relatively weak interaction with a single boron nucleus. Griscom et al. (1968) simulated this spectrum numerically under the assumption of orthorhombic g and hyperfine tensors with a skewed distribution of g_3 values. Their results for a glass containing 20 mole % K_2O at 9 GHz are shown in Fig. 13. Similar results were obtained at 35 GHz and for samples enriched in the isotope ^{10}B, where $I = 3$.

Recent photoemission results on SiO_2 (DiStefano and Eastman, 1971) and on various chalcogenide glasses (Bishop and Schevchik, 1975) indicate that the upper valence band energy levels for the oxygen and chalcogen (S, Se, or Te) atoms are entirely p-like with essentially no s–p hybridization. These results are not entirely unexpected for the valence band energy level which is a nonbonding lone pair orbital, but the lack of hybridization in the two bonding orbitals has not generally been appreciated. These photoemission data strongly imply both that the unpaired spin of the 1:3 type hole center is predominantly localized on an oxygen p wavefunction and that the excited states admixed to the ground state by the spin orbit interaction (which determine the departures of g_2 and g_3 from 2.00) are also predominantly p states.

Similar hole centers have been observed in a variety of oxide glasses (See Table I for a partial listing) including an interesting nitrogen-related center

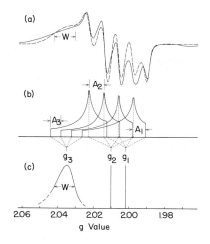

Fig. 13. (a) Experimental ESR derivative spectrum of the 1:3 type hole center in an irradiated potassium borate glass with 20% K_2O (solid line). The dashed line represents a computer simulation using the most probable powder patterns of (b) and the g value distributions of (c). (After Griscom, 1973/74.)

TABLE I

Most Probable Spin–Hamiltonian Parameters for Hole-Type Centers in a Variety of Oxide and Chalcogenide Glasses[a]

Material	Nucleus	g_1	g_2	g_3	$\langle A \rangle / A_0$	Reference
Oxide glasses						
Silicate glass	^{29}Si	2.003	2.009	2.019	0.01	Schreurs (1967)
Borate glass	$^{11}B, ^{10}B$	2.002	2.010	2.035	0.01	Griscom *et al.* (1968)
Titanate glass	$^{47}Ti, ^{49}Ti$	2.003	2.010	2.022	—	Kim and Bray (1970)
Germanate glass	^{37}Ge	2.002	2.008	2.051	—	Purcell and Weeks (1969)
Selenide glass						
Arsenate glass	^{75}As	2.00	2.03	2.17	$\lesssim 0.01$	Taylor *et al.* (1976b).

[a] Oxide glass data adapted from Griscom, (1976) and selenide glass data from Taylor *et al.* (1976b).

observed in irradiated sodium silicate glasses containing nitrogen (Mackey *et al.*, 1969 and 1970). Figure 14 shows that the essential features of the center including the broad distribution of g_3 values, are strikingly similar to those already described for the 1:3 type oxygen center of Fig. 13. There is one important difference between these two hole centers: in the nitrogen center the hyperfine interaction is with the atom on which the paramagnetic spin is localized and one may thus calculate directly the s and p character of the wavefunction on the nitrogen. A simple estimate yields about 75% p character and only about 2% s character which is in agreement with the conclusions drawn for the 1:3 type oxygen center by less direct means.

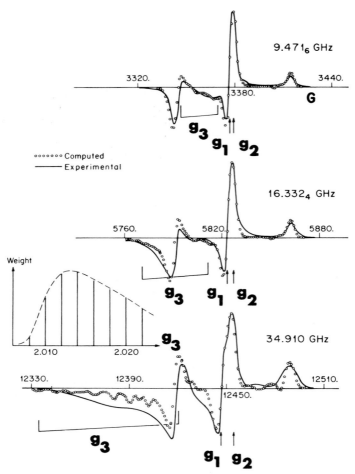

Fig. 14. Experimental (solid lines) and computer-evaluated (open circles) ESR derivative spectra of nitrogen-related hole center in x-irradiated sodium silicate glasses at 9, 16, and 35 GHz. The three singularities in the powder pattern neglecting hyperfine effects $[H(g_i)]$ are denoted as g_i in this figure. The inset denotes the assumed distribution function in g_3 as described in the text. (After Mackey *et al.*, 1969.)

Despite the successful analyses of the spectra such as the ones presented in Figs. 13 and 14 and the observation of a similar center in several crystalline compounds, a unique model for the 1:3 type hole center does not yet exist. It has been established that the hyperfine structure probably results from the mechanism of core polarization rather than direct overlap of the wave-function of the unpaired spin on the boron atom (Symons, 1970, and Taylor *et al.*, 1971). But it is still undecided whether the hole resides on a nonbonding p orbital of a bridging oxygen (Griscom *et al.*, 1968; Taylor and Griscom, 1971; Griscom, 1975), or on an oxygen dangling bond (nonbonding p orbital of a nonbridging oxygen) (Symons, 1970). The details of this debate over models are unimportant for the purposes of the present discussion, but two interesting points can be made by examining the principal components of the g tensors and hyperfine tensors for several hole centers occurring in irradiated oxide and chalcogenide glasses which are listed in Table I. First, the isotropic components of the hyperfine interaction $\langle A \rangle$ normalized to the atomic isotropic coupling constants A_0 are all of the order of 10^{-2} and indicate that the paramagnetic centers are either localized on nonbonding or dangling bond oxygen orbitals. Second, the principal components of the g tensors in the oxides are all quite similar (including broad distributions in g_3 values in all cases) which implies either that all of the centers are on nonbridging oxygens, as is the case described above for HCl in the silicate glasses, or that the electronic Zeeman effect is essentially independent of the two bonding configurations (bridging versus nonbridging oxygens).

The analog on a boron atom of the silicon E' center has been identified in several borate glasses (Griscom *et al.*, 1976). The ESR spectrum of the E' center in a borosilicate glass ($B_2O_3 \cdot 3SiO_2$) is shown in Fig. 15a. In Fig. 15b

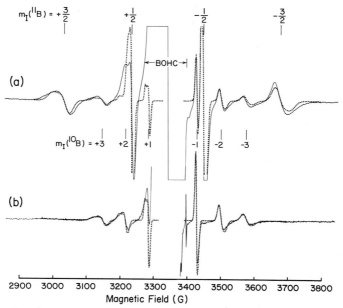

Fig. 15. Experimental ESR derivative spectra (heavy lines) of irradiated $B_2O_3 \cdot 3SiO_2$ glass for (a) a naturally abundant mixture of ^{11}B and ^{10}B and (b) an enrichment of ^{10}B to 95%. The dotted curves are computer simulations of the boron E' center contributions to the spectra as described in the text. (After Griscom *et al.*, 1976.)

a similar trace is displayed for a glass enriched to 95% in ^{10}B. Because the same parameters were used to fit both spectra (dotted curves), one may conclude that the hyperfine interaction with a single boron nucleus is well established. Furthermore, the mean value spin-Hamiltonian parameters and statistical distributions used in the computer simulations of Fig. 15 are the direct analogs for boron of those used in fitting the silicon E' center described above. We conclude from this rather convincing model calculation that the E' centers on boron atoms in borate glasses are similar in every respect, including the bond angle distortions, to those observed on silicon atoms in silicate glasses.

The intensity of the boron E' ESR response in borate glasses is generally insufficient to account for all of the trapped electron density inferred from the intensity of the 1:3 type trapped hole resonance. Most of the electrons in alkali borate glasses are trapped on interstitial alkali atoms or on clusters of alkali atoms (Griscom, 1971) which form as a consequence of irradiation. These centers are difficult to analyze in detail because of a large hyperfine interaction with the alkali atoms which generates a broad featureless ESR response.

In irradiated alkali borate glasses doped with halogen atoms ESR responses have been observed from atomic halogens and doubly charged dihalogen complexes (Griscom, 1972; Griscom *et al.*, 1969). The analysis of the atomic halogen spectrum, which is only stable at low temperatures, is complicated by the presence of several underlying resonances, but the dihalogen complex can be isolated in many borate glasses. The ESR spectrum of Cl_2^- in $NaCl–Na_2I–B_2O_3$ glass is shown in Fig. 16 together with the

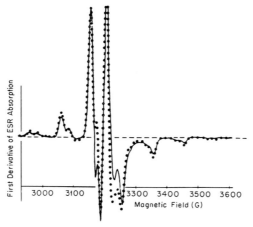

Fig. 16. Experimental (solid curve) and computer simulated (open circles) ESR derivative spectra observed in irradiated $NaCl–Na_2O–B_2O_3$ glasses. The response is due to Cl_2^- ions as described in the text. (After Taylor and Bray, 1970.)

computer simulation (Taylor and Bray, 1969). In the computer simulation of this interstitial center the parameters found in KCl and NaCl crystals were employed and no distributions of spin-Hamiltonian parameters were necessary to fit the data. This result illustrates the fact that interstitial molecular complexes incorporated into a glassy matrix are usually not greatly affected by the random distortions in the glass network.

3. Se, As$_2$Se$_3$, AND CHALCOGENIDE GLASSES

Chalcogenide glasses (i.e., those containing the elements S, Se, or Te) do not generally display radiation-damaged ESR responses at 300 K, but very recent measurements below 80 K indicate that analogs of some of the same centers observed in the oxide glasses are induced by electron irradiation in the chalcogenides at low temperatures (Taylor *et al.*, 1976b). In Fig. 17 the

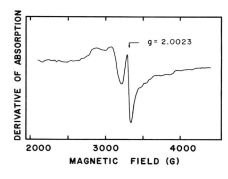

Fig. 17. Experimental ESR derivative trace of irradiated As$_2$Se$_3$ glass *at* 4.2 K. (After Taylor *et al.*, 1976b.)

ESR response recorded in glassy As$_2$Se$_3$ at 4.2 K is presented. The predominant features of this spectrum can be separated into two distinct centers, a hole center with *g* values listed in Table I and a narrow response at $g = 2.0023$. The narrow resonance is of unknown origin, but the hole center can be identified by comparison with the hole centers observed in irradiated oxide glasses. The spin–orbit coupling constant λ for selenium is about an order of magnitude larger than λ for oxygen. Thus for equivalent bonding configurations one would expect the deviations of the principal components of the g tensor from the free electron *g* value to scale with λ for oxygen and selenium if the two types of center are similar. Photoemission results (DiStefano and Eastman, 1971; Bishop and Schevchik, 1975) indicate that the bonding configurations for oxygen and selenium are quite similar, and the *g* value deviations of Table I do scale roughly with λ. We thus conclude that the hole center in electron-irradiated As$_2$Se$_3$ is localized on either a nonbonding or a dangling bond selenium p orbital. In selenium a similar hole center is also observed, but with greater resolution. We take this fact to indicate that there is possibly a small hyperfine interaction of the hole center in As_2Se_3 with a neighboring arsenic atom just as occurs with ^{11}B and ^{29}Si in borate and silicate glasses.

The results just described for the chalcogenide glasses are minimal compared to the wealth of detailed information which has been determined over the years for the oxide glasses. Nonetheless, these preliminary measurements indicate that more detailed studies of chalcogenide glasses will prove fruitful, especially when comparisons can be made to the more familiar results in the oxide glasses.

B. Paramagnetic Impurities

For several reasons ESR studies of paramagnetic impurities in glasses have not yielded the rich structural information that the ESR studies of irradiated glasses have. First, most paramagnetic impurities have spins I greater than $\frac{1}{2}$ and undergo strong interactions with the electric fields present in the glass. This so-called fine structure interaction [the second term in Eq. (1)] is often strong enough with such transition metal ions as Mn^{2+}, Fe^{3+}, Cu^{2+}, and Cr^{3+} that the electronic Zeeman term no longer dominates the spin Hamiltonian, in which case the perturbation calculations which yield resonance conditions such as Eq. (4) are no longer valid. In this situation the spin Hamiltonian of Eq. (1) must be numerically diagonalized for each orientation of the magnetic field with respect to the principal axes of the system, and the calculation of the absorption envelope often becomes prohibitively expensive. In some advantageous situations a mixture of perturbation calculations and exact diagonalization techniques can be employed effectively as will be demonstrated in an example to follow. The ESR of transition metal ions has been reviewed by Low (1960).

A second reason for the dearth of structural information which has been obtained from ESR of paramagnetic impurities is that the statistical distribution of spin-Hamiltonian parameters, in particular the fine-structure constants D and E, are often very broad and they smear out many of the details which would otherwise be present in the ESR spectra. In addition, the observed ESR spectra for a given paramagnetic impurity are often insensitive to the host glass.

As will be illustrated below, this situation results not because the paramagnetic sites are identical in all glasses, but rather because the statistical distributions are so broad that quite different average sites can yield similar envelopes. Here again, it is imperative that the glass spectra be compared with spectra occurring in related crystalline compounds if there is to be any progress toward extracting meaningful structural information from the ESR results. Of course, the structural information which is obtained relates to the bonding of the paramagnetic impurity in the host glass and is only indirectly related to the structure of the glass itself.

The paramagnetic impurity most often studied by ESR techniques in both oxide and chalcogenide glasses is Mn^{2+} and we discuss these results in some detail. We first briefly mention the pioneering experiments involving Fe^{3+} and Cu^{2+} in oxide glasses.

The first experiments on ESR of transition metals doped into glassy hosts were performed on Cu^{2+} and other ions by Sands (1955) and on Fe^{3+} by Castner et al. (1960). Sands observed features near $g = 6$ and $g = 4$ as well as $g = 2$ in most of his samples, and Castner et al. successfully explained these features by examining the details of the fine structure spin Hamiltonian for large values of the fine structure constant. In nearly every glassy host the spectrum of Fe^{3+} shows a sharp feature near $g = 4.3$. This feature occurs because the diagonalization of the fine structure spin Hamiltonian yields an allowed transition at this g value which is essentially independent of the fine structure constants for large values of D. Diagrams of the allowed transitions for $S = \frac{5}{2}$ ions such as Fe^{3+} and Mn^{2+} are available from Aasa (1970), Dowing and Gibson (1969) and Barry (1967). Details concerning the application of these diagrams to the analysis of polycrystalline and glassy spectra can be found in Taylor et al. (1975).

A refinement of the original analysis by Sands of Cu^{2+} in various oxide glasses has been performed by Imagawa (1968) who has obtained semi-quantitative fits to the Cu^{2+} spectra in borate, silicate and phosphate glasses and extracted rough distribution functions for the spin-Hamiltonian parameters. These calculations are based on a model for the Cu^{2+} center which assumes that the dominant fluctuations in local environment involve only one of the copper ligand orbitals. There is unfortunately no crystalline data with which to compare the results of this model calculation.

Several authors have investigated the ESR of Mn^{2+} in borate glasses (de Wijn and van Balderen, 1967; Griscom and Griscom, 1967; Taylor and Bray 1972b). The ESR spectra are quite similar in all borate glasses, and Fig. 18 presents two representative traces from the strontium borate system.

Although the Mn^{2+} ESR spectra in borate glasses are all similar, different explanations of the Mn^{2+} spectra observed in alkali borate glasses have been given. De Wijn and van Balderen (1967) describe the glass spectrum in potassium borate glasses as arising from one site with $E = 0$. This single site is assumed to give rise both to the features near $g = 2$ and to those near $g = 4.3$. Tucker (1962) in his discussion of borate glasses, considers two types of sites: type I giving rise to the features near $g = 4.3$ in the spectrum, and type II giving rise to the features near $g = 2.0$ in the spectrum. Griscom and Griscom (1967) studying lithium borate glasses also consider two types of sites where the ratio E/D remains $\frac{1}{3}$ for both types. Sites of

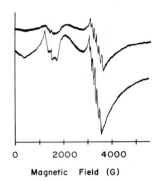

Fig. 18. Experimental ESR derivative spectra observed in Mn^{2+}-doped strontium borate glasses containing 20 (*upper trace*) and 40 (*lower trace*) mole % SrO. (After Taylor and Bray, 1972b.)

type II give rise to features near $g = 2.0$ at X band and those of type I to features near $g = 4.3$ and $g = 3.3$. In addition, Griscom and Griscom (1967) consider a continuous distribution of sites between these two types where D/h ranges from near zero to at least 3.7 GHz and possibly up to 7 GHz. From parallel investigations of Mn^{2+} in the 1:4 lithium borate polycrystalline compound these authors also conclude that many of the Mn^{2+} sites in the glasses are probably distorted versions of the site occupied in the crystalline material.

The results of Taylor and Bray (1972b) support the basic conclusions of Tucker (1962) and of Griscom and Griscom (1967) where in the strontium borate system the type I and type II sites are those of the 1:3 and 1:2 strontium borate compounds, respectively. The $E/D = \frac{1}{3}$ restriction of Griscom and Griscom is found not to hold in this case. Figures 19 and 20 show the Mn^{2+} ESR spectra observed in the 1:3 and 1:2 strontium borate compounds, respectively. The singularities marked for the 1:3 spectrum of Fig. 19 were determined by diagonalization of the spin Hamiltonian along the three principal axis directions because of the large value of D. The analysis of the

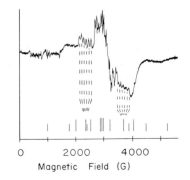

Fig. 19. ESR derivative spectrum of Mn^{2+}-doped $SrO\cdot3B_2O_2$ compound. Solid lines indicate the positions of singularities in the powder patterns assuming no hyperfine interaction. Height of each solid line is proportional to the transition probability for the singularity. Dashed lines mark the six hyperfine lines for two representative singularities. (After Taylor and Bray, 1972b.)

1:2 spectrum of Fig. 20 uses a computer simulation which employs the usual perturbation expression for the resonance condition. The salient features of the glass spectra (Fig. 18) at $g = 4.3, 3.3$, and 2.0 as well as the six hyperfine lines centered at $g = 2.0$ can be reproduced by a suitable addition of the 1:3 and 1:2 compound spectra under the general assumption of a broad distribution of both D and E values.

Although there are no crystal field splittings in the strontium borate compounds nearly as large as those observed in the 1:4 lithium borate compound, the glass spectra for these two systems are virtually identical. The silicate glasses are structurally quite different from the borate systems; however, the Mn^{2+} spectrum in silicate glasses is again virtually identical with the borate glasses. The ubiquitous nature of the glass spectrum results not from any unique site found in all of the glass systems but rather from the fact that distributions of nearly any sites where a substantial fraction of the sites have $D/h > 1.5$ GHz yield the appropriate envelope. The Mn^{2+}

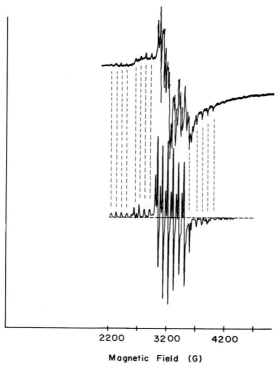

Magnetic Field (G)

Fig. 20. ESR derivative spectra of Mn^{2+}-doped $SrO\cdot2B_2O_2$ compound. Upper trace is experimental; lower trace represents a computer simulation. (After Taylor and Bray, 1972b.)

glass spectrum arises from singularities of powder patterns which are stationary with respect to large changes in both D and E/D. These results indicate that great caution must be exercised in interpreting the spectra of Mn^{2+} or Fe^{3+} ions in glasses without due consideration of the corresponding spectra occurring in the crystalline compounds of the system where unique spin-Hamiltonian parameters can usually be obtained.

In the chalcogenide glasses the Mn^{2+} spectrum is again essentially independent of the host glass but the spectrum is characteristically different from that observed in the oxide glasses (Kumeda et al., 1975; Watanabe et al., 1975). Features are again observed at $g = 4.2$ and $g = 2.0$ with the feature at 4.2 showing a resolved hyperfine structure and the one at 2.0 being motionally narrowed with no resolved structure. The 4.2 feature is probably characteristic of Mn^{2+} in the chalcogenide glasses but the feature at 2.0 has been attributed to polycrystalline MnO precipitates in the glassy matrix. No detailed model calculations or studies of relevant polycrystalline chalcogenide compounds have been performed.

C. Optically Induced Paramagnetic States

We end this section on ESR in glasses with a discussion of a class of paramagnetic center which appears to have no counterpart in crystalline materials. These centers are similar to, but characteristically distinguishable from, the centers produced by radiation damage in oxide and chalcogenide glasses. The characteristic ESR signals observed at 4.2 K in As_2S_3, As_2Se_3, amorphous As and Se are shown in Fig. 21 (Bishop et al., 1975). Comparison of the As_2Se_3 trace of this figure with the response induced by electron irradiation (Fig. 17) in the same material indicates that, although the overall shapes are similar, there is less resolved structure in the spectrum of Fig. 21.

The curves of Fig. 21 were produced by low-temperature irradiation of these chalcogenide glasses with low-intensity (~ 1 mW/cm^2) light whose energy corresponds to the Urbach tail in the fundamental absorption edge (absorption coefficient $\simeq 100$ cm^{-1}). An associated induced midgap optical absorption has demonstrated that the paramagnetic states are located in the forbidden energy gap of these glasses (Bishop et al., 1975). These centers are unstable above about 100 K and can be bleached optically at low temperature with low-intensity light whose energy lies below the bandgap energy. The ability to create and bleach these ESR centers with low-intensity light distinguishes them from the radiation-induced centers in the chalcogenides which are not optically reversible (Taylor et al., 1976b). Similar optically induced ESR centers should be observable in the oxide glasses but their production requires ultraviolet light rather than the infrared light used

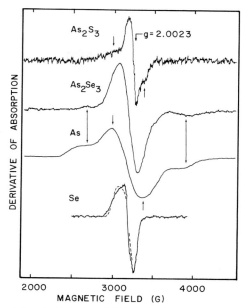

Fig. 21. Experimental optically induced ESR derivative spectra observed in several chalcogenide glasses and amorphous arsenic. The dashed line represents a computer simulation of the selenium hole center. The arrows illustrate the presence of an As-related line in As_2Se_3 and As_2S_3 as described in the text. (After Bishop *et al.*, 1976a).

for the semiconducting chalcogenide glasses. The facts that these optically induced ESR centers can be produced and destroyed with very low-intensity light and that these centers occur only in glassy solids suggest that these paramagnetic "defects" are perhaps more "intrinsic" to the glassy phase than those created by radiation damage.

The ESR spectra of Fig. 21 can be analyzed using procedures similar to those described in detail for the irradiated oxide glasses. The sharp features in the Se, As_2Se_3, and As_2S_3 spectra have been identified as holes on chalcogen atoms which are localized either on a lone-pair p orbital or on a dangling-bond p orbital. This identification was made by an analysis of the g tensor components and the spin–orbit interaction which is strictly analogous to that described in detail for the hole centers created by radiation damage. The spin-Hamiltonian parameters were extracted using computer simulation techniques (see dashed line on Se trace) but no distributions of these parameters were determined.

The spectrum observed in amorphous As can be identified by an analysis of the hyperfine tensor which is similar to that described above for the center HCl in SiO_2 enriched in ^{29}Si (Bishop *et al.*, 1976b). The As center is due to an unpaired spin (probably an electron center) which is localized predominately on a p orbital of a single As atom ($\simeq 5\%$ S spin density). There is a similarity between the optically induced As center and the E' center, but the As center does not posses sp^3 hybridization and probably is not situated in an atomic vacancy. In As_2Se_3 there are broad features (indicated by arrows in Fig. 21) which are due to the presence of an As associated ESR center in these glasses.

Although no unique model has yet been accepted to explain all of these optically induced ESR centers, the current knowledge of the conjugate localized hole and electron centers in the chalcogenide glasses is sufficient to allow one to categorize those simple models which will explain the data. All such models must incorporate the restriction that there be essentially no overlap between the wavefunctions of the paramagnetic hole and the As atoms. The simplest model assumes an "ideal" glass structure with no dangling bonds. In this picture the hole center is formed by optically removing an electron from the normally filled nonbonding chalcogen orbital, and the electron center in As_2Se_3 and As_2S_3 is formed by adding an electron to a fourth antibonding As orbital. A second simple interpretation allows for the existence of dangling bonds. In this model the dangling bonds are either doubly occupied or unoccupied in the cold dark in order to be consistent with the observed diamagnetism of the glasses. The most likely optically induced ESR centers (consistent with the data) are in this case holes on chalcogen dangling-bond orbitals and electrons on As dangling bonds. Here again one must make the logical assumption that the singly occupied dangling-bond orbital, which is normally bonding, is similar to a nonbonding orbital in order for the g tensor scaling arguments described above to remain valid.

Street and Mott (1975) have recently proposed a more detailed model of localized gap states in chalcogenide glasses based on specific dangling-bond configurations. A positively charged recombination center consists of an unoccupied dangling bond which interacts with the nonbonding lone pair of a neighboring fully bonded chalcogen; the negatively charged center consists of a doubly occupied dangling bond which forms a valence band like lone pair. The optically induced, metastable ESR centers are explained by these authors as the trapping of free electrons or holes by these charged centers. This model considers only one specific type of defect which can be unoccupied, singly occupied, or doubly occupied, while it is now obvious that the observed ESR spectra in As_2Se_3 and As_3S_3 require two rather specific types of paramagnetic center. It is also unclear how the model relates to amorphous As where there is no conclusive evidence for the existence of nonbonding p type lone-pair wavefunctions.

IV. NMR in Glasses

Three general types of NMR experiments have provided useful information concerning the physical and chemical properties of glasses. Local structural properties of glasses can be probed through either the quadrupolar or chemical shift terms of Eq. (2). In the oxide glasses the most complete studies of local structural order using the quadrupolar interaction have been the ^{11}B and ^{10}B NMR experiments in borate glasses, although a number of results on other nuclei have been reported. Investigations of As quadrupole resonance have provided useful structural information in the chalcogenide glasses. The information provided by the chemical shift interaction is generally not as easily relatable to structural properties as that provided by the quadrupolar interaction. This difficulty arises because tractable models do not generally exist which enable one to calculate an observed chemical shift from an assumed local bonding configuration. Nonetheless some useful qualitative results have been obtained for ^{31}P and ^{51}V in oxide glasses and for ^{203}Tl or ^{205}Tl in chalcogenide glasses.

A second type of NMR experiment which is capable of providing useful information in glasses is the study of diffusion of ions through the motional narrowing of the NMR lineshape. The most successful studies of this type concern the motion of various alkali ions in alkali oxide glasses.

With the advent of pulsed NMR some detailed investigations of elementary excitations in glasses have recently been performed. These studies use measurements of the spin–lattice relaxation time T_1 in glasses and related crystals which attempt to determine the type of vibrational modes which are effective in relaxing the nuclear spins. The most detailed T_1 measurements to date have been performed on ^{11}B in B_2O_3 glass and ^{75}As in As_2S_3 glass.

A. Structural Studies—Quadrupolar Interaction

Nuclei with spin I greater than $\frac{1}{2}$ possess nonspherical symmetry and hence a quadrupole moment. This quadrupole moment interacts with the gradient of the electric field present at the nuclear site to produce a shift in the NMR frequency. Electric field gradients in solids arise from either covalent bonding or from ionic charges, and several methods exist to calculate these effects. The simplest and easiest molecular orbital model calculation of the electric field gradient is due to Townes and Dailey (1949), but numerous much more sophisticated linear combination of atomic orbital (LCAO) methods are now available (see for example Snyder and Basch, 1972). When possible in the examples to follow we will use the most simple and transparent model calculations to illustrate the information one can obtain, but deficiencies in these calculations will be pointed out and references to more sophisticated and complicated calculations will be made when available.

1. B_2O_3 AND BORATE GLASSES

The first systematic quadrupolar studies of ^{11}B in borate glasses were performed by Silver and Bray (1958) who observed a broad resonance which they attributed to boron in three coordination and a narrow resonance which they attributed to boron in four coordination. Although many refinements have been published since 1958 (see for example the reviews of Muller-Warmuth, 1965a,b; Bray, 1967, 1970), the basic conclusions of this pioneering study have remained unchanged. One important refinement in alkali borate glasses of high alkali–oxide content is the separation of ^{11}B NMR responses due to two distinct 3-coordinated boron sites (one with three bridging oxygens and one with one nonbridging oxygen) by Taylor and Bray (1970) and Kriz et al., (1971). A second important refinement is the determination in B_2O_3 of the distribution of Hamiltonian parameters for the 3-coordinated boron sites in the glass (Kriz and Bray, 1971a; Jellison and Bray, 1976). Model calculations which yield the observed distribution of coupling constants e^2qQ/h and asymmetry parameters η for B_2O_3 glass have been attempted (Taylor and Friebele, 1974; Snyder et al., 1976) with varying degrees of success.

In this section we first describe the behavior of ^{11}B and ^{10}B NMR in B_2O_3 and the successes to date of model calculations which attempt to extract structural information. We then summarize the structural information extracted from ^{11}B and ^{10}B NMR in glasses of the alkali borate system.

As usual, it is extremely useful to examine the ^{11}B NMR response in crystalline B_2O_3 before proceeding to discuss the glass. Figure 22 shows the ^{11}B NMR spectra observed by Kline et al. (1968) in crystalline and glassy B_2O_3. The spectrum for crystalline B_2O_3 can be fit with one unique site where $e^2qQ/h = 2.76 \pm 0.1$ MHz and $\eta = 0.0 \pm 0.1$. This spectrum is due to boron atoms in planar BO_3 triangular units where all three oxygens are bridging and agrees with the most recent x-ray structural determination of crystalline B_2O_3 (Strong and Kaplow, 1968).

The glass spectrum of Fig. 22 can also be fit with a single site ($e^2qQ/h = 2.66$ MHz and $\eta = 0.1$) and no statistical distribution of Hamiltonian parameters (solid line), but spectra observed at lower frequencies require distributions in e^2qQ/h and η. Kriz and Bray (1971a) made a first attempt at calculating these distributions under the reasonable assumption that the most probable site in the glass was the one to which the glass spectrum collapsed at the highest frequencies (where the magnitude of the quadrupolar interaction with respect to that of the Zeeman interaction is the smallest).

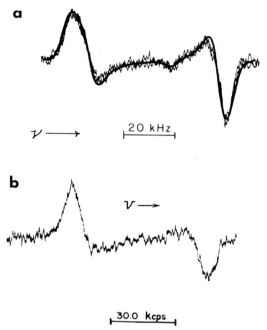

Fig. 22. (a) Experimental (noisy trace) and computer simulated (solid curve) ^{11}B NMR derivative spectra in glassy B_2O_3. The computer simulation assumes one unique site. (After Kriz and Bray, 1971a.) (b) Experimental ^{11}B NMR derivative spectrum observed in crystalline B_2O_3. (After Kline *et al.*, 1968.)

For the sake of simplicity these authors also assumed that the asymmetry as indicated by η did not change from site to site and that the distribution was entirely due to changes in the coupling constant. The fits obtained by Kriz and Bray at 8 and 16 MHz are shown in Fig. 23. The distribution function in e^2qQ/h in these fits is slightly skewed toward lower coupling constants. These authors were quick to point out that without a physical model of the field gradient distribution it is not possible to choose a unique distribution of parameters.

Taylor and Friebele (1974) used a simple model as a first attempt toward fitting the observed spectrum in B_2O_3 and other borate glasses from an assumed distribution in bonding configurations. This model calculation assumed axial symmetry which implies that $\eta = 0$ and that the distortions of the BO_3 units are departures from planarity given by statistical distributions in the O–B–O bond angle about the planar value of 120°. The Townes and Dailey (1949) theory was employed to calculate the coupling constants in this simple model. The fit to the ^{11}B NMR spectrum in B_2O_3 glass using

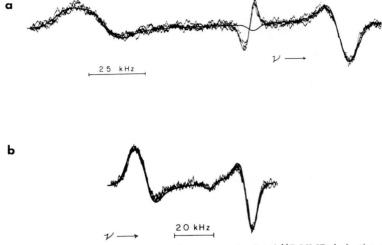

Fig. 23. Experimental (noisy traces) and computer simulated ^{11}B NMR derivative spectra at (a) 8 MHz and (b) 16 MHz in glassy B_2O_3. The computer simulations employed a distribution of quadrupole coupling constants as described in the text. (After Kriz and Bray, 1971a.)

this model is shown in Fig. 24. The width of the distribution of bond angles used in Fig. 24 is $\pm 2°$.

Although the approximations involved in the Taylor and Friebele approach are significant, two valid conclusions can be drawn from their results. First, the calculation shows how a distribution in coupling constants which is skewed toward lower values arises quite naturally out of a normal (Gaussian) distribution of bond angles. This behavior is reminiscent of similar effects observed in the model calculations of ESR centers described above. A second conclusion which can be inferred from these calculations is that, whatever

Fig. 24. Distribution function $F(v_Q)$ as a function of $v_Q = [3e^2qQ/h]/2I(2I - 1)$ and the resulting fit (smooth line) to the ^{11}B NMR experimental derivative spectrum (noisy curve) in glassy B_2O_3 using the model described in the text. (After Taylor and Friebele, 1974.)

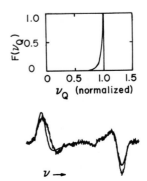

the actual distortions turn out to be, only small distortions in the BO_3 units are necessary to account for the observed ^{11}B NMR spectrum in glassy B_2O_3.

More detailed self-consistent field (SCF) molecular orbital calculations of axially symmetric $B(OH)_3$ molecules have been performed by Snyder et al. (1976). These calculations do not predict any appreciable change in the coupling constant for large changes in the bonding angle, but they do predict a rapid change in the energy of the molecule with distortion. On the basis of the energy calculations the authors conclude that the BO_3 units in B_2O_3 glass are not highly distorted in agrrement with the results of Taylor and Friebele. However, Snyder et al. also suggest that departures from planarity cannot account for the shape of the glass spectrum. Although this conclusion may in fact turn out to be correct, the calculations as performed do not provide a definitive test because of the rapid increase in energy with distortions under the given set of simplifying assumptions.

Perhaps the most appropriate conclusions to be drawn from these two model calculations are that the BO_3 structural units remain very nearly planar in the glass and that the detailed fitting of the observed NMR spectrum will eventually involve the consideration of both departures from acial symmetry and larger molecular clusters. There is in fact evidence from recent x-ray studies of glassy B_2O_3 (Mozzi and Warren, 1970) that, although the BO_3 triangular units are similar in the crystalline and glassy phases, the glass may have a different longer range order than the crystal. In particular, Mozzi and Warren suggest that glassy B_2O_3 is composed primarily of six-membered boroxol rings which are linked together randomly by the three bridging oxygens. Future model calculations should consider both the effects of this ring structure and the effects of in-plane changes in bond lengths and bond angles.

The interpretation of ^{11}B NMR measurements in B_2O_3 and other borate glasses is complicated by the inability to separate the statistical distributions in e^2qQ/h and η. Recent measurements of Jellison and Bray (1976) have effectively removed this ambiguity through the use of [10]B isotopic enrichment. In [10]B NMR the dipolar broadening is much less important than it is for [11]B and hence the spectral resolution is greatly improved. In addition, there exists a feature in the [10]B powder envelope that depends solely on the quadrupolar coupling constant.

Figure 25 shows the [10]B NMR derivative spectrum observed in B_2O_3 glass and a schematic powder pattern for the most probable boron site. Perturbation calculations of the resonance condition similar to those which yielded Eq. (5), but which include terms up to third order, show that the features labeled a and f in Fig. 25 are independent of η to within 1% for the values of e^2qQ/h appropriate to B_2O_3 glass. Since the dipolar broadening is much less than the widths of the features a and f in Fig. 25, these widths

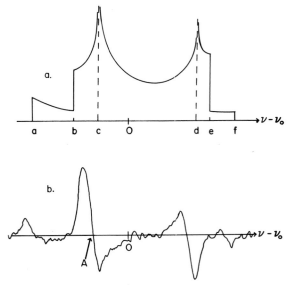

Fig. 25. (a) Powder pattern for ^{10}B NMR. (b) Experimental ^{10}B NMR derivative spectrum observed in glassy B_2O_3. (After Jellison and Bray, 1976.)

provide an accurate measurement of the distribution in the quadrupole coupling constants independent of changes in the asymmetry parameter.

From an analysis of the spectrum in the region of point f in Fig. 25 Jellison and Bray calculate an average value for the ^{10}B coupling constant of 5.51 MHz with a halfwidth for the assumed Gaussian distribution of 0.21 MHz. This halfwidth corresponds to an average deviation in coupling constants on the order of $\pm 5\%$ which compares favorably with the deviation calculated for ^{11}B by Kriz and Bray of $\pm 3\%$.

The signal-to-noise ratio of the experimental derivative trace in Fig. 25 is such that only limited information can presently be obtained about the distribution of coupling constants, and there are significant experimental difficulties involved in obtaining a more accurate line shape. Nonetheless if improvements in the signal-to-noise ratio can be made, then one can in principle determine the distribution in η from the distribution in e^2qQ/h and a fit to the entire ^{10}B NMR derivative spectrum.

In the alkali borate glass systems there is good evidence from ^{11}B NMR for at least three types of rather discrete boron sites whose probability of occurrence in glass depends on composition in a chemically predictable fashion. The initial evidence for these distinct sites comes from detailed investigations of ^{11}B NMR in alkali borate polycrystalline compounds by

Kriz and Bray (1971b,c). It has long been known that in crystalline materials boron in four coordination yields coupling constants less than about 0.85 MHz while boron in three coordination yields values that range from 2.45 to 2.81 MHz. Kriz and Bray have in addition determined the effects on the ^{11}B NMR derivative spectra for BO_3 planar units known to contain zero, one and two nonbridging oxygen atoms. For example, the case of one nonbridging oxygen in a BO_3 unit has been studied in calcium metaborate where each boron is bonded to one nonbridging and two bridging oxygen atoms. The upper part of Fig. 26 shows the ^{11}B NMR spectrum observed in calcium metaborate while the lower portion shows the calculated theoretical line shape (Bray, 1970). The coupling constant for these boron sites is essentially the same as that for the sites with three bridging oxygens but the interaction is no longer axially symmetric. In this case $\eta = 0.54$. The difference in lineshapes is apparent on comparison of the B_2O_3 and calcium metaborate compound traces of Figs. 22 and 26. One can thus separate quite easily three different ^{11}B NMR responses for boron in polycrystalline samples—a narrow line due to 4-coordinated boron and two broad lines due to 3-coordinated boron with three bridging and two bridging oxygens, respectively.

The hypothesis of Krogh-Moe (1962, 1965) that the structural units in the glass are made up of those appearing in corresponding crystalline compounds

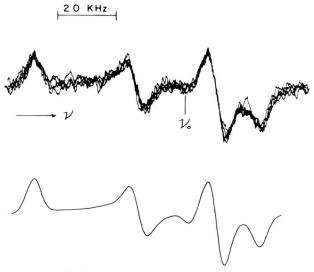

Fig. 26. Experimental ^{11}B NMR derivative spectrum observed in calcium metaborate (*top trace*). Computer simulation of the experimental spectrum with $e^2qQ/h = 2.56$ MHz and $\eta = 0.54$. (After Bray, 1970.)

of the system can be combined with some simple chemical bonding considerations (Bray and O'Keefe, 1963; Taylor and Bray, 1970) to predict the relative contributions due to the three types of boron site from the glass composition. The fraction N_4 of boron atoms in four coordination in alkali borate glasses is shown in Fig. 27 as determined from the observed ^{11}B NMR spectra. Up to around 30 mole % alkali oxide N_4 increases at a rate (solid line of Fig. 27) which indicates that each oxygen added to the glass in the form of alkali oxide converts two boron atoms from BO_3 units to BO_4 units. Above this composition the fall off in N_4 is due to the formation of BO_3 units with at least one oxygen nonbridging and may also be predicted from simple chemical bonding considerations (Taylor and Bray, 1970).

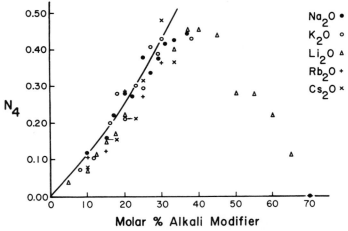

Fig. 27. The fraction N_4 of boron atoms in BO_4 configurations in alkali borate glasses plotted as a function of the molar percent of alkali oxide. (After Bray, 1970.)

In Fig. 28 a compendium of traces is presented (Taylor and Bray, 1970; Kriz and Bray, 1971a; Kim and Bray, 1974; Rhee and Bray, 1971) which illustrates the agreement typically obtained between experimental and computer simulated ^{11}B NMR derivative spectra for representative borate glasses whose values of N_4 range from 0 to 0.45. The average values of the Hamiltonian parameters used in fitting these spectra are dictated by the constraints of the simple chemical model described above and are listed in Table II. Similar results are obtained in the borosilicate system (Milberg et al., 1972).

It has been suggested that perhaps the sites in the glasses are not discrete (4-coordinated, 3-coordinated) but rather that changing features in the ^{11}B NMR spectra with composition represent changes in very broad featureless

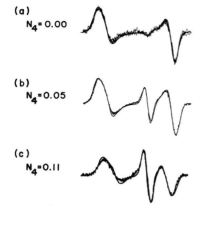

(a) $N_4 = 0.00$

(b) $N_4 = 0.05$

(c) $N_4 = 0.11$

(d) $N = 0.45$

Fig. 28. Experimental and computer-simulated ^{11}B NMR derivative spectra at $v_0 = 16$ MHz for various borate glasses whose fractions of four-coordinated boron atoms N_4 vary from 0 to 0.45; (a) B_2O_3 glass: (b) $(Ag_2O)_{0.05}$ $(B_2O_3)_{0.95}$ glass; (c) $(Cs_2O)_{0.1}$ $(B_2O_3)_{0.9}$ glass; (d) $(Cs_2O)_{0.4}$ $(B_2O_3)_{0.6}$ glass. The simulation parameters are listed in Table II and explained in the text. (After Taylor and Friebele, 1974.)

TABLE II

PARAMETERS OF COMPUTER-SIMULATED ^{11}B NMR SPECTRA FOR VARIOUS GLASSES [a]

Glass	Site	e^2qQ/h (MHz)	η	σ (kHz)	Fractional intensity
B_2O_3	BO_3	2.66	0.11	2.9	1.0
$(Ag_2O)_{0.05}(B_2O_3)_{0.95}$	BO_3	2.65	0.13	3.0	0.95
	BO_4	0.39	0.00	2.9^b	0.05
$(Cs_2O)_{0.1}(B_2O_3)_{0.9}$	BO_3	2.60	0.15	2.0	0.89
	BO_4	0.40	0.00	3.2^b	0.11
$(Cs_2O)_{0.4}(B_2O_3)_{0.6}$	BO_3	2.56	0.13	3.9	0.33
	$BO_3{}^c$	2.45	0.59	2.5	0.22
	BO_4	0.44	0.0	2.4^b	0.45

[a] Adapted from Taylor and Friebele (1974).
[b] Includes the effects of over modulation.
[c] BO_3 unit with one nonbridging oxygen.

distributions in coupling constants and asymmetry parameters (Peterson *et al.*, 1974a,b). As was stressed in the introduction to this chapter, it is impossible to refute a contention of this sort without recourse to some physically and chemically reasonable model because in the absence of a comprehensive theory of the structure of glasses, all experimental data extracted from glassy solids require an assumed model in order to be interpreted. However, detailed comparisons of computed and experimental line shapes using the simple molecular model described above do not support this broad, featureless distribution hypothesis. These calculations, which assume axial symmetry, indicate that a bimodal distribution in e^2qQ/h with peaks near 0.5 and 2.5 MHz is necessary to fit the experimental spectra of borate glasses (Taylor and Friebele, 1974).

The foregoing discussion has illustrated the substantial contributions which the results of ^{11}B NMR have made to an understanding of the details of the local structural order in alkali borate glasses. The analysis presented has successfully used the unique boron sites in corresponding crystalline solids as a starting point toward understanding the spectra of the glasses.

2. As$_2$S$_3$ and Chalcogenide Glasses

The only detailed studies of network-forming quadrupolar nuclei in chalcogenide glasses have been the investigations of pure nuclear quadrupole resonance (NQR) of ^{75}As in glassy As$_2$S$_3$ and As$_2$Se$_3$ (Rubinstein and Taylor, 1974; Taylor *et al.*, 1976a). In NQR there is no applied magnetic field and only the quadrupolar term of Eq. (2) is important. (For details concerning NQR spectroscopy see Das and Hahn, 1958).

In both crystalline and amorphous As$_2$S$_3$ the basic structural unit is an AsS$_3$ pyramid with the arsenic atom at the apex and three sulfurs at the base. In the crystalline phase there are two inequivalent As sites in the unit cell whose average pyramidal apex bonding angles differ from one another by only 1.3°. The experimentally observed NQR absorption spectra for both crystalline and glassy As$_2$S$_3$ are shown in Fig. 29. Note that these measurements, which were performed using pulsed techniques, yield the absorption spectrum directly. The two sharp lines are spectra observed in the crystal and the circles represent the data observed in the glass.

Through the use of a model to calculate the field gradient, one can estimate the distortions present in the glass from the observed crystal and glass spectra. As we did in the previous section, we use the simple Townes and Dailey model and assume axial symmetry in order to illustrate the calculation of the field gradients while we recognize that more detailed calculations are necessary to obtain exact numerical results. Although it is not possible from the data of Fig. 29 to tell whether the distribution of sites in the glass is single peaked or bimodal, the Townes and Dailey calculations predict that the

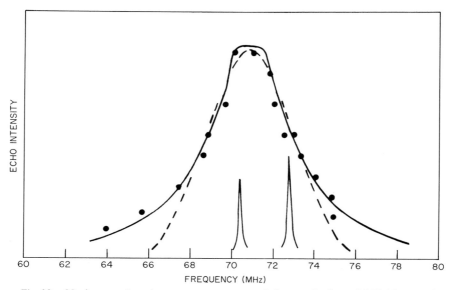

Fig. 29.   Nuclear quadrupole resonance spectrum of vitreous As_2S_3 at 4.2 K. The experimental data are shown by the points. The solid and dashed lines are fits to the data by Lorentzian and Gaussian distributions, respectively, of resonance frequencies. The spectrum of crystalline As_2S_3 is also displayed for purposes of comparison. (After Rubinstein and Taylor, 1974.)

pyramidal units in the glass are very well defined, having a distribution width of average apex pyramidal bonding angles of around $\pm 1°$. Note that these Townes and Dailey calculations accurately predict the observed changes in average apex bonding angle in crystalline As_2S_3 of $1.3°$.

B.   Structural Studies—Chemical Shift Interaction

There have been few systematic studies of the chemical shift interaction in glasses because of the basic inability to relate observed experimental results to local bonding configurations even in crystalline materials. Some success has been obtained in selected glass systems where empirical comparisons can be made with appropriate crystalline materials without resort to model calculations. We illustrate the kinds of results which have been obtained with two examples—studies of ^{31}P and [51]V NMR in vanadium phosphate glasses and studies of [205]Tl in thallium–arsenic–selenium glasses.

The chemical shift is due to several different components, one of which arises because the resonant nucleus is affected not only by the externally applied magnetic field but also by the local magnetic field produced at the nuclear site by the diamagnetic response of the orbital electrons to the applied field. The second component is a second-order paramagnetic contribution which arises from the interaction of the applied field with excited paramagnetic states of the free ion mixed into the ground state by the electrostatic interaction with neighboring ions. It is this paramagnetic contribution to the chemical shift which provides a measure of the relative covalency of the chemical bonds.

The vanadium phosphate glass system has been studied using NMR by France and Hooper (1970) and by Landsberger and Bray (1970). The latter authors have determined the average values of the two principal components of the chemical shift tensor $\sigma_{\parallel}$ and $\sigma_{\perp}$ for ^{51}V as a function of glass com-composition using computer simulation techniques. The average values of $\sigma_{\parallel}$ and $\sigma_{\perp}$ in glasses of low P_2O_5 content approach those determined in crystalline V_2O_5. From the behavior of $\sigma_{\parallel}$ and $\sigma_{\perp}$ with composition Landsberger and Bray suggest that a PO_4 tetrahedron replaces the apex oxygens of VO_5 units as one increases the phosphorous pentoxide content of the glass. Qualitative agreement is obtained with some basic chemical expectations through the use of this model, but the analysis is not precise enough to yield any quantitative details. The ^{31}P results are even less quantitative (France and Hooper, 1970).

Similar studies of ^{205}Tl in several chalcogenide glasses have been published (Bishop and Taylor, 1973; Baidakov and Borisova, 1974). In the pseudobinary system $(Tl_2Se)_x(As_2Se_3)_{1-x}$ the measured values of the isotropic chemical shift exhibit the behavior illustrated in Fig. 30 as a function of x (Bishop and Taylor, 1973; Taylor et al., 1973). Two facts are evident from this figure. First, since most chemical shifts are of the order of 10^{-4}, the paramagnetic shifts displayed in Fig. 30 are large. These large shifts indicate that the thallium atoms are strongly covalently bonded to the network. Second, the abrupt change in the magnitude of the chemical shift at $x = 3$ indicates that the thallium atoms are undergoing an abrupt change in local environment near this composition. The thallium sites in glasses with $x \leq 0.3$ are somewhat less covalent than those for $x > 0.3$.

Fig. 30. Magnitude of the ^{205}Tl NMR isotropic chemical shift in Tl–As–Se glasses plotted as a function of Tl_2Se content. (After Bishop and Taylor, 1973.)

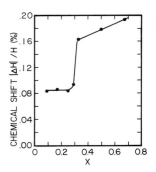

Additional evidence for identifying the chemical shifts for $x < 0.3$ as somewhat less covalent is provided by a comparison with chemical shifts measured in thallium halides. In largely covalent thallic chloride ($TlCl_3$) the chemical shift is nearly identical with that observed in the $(Tl_2Se)_x(As_2Se_3)_{1-x}$ glasses with $x > 0.3$. In ionic thallous chloride ($TlCl$) the absolute value of the chemical shift is only slightly less than that in the glasses for $x < 0.3$. One might conjecture from these comparisons that the thallium atoms enter the glass as Tl^+ ions for low thallium content ($x < 0.3$) and as Tl^{3+} ions for high thallium content ($x > 0.3$).

C. Diffusion and Motional Narrowing

In the previous sections of this chapter much has been made of the fact that the local structural order in glasses is often quite similar to that which exists in corresponding crystalline compounds. A similar parallel between glasses and crystals does not exist when one considers either the motion of ions to be discussed in this section or the elementary excitations to be considered in the next section. In these two cases glasses and crystals usually display markedly different behavior.

Bishop and Bray (1968) first investigated the diffusion of alkali ions in alkali borate glasses at elevated temperatures. Since then several other authors have studied diffusion in both borate and silicate glasses (Kramer *et al.*, 1973; Svanson and Johansson, 1970; and Hendrickson and Bray, 1974). The only NMR studies of diffusion in chalcogenide glass are those of ^{77}Se in liquid Se (Bishop and Taylor, 1972) and ^{11}B in B_2S_3 (Hendrickson and Bishop, 1975).

In this section we briefly describe the NMR measurements of diffusion of alkali ions in borate and silicate glasses and corresponding crystals. The diffusion process is manifested in a narrowing of the broad NMR line shape with increasing temperature due to the fact that, as the ions move in the solids, the static interactions which cause the broad linewidth are averaged to zero. The diffusion and hence the line narrowing is a thermally activated process. In general crystals display higher activation energies for diffusion than do glasses. This fact has often been cited as evidence that the line narrowing in glasses is dominated by some type of local diffusion in contrast to the long-range diffusion in crystals.

The exact activation energies and other parameters extracted from the line narrowing depend on the model which is used to reduce the data. The original interpretation of Bloembergen *et al.*, (1948) has been modified for glasses by Resing (1972) and by Hendrickson and Bray (1974). In Fig. 31 the data for ^7Li NMR in a borate glass are presented as they were reduced using the BPP approach (triangles) (Bloembergen *et al.*, 1948) and the approach

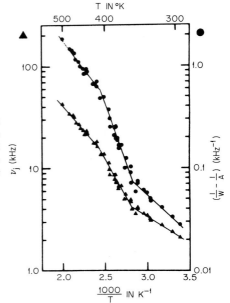

Fig. 31. Graphical analysis of measured ^{7}Li linewidth data in a mixed-isotope borate glass. A comparison of the BPP method (triangles) and the Hendrickson and Bray approach (circles) to reducing the data. (After Hendrickson and Bray, 1974.)

of Hendrickson and Bray (circles). Note that the slopes in $1/T$, which determine the activation energies, differ slightly in these two approaches. In either case, however, one obtains regions where the activation energies for diffusion are less than those obtained for bulk diffusion in the crystalline phase.

The range of localization of the diffusion in glasses is still an open question as is the exact mechanism. Both hopping of alkali ions around a tetrahedral boron site and diffusion of alkali ions in an alkali-rich phase-separated region have been suggested as possible mechanisms for short- or intermediate-range diffusion in these glasses.

NMR studies of diffusion in alkali borate and silicate glasses have also been useful probes of the mixed alkali effect in glasses. Both glass systems display an increase of the activation energy for long-range ionic motion upon partial substitution of lithium by a second type of alkali. In the silicate glasses this increase becomes greater with increasing alkali oxide content and conforms to a trend established from other types of measurements (Isard, 1969).

D. Spin–Lattice Relaxation and Elementary Excitations

In crystalline solids with nuclei which possess quadrupole moments, nuclear spin–lattice relaxation results from first-order Raman processes involving the inelastic scattering of a phonon by the spin system. The relaxation can proceed via either acoustical or optical phonons, and the theory of these two processes is well understood in crystals (Van Kranendonk, 1954; Jeffrey and Armstrong, 1968). The problem of spin–lattice relaxation in amorphous materials has only recently been addressed (Rubinstein and Taylor, 1974; Rubinstein et al., 1975; Reinecke and Ngai, 1975; Szeftel and Alloul, 1975), and only a basic understanding of the phenomenon exists at the present time.

A host of experimental measurements on glasses at low temperature (specific heat, thermal conductivity, microwave absorption and others) have indicated the existence of an enhanced density of low-frequency vibrational modes which appear to be characteristic of the amorphous state. Since nuclear spin–lattice relaxation proceeds via vibrational modes, one might expect greater low-temperature spin–lattice relaxation rates in glasses over those in the corresponding crystalline materials. In fact, increased nuclear spin–lattice relaxation rates are a general feature of disordered materials. Greatly increased spin–lattice relaxation rates have been observed for protons in organic glasses as compared to the rates in crystalline counterparts (Haupt and Müller-Warmuth, 1968, 1969; Haupt, 1970). A similar increase in the relaxation rate of ^{11}B in glassy B_2O_3 over the crystalline material has also been observed (Rubinstein et al., 1975; Rubinstein and Resing, 1976). In addition, measurements of the temperature dependence of ^{75}As spin–lattice relaxation in glassy and crystalline As_2S_3 also show greater relaxation rates in the glass. In this section we briefly discuss the NMR results on crystalline and glassy B_2O_3 and the NQR results on crystalline and glassy As_2S_3.

In B_2O_3, Szeftel and Alloul (1975) first examined the nuclear spin–lattice relaxation in both the glass and crystal and found that the relaxation time T_1 in the crystal was much longer than in the glass. The temperature dependence of the relaxation time in the glass varied as $T_1 \propto T^{-1.3}$ from 1.2 to 90 K. Rubinstein et al. studied the spin–lattice relaxation in crystalline and glassy B_2O_3 from 150 to 500 K as a function of frequency. Data by these authors and Szeftel and Alloul agreed in the temperature region where the data overlapped, but at higher temperatures Rubinstein et al. observed a minimum in T_1 as a function of temperature near 300 K as shown in Fig. 32. It can be seen from Fig. 32 that this minimum is nearly independent of the resonance frequency which is in contrast to the prediction of the BPP theory (Bloembergen et al., 1948) of motional narrowing which works well for crystalline materials. Such a frequency independent minimum in T_1 has recently been observed in several other amorphous materials and may well be a universal property of this class of materials. For this reason even a qualitative understanding of this feature is important.

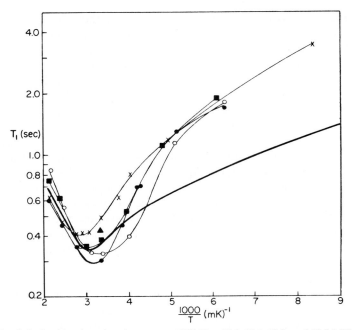

Fig. 32. Spin–lattice relaxation time versus $1000/T$ at 10.2, 13.1, 19.3, and 28.9 MHz. Heavy solid line is the theoretical fit to the data. ● = 13.661 MHz; ■ = 19.344 MHz; ○ = 10.200 MHz; x = 28.88 MHz. Solid triangle is a point from Szeftel and Alloul (1975). (After Rubinstein *et al.*, 1975.)

Rubinstein *et al.*, (1975) have developed a theoretical model which explains this frequency-independent T_1 minimum using ideas introduced by Anderson *et al.* (1972) and by Phillips (1972). Anderson *et al.* and Phillips successfully explained many of the anomalous thermal and ultrasonic properties of glasses by introducing a series of two level systems with a broad and continuous distribution of energy splittings. These two-level systems, called disorder modes or tunneling modes, arise from an atom or a group of atoms which can sit in two near-equilibrium configurations which are separated by a small energy difference.

For low temperatures, Reinecke and Ngai (1975) consider three possible Raman processes to account for the spin–lattice relaxation in B_2O_3: (1) excitation and de-excitation of two disorder modes of energies E and E, respectively, (2) excitation and de-excitation of a phonon and a disorder mode of energies E and E', respectively, and (3) excitation and de-excitation of two phonons of energies E and E', respectively. The appropriate process in the glasses at low temperatures is process (1) because it alone gives the correct temperature dependence at low temperatures ($T^{-1.3}$). The fit to the data assuming a uniform energy density of tunneling modes is given by the heavy solid line in Fig. 32. This fit is certainly not quantitatively correct but, considering that the detailed nature of the tunneling modes is as yet completely unknown, the qualitative agreement is encouraging. At least one can have some confidence that the correct Raman process at low temperatures has been isolated.

In As_2S_3 the results are obtained by NQR where, as mentioned previously, no external magnetic field is applied. For this reason the ^{75}As spin–lattice relaxation time in this material cannot be studied as a function of frequency as was done for ^{11}B in B_2O_3. Nonetheless some informative results have also been obtained in this system (Rubinstein and Taylor, 1974 and Taylor *et al.*, 1976a). The normalized relaxation curves (where the temperature dependence has been divided out), for glassy As_2S_3 are shown in Fig. 33 at three temperatures. One difference between spin–lattice relaxation in glasses and the crystals which is apparent from this figure is that the decay cannot be represented by a single exponential characterized by one relaxation time. A second difference is the rate of decay, which is ~ 15 sec to reach $1/e$ in As_2S_3 glass, compared to ~ 2 hr in As_2S_3 crystal at 4.2 K.

Even though the decay curve in the glass is not exponential, one may still define a characteristic relaxation time as the time for the nuclear spin system to attain $(1 - 1/e)$ of its equilibrium magnetization after the application of a saturating comb of rf pulses (i.e., the time at which the decay curve of Fig. 33 falls to $1/e$). This definition is particularly meaningful because the shape of the decay curves is independent of temperature, as shown in Fig. 33. Since the distribution of relaxation rates is governed by the slope of the decay curve, all relaxation rates in the distribution have the same temperature dependence. This temperature dependence was determined by

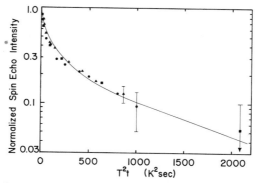

Fig. 33. Normalized spin–lattice decay for vitreous As_2S_3 taken at 4.2, 14.8, and 77 K vs T^2t, where T is temperature and t is the time. The solid line is a model fit to the data. ● = 77 K; ■ = 14.8 K; ▲ = 4.2 K. (After Rubinstein and Taylor, 1974.)

Rubinstein and Taylor to be proportional to T^{-2}, but the data are probably not accurate enough to rule out a slightly slower temperature dependence such as the $T^{-1.3}$ which is observed in B_2O_3 and various proton-containing glasses (Haupt, 1970). It is to be hoped that more detailed model calculations not only will provide a more qualititative fit to the T_1 minimum but will also produce a fit to the observed distribution of relaxation times.

V. Mössbauer Spectroscopy in Glasses

Of the three techniques described in this chapter Mössbauer spectroscopy presents the most formidable experimental difficulties and has to date provided the least structural information. For example, calculations of the Mössbauer effect probability (or recoil-free fraction) f' of Eq. (3) could give information about the vibrational properties of glasses, but they have as yet not been performed on any glassy materials. We do not mean to imply that this technique is inherently less appropriate to the study of glasses but rather that the experimental effort required is perhaps greater than that required in NMR and ESR to obtain roughly comparable information. One important exception to this assessment must be made—Mössbauer spectroscopy is the only technique of the three which gives information concerning the excited states of nuclei in glasses.

The first measurements using the Mössbauer effect in glasses were performed on [57]Fe in silica glasses by Pollak et al., (1962). Since that time a great many oxide glasses containing iron have been investigated using the Mössbauer technique. Other elements observed in oxide glasses include [119]Sn, [169]Tm, and [151]Eu. Since most of the systematic studies have been performed using [57]Fe, we use this nucleus as an example of the results which can be obtained in the oxide glasses.

In the chalcogenide glasses [125]Te is by far the most important nucleus because it is to date the only glass forming element which can be directly studied using the Mössbauer effect. Other nuclei which have been studied include [119]Sn and [121]Sb. Examples of the [125]Te and [121]Sb Mössbauer isotopes in chalcogenide glasses will be discussed.

A. Structural Studies—Electric and Magnetic Hyperfine Interactions

In crystalline compounds containing Fe^{3+}, a quadrupole splitting (electric hyperfine splitting) occurs as a result of distortions from cubic symmetry. In glasses all Fe^{3+} spectra display quadrupolar splittings. Although a great deal of effort has been expended in studying Fe^{3+} in glasses using Mössbauer spectroscopy, it is still difficult to define any trends in the Fe^{3+} quadrupolar splittings or to identify changes in quadrupolar splitting with changes in the Fe^{3+} local bonding configurations. In silicate glasses the quadrupolar splittings range from 0.070 to 0.100 cm/sec. (Kurkjian, 1970). For Fe^{2+} even fewer systematic trends are apparent and values for the splittings range from 0.200 to 0.209 cm/sec.

Figure 34 shows ^{57}Fe Mössbauer spectra in sodium and potassium silicate glasses doped with iron (Kurkjian and Sigety, 1965). In this figure the number of counts recorded per channel is plotted as a function of the relative velocity of the Co^{57} source and the alkali borate glass absorber. The velocity difference between the two peak positions in each trace is a measure of the quadrupolar splitting. The quadrupolar splittings and isomer shifts (to be discussed below) of the top two traces are representative of Fe^{3+} and those of the bottom trace are representative of Fe^{2+}. Although the Mössbauer spectra can unambiguously distinguish between Fe^{3+} and Fe^{2+} little else

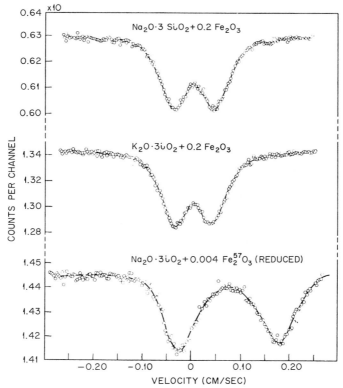

Fig. 34. Mössbauer spectra of ^{57}Fe for alkali silicate glasses containing iron. (After Kurkjian and Sigety, 1965.)

concerning the local bonding of the iron in the glasses has yet been extracted using the quadrupolar splittings alone. Magnetic hyperfine effects can be observed for Fe^{3+} and Fe^{2+} in oxide glasses for low concentrations of iron at low temperatures (Kurkjian and Buchanan, 1964). In some cases magnetic iron oxide precipitates can also be detected in iron-doped oxide glasses.

Several chalcogenide glasses doped with ^{125}Te have been investigated using Mössbauer spectroscopy including films of Te (Blum and Feldman, 1974), films of $GeSe_xTe_{1-x}$ (Boolchand et al., 1974) and Se doped with Te (Boolchand, 1973; Henneberger and Boolchand, 1973). In what follows we present the amorphous Te results as a useful illustrative example of the data obtained on chalcogenide glasses.

The ^{125}Te Mössbauer spectra in amorphous and crystalline Te at 80 K are shown in Fig. 35. These data were taken on the same 2-μm-thick film before and after crystallization. The noise on the amorphous trace is such that no significant least squares fit to the data could be performed, but in the crystalline data the least squares fit is indicated by the solid line. The reasons for the low signal-to-noise ratio in the amorphous material are the low recoil-free fraction f' and a possible distribution of quadrupolar interactions in the amorphous material. The value of f' is a factor of three smaller in the amorphous than in the crystalline phase at 4.2 K, but no details concerning the distribution of quadrupole coupling constants can be obtained due to the low signal-to-noise ratio. The average coupling constant in the

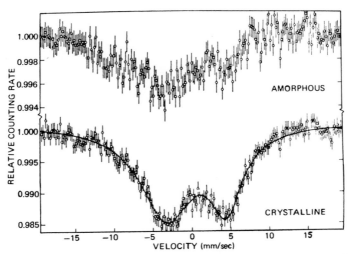

Fig. 35. Mössbauer absorption spectra of the same tellurium film (65 percent enriched ^{125}Te and ~2-μm thick) as deposited (amorphous, above) and after annealing for a few minutes at room temperature (crystalline, below). (After Blum and Feldman, 1974.)

amorphous phase appears to be slightly larger than that observed in the crystalline phase.

As discussed in the NMR section above in some detail, the electric field gradient, which determines the magnitude of the quadrupole coupling constant, depends primarily on the covalent bonding of the Te atoms in both the crystalline and the amorphous phases. In crystalline hosts, Boolchand *et al.* (1973) observed that the magnitude of e^2qQ/h varied rapidly with the inverse of the covalent bond distance. The slight increase in e^2qQ/h in the glass is thus consistent with a slight increase which has been observed in the first nearest neighbor distance in the amorphous material over that measured in the crystalline phase. Of course the poor signals in the amorphous films preclude any quantitative determination of relative bond lengths in the crystalline and amorphous phases.

The fact that the Mössbauer effect probability f' is smaller by a factor of three in the amorphous phase indicates qualitatively that there is an increased density of low-frequency vibrational modes in the amorphous phase, such as those considered in some detail in the NMR section above. The recoil energies of the Mössbauer γ quanta are below about 100 cm^{-1} in this system so that the increased density of vibrational modes which is important in the ^{125}Te Mössbauer measurements lies below about 100 cm^{-1} which is consistent with what is currently known about the dominance of low-frequency vibrational modes in glasses. Beyond these qualitative consistency arguments, no more detailed or quantitative information can presently be extracted from the limited data which are available.

In Te-doped crystalline Se the ^{125}Te spectra for the monoclinic and trigonal forms exhibit slighly different quadrupolar splittings (Boolchand, 1973). In glassy Se a splitting intermediate to these two crystalline splittings is observed which has been interpreted as evidence that both monoclinic rings and trigonal chains of Se atoms exist in the glass. Lack of knowledge as to the relative probabilities of substitution of ^{125}Te in rings and chains and the lack of resolution in the glass spectrum once again preclude any more quantitative conclusions.

B. *Structural Studies—Isomer Shifts*

The isomer shifts of ^{57}Fe in the alkali borate glasses of Fig. 34 are obtained from the velocity difference between the center of gravity of the two quadrupolar peaks and the point where the velocity equals zero. One would expect that as the coordination number of either the Fe^{3+} or the Fe^{2+} changed from six to four that the isomer shift would decrease because of a resultant decrease in the charge density at the nucleus due to increased screening by the 3d electrons. For Fe^{3+} in silicate and phosphate glasses, the isomer shifts fall into two distinct ranges which may be associated uniquely with Fe^{3+} in octahedral and tetrahedral sites. The situation for Fe^{2+} in oxide glasses is more ambiguous because the ranges of isomer shifts tend to overlap and there is little quantitative agreement between the shifts observed in tetrahedrally and octahedrally coordinated crystals and the glasses.

In the chalcogenide glasses the ^{125}Te isomer shift is too small to be observed given the present resolution, but ^{121}Sb isomer shifts have been successfully studied in a series of antimony chalcogenide glasses and crystals (Long et al., 1969a,b; Ruby et al., 1971). There is an observed increase of the isomer shift for ^{121}Sb in the series of glasses Sb_2O_3, Sb_2S_3, Sb_2Se_3 which corresponds to an increase in the s-electron density at the Sb nucleus as one goes from O to S to Se. This trend is consistent with the decreasing trend in the electronegativity differences among Sb–O, Sb–S, and Sb–Se bonds, but a more quantitative correlation is difficult to establish because of the many difficulties previously discussed.

The isomer shift can thus be used successfully in glasses to distinguish between valence states such as between Fe^{3+} and Fe^{2+} or to predict certain qualitative bonding trends such as changes in covalency, but like the chemical shift measurements of NMR, more quantitative conclusions are extremely difficult to draw.

VI. Concluding Remarks

In this chapter the versatility and the limitations of magnetic resonance and Mössbauer techniques in the study of glasses have been discussed. The utility of these techniques for probing local structural order and bonding, elementary excitations, ionic motion, and the nature of impurities and defects in glasses has been illustrated. Just as any other probe of the glassy state, NMR, ESR, and Mössbauer measurements all require the framework of chemically and physically reasonable models in which to interpret the observed results. Some of the general successes of these techniques as applied to glasses can now be summarized.

Systematic and detailed ESR measurements of irradiated oxide glasses have observed and identified the two prevalent and stable radiation induced defects as (1) an electron trapped on a dangling-bond orbital (on Si in silicate and on B in borate glasses, respectively) at an atomic vacancy, and (2) a hole trapped predominantly on a nonbonding lone-pair or dangling-bond orbital on an oxygen atom. The unpaired electron is possibly stabilized by the presence of a positively charged modifier ion and the hole center may be stabilized by a net negative charge on nearby glass network atoms such as 4-coordinated borons. Preliminary estimates of the distortions of these two, and other defects in oxide glasses have been made. Initial studies in chalcogenide glasses have found similar radiation-induced defects stable at low temperatures and also a new optically reversible paramagnetic center which is distinct from the radiation-induced defects.

In NMR, a series of comprehensive studies has enabled one to identify quantitatively the relative numbers of 4-coordinated and 3-coordinated boron atoms in borate glasses. In addition, those 3-coordinated boron atoms which are bonded to three bridging oxygen atoms can be distinguished from those which are bonded to one or more nonbridging oxygens. Even greater detail in the separation of the types of boron structural units should be possible in the future. Like the progress in ESR in glasses, preliminary estimates of the distortions of boron sites in borate glasses are also currently available.

There has been far less NMR research on other network former nuclei because of a host of experimental difficulties. But with the advent of signal averaging and pulsed techniques, additional nuclei should become accessible for more detailed study. Both ^{17}O and ^{29}Si are network forming nuclei which hold great promise as future useful structural probes in oxide glasses.

An unusually weak temperature dependence has been established for the spin-lattice relaxation time ($T_1 \propto T^{-(1+\delta)}$ where $\delta < 1$) in all glasses studied to date. These enhanced relaxation rates in glasses are probably a universal feature of disordered materials. Measurements to date have established that local atomic tunneling modes are important in the spin–lattice relaxation in glasses, and detailed T_1 measurements in the future should provide an extremely useful probe of these modes.

The Mössbauer effect has not yet provided as detailed information about glasses as that extracted from both NMR and ESR measurements, but some success has been obtained using ^{57}Fe in oxide glasses. One can distinguish between Fe^{3+} and Fe^{2+} in glasses using Mössbauer spectroscopy, and sometimes one can distinguish the difference between tetrahedral and octahedral symmetry by comparison with spectra observed in crystals. Studies of other Mössbauer nuclei in glasses such as ^{121}Sb, ^{119}Sn, and ^{125}Te, have not been investigated as thoroughly as ^{57}Fe, but these nuclei, especially Te, should yield significant information about the structure of glasses in the future.

The major future developments in NMR and ESR of glasses will come from measurements which employ either signal averaging techniques or pulsed techniques. Point-by-point pulsed measurements of lineshapes which saturate easily should be most useful (for example, ^{29}Si in pure SiO_2 or ^{77}Se in the chalcogenide glasses). The need for pulsed measurements of T_1 in ESR is apparent. Similar anomalous behavior is to be expected for the spin–lattice relaxation of paramagnetic electrons as has been found for nuclei in glasses. Comparisons of both nuclear and electronic spin–lattice relaxation data with other probes of low-energy modes, such as ultrasonic attenuation and Raman scattering, should prove useful.

Systematic Mössbauer measurements of the recoil-free fraction f' of ^{57}Fe as a function of temperature in simple oxide glasses should also provide information on the tunneling modes. Since these measurements are difficult to perform, the best chance for success is to compare the results on glasses directly with those obtained in related crystalline materials.

There is a need for more sophisticated, yet physically plausible and tractable, model calculations to provide a framework in which to evaluate better the distributions of Hamiltonian parameters extracted from NMR and ESR spectra of glasses. The consequences on ESR and NMR data of the presence of larger structural units, such as boroxal rings in glassy B_2O_3, should be investigated.

The future utility of NMR, ESR, and Mössbauer spectroscopy in studies of glasses depends upon experimental ingenuity with simple glass systems. We must avoid the temptation to perform, exclusively, routine measurements on ever more complex glass systems.

References

Aäsa, R. (1970). *J. Chem. Phys.* **52**, 3919–3930.
Abragam, A. (1961). "The Principles of Nuclear Magnetism." Oxford Univ. Press (Clarendon), London and New York.
Anderson, P. W., Halperin, B. I., and Varma, C. M. (1972). *Phil. Mag.* **25**, 1.
Baidakov, L. A., and Borisova, Z. U. (1974). *In* "Amorphous and Liquid Semiconductors" (J. Stuke and W. Brenig, eds.), pp. 1036–1042. Taylor and Francis, London.
Barry, T. I. (1967). Nat. Lab. Rep. (unpublished), Teddington, Middlesex, England.
Bishop, S. G. (1974). *In* "Amorphous and Liquid Semiconductors" (J. Stuke and W. Brenig, eds.), pp. 997–1014. Taylor and Francis, London.
Bishop, S. G., and Bray, P. J. (1968). *J. Chem. Phys.* **48**, 1709–1717.
Bishop, S. G., and Schevchik, N. J. (1975). *Phys. Rev. B* **12**, 1567–1578.
Bishop, S. G., and Taylor, P. C. (1972). *Solid State Commun.* **11**, 1323–1326.
Bishop, S. G., and Taylor, P. C. (1973). *Phys. Rev. B* **7**, 5177–5183.
Bishop, S. G., Strom, U., and Taylor, P. C. (1975). *Phys. Rev. Lett.* **34**, 1346–1350.
Bishop, S. G., Strom, U., and Taylor, P. C. (1976a). *Phys. Rev. Lett.* **36**, 543–547.
Bishop, S. G., Strom, U., and Taylor, P. C. (1976b). *Solid State Commun.* **18**, 573–576.
Bloembergen, N., Purcell, E. M., and Pound, R. V. (1948). *Phys. Rev.* **73**, 679.
Blum, N. A., and Feldman, C. (1974). *Solid State Commun.* **15**, 965–968.
Boolchand, P. (1973). *Solid State Commun.* **12**, 753–756.
Boolchand, P., Triplett, B. B., Hanna, S. S., and de Neufville, J. P. (1974). *In* "Mossbauer Effect Methodology" (I. J. Graverman, C. W. Seidel, and D. K. Dieterly, eds.), Vol. 9, pp. 53–80. Plenum Press, New York.
Bray, P. J. (1967). *In* "Interaction of Radiation with Solids" (A. Bishay, ed.), pp. 25–54. Plenum Press, New York.
Bray, P. J. (1970). *In* "Magnetic Resonance" (C. K. Cougan, N. S. Ham, S. N. Stewart, J. R. Pilbrow, and G. V. H. Wilson, eds.), pp. 11–39. Plenum Press, New York.
Bray, P. J., and O'Keefe. (1963). *Phys. Chem. Glasses* **4**, 37–46.

Castner, Th., Newell, G. S., Holton, W. C., and Slichter, C. P. (1960). *J. Chem. Phys.* **32**, 668–673.

Coey, J. M. D. (1974). *J. Phys.* (*Paris*) **35**, C6-89–C6-105.

Das, T. P., and Hahn, E. L. (1958). *Solid State Phys. Suppl. 1.*

de Wijn, H. W., and van Balderen, R. F. (1967). *J. Chem. Phys.* **46**, 1381–1387.

DiStefano, T. H., and Eastman, D. E. (1971). *Phys. Rev. Lett.* **27**, 1560–1562.

Dowsing, R. D., and Gibson, J. F. (1969). *J. Chem. Phys.* **50**, 294–303.

Edwards, J. O., Griscom, D. L., Jones, R. B., Walters, K. L., and Weeks, R. A. (1969). *J. Am. Chem. Soc.* **91**, 1095.

Feigl, F. J. (1970). *J. Phys. Chem. Solids.* **31**, 575.

France, P. W., and Hooper, H. O. (1970). *J. Phys. Chem. Solids* **31**, 1307–1315.

Friebele, E. J., Griscom, D. L., Ginther, R. J., and Sigel, G. H. (1974). *Proc. Int. Congr. Glass, 10th, Kyoto, Jpn.* pp. 6–16.

Goldanskii, V. I., and Makarov, E. F. (1968). *In* "Chemical Applications of Mössbauer Spectroscopy" (V. I. Goldanskii and R. H. Herber, eds.), pp. 1–113. Academic Press, New York.

Griscom, D. L. (1971). *J. Non-Cryst. Solids* **6**, 275–282.

Griscom, D. L. (1972). *Solid State Commun.* **11**, 899–902.

Griscom, D. L. (1973/74). *J. Non-Cryst. Solids* **13**, 251–285.

Griscom, D. L. (1976). *Proc. Exeter Summer School on Defects in Solids.* Plenum, New York.

Griscom, D. L., and Griscom, R. E. (1967). *J. Chem. Phys.* **47**, 2711–2722.

Griscom, D. L., Taylor, P. C., Ware, D. A., and Bray, P. J. (1968). *J. Chem. Phys.* **48**, 5158–5173.

Griscom, D. L., Taylor, P. C., and Bray, P. J. (1969). *J. Chem. Phys.* **50**, 977–983.

Griscom, D. L., Friebele, E. J., and Sigel, G. H. (1974). *Solid State Commun.* **15**, 479–483.

Griscom, D. L., Sigel, G. H., and Ginther, R. J. (1976). *J. Appl. Phys.* **47**, 960–967.

Haupt, J. (1970). *Proc. Colloq. Ampere, 16th, Bucarest,* pp. 630–634.

Haupt, J., and Müller-Warmuth, W. (1968). *Z. Naturforsch.* **23a**, 208–216.

Haupt, J., and Müller-Warmuth, W. (1969). *Z. Naturforsch.* **24a**, 1066–1074.

Hendrickson, J. R., and Bishop, S. G. (1975). *Solid State Commun.* **17**, 301–304.

Hendrickson, J. R., and Bray, P. J. (1974). *J. Chem. Phys.* **61**, 2754–2764.

Henneberger, T., and Boolchand, P. (1973). *Solid State Commun.* **13**, 1619.

Imagawa, H. (1968). *Phys. Status Solidi* **30**, 469–478.

Isard, J. O. (1969). *J. Non-Cryst. Solids* **1**, 235.

Jeffrey, K. R., and Armstrong, R. L. (1968). *Phys. Rev.* **174**, 359–369.

Jellison, G. E., and Bray, P. J. (1976). *Solid State Commun.* **19**, 517–520.

Jellison, G. E., Bray, P. J., and Taylor, P. C. (1976). *Phys. Chem. Glasses* **17**, 35–37.

Känzig, W., and Cohen, M. H. (1959). *Phys. Rev. Lett.* **3**, 509–511.

Karapetyan, G. O., and Yudin, D. M. (1963). *Sov. Phys.-Solid State* **4**, 1943–1949.

Kim, K. S., and Bray, P. J. (1974). *J. Non-Met.* **2**, 95–101.

Kim, Y. M., and Bray, P. J. (1970). *J. Chem. Phys.* **53**, 716.

Kline, D., Bray, P. J., and Kriz, H. M. (1968). *J. Chem. Phys.* **48**, 5277–5278.

Konnert, J. H., and Karle, J. (1972). *Nature* (*London*) *Phys. Sci.* **236**, 92.

Kopp, M., and Mackey, J. H. (1969). *J. Comput. Phys.* **3**, 539.

Krämer, F., Müller-Warmuth, W., Schoerer, J., and Dutz, H. (1973). *Naturforsch.* **28a**, 1338–1350.

Kriz, H. M., and Bray, P. J. (1971a). *J. Non-Cryst. Solids* **6**, 27–36.

Kriz, H. M., and Bray, P. J. (1971b). *J. Magn. Res.* **4**, 69–75.

Kriz, H. M., and Bray, P. J. (1971c). *J. Magn. Res.* **4**, 76–84.

Kriz, H. M., Park, M. J., and Bray, P. J. (1971). *Phys. Chem. Glasses* **12**, 45–49.

Krogh-Moe, J. (1962). *Phys. Chem. Glasses* **3**, 101–110.

Krogh-Moe, J. (1965). *Phys. Chem. Glasses* **6**, 46–54.

Kumeda, M., Kobayashi, N., Suzuki, M., and Shimizu, T. (1975). *Jpn. J. Appl. Phys.* **14**, 173–180.

Kurkjian, C. R. (1970). *J. Non-Cryst. Solids* **3**, 157–194.

Kurkjian, C. R., and Buchanan, D. N. E. (1964). *Phys. Chem. Glasses* **5**, 63.

Kurkjian, C. R., and Sigety, E. A. (1965). *Proc. Int. Congr. Glass, 7th* Vol. 1, Brussels paper #39. Gordon and Breach, New York.

Landsberger, F. R., and Bray, P. J. (1970). *J. Chem. Phys.* **53**, 2757–2768.

Lee, S., and Bray, P. J. (1963). *J. Chem. Phys.* **39**, 2963–2973.

Lee, S., and Bray, P. J. (1964). *J. Chem. Phys.* **40**, 2982–2988.

Lefebvre, R., and Maruani, J. (1965a). *J. Chem. Phys.* **42**, 1480–1496.

Lefebvre, R., and Maruani, J. (1965b). *J. Chem. Phys.* **42**, 1496–1502.

Long, G. C., Stevens, J. G., Bowen, L. H., and Ruby, S. L. (1969a). *Inorg. Nucl. Chem. Lett.* **5**, 21.

Long, G. C., Stevens, J. G., and Bowen, L. H. (1969b). *Inorg. Nucl. Chem. Lett.* **5**, 799.

Low, W. (1960). *Solid State Phys. Suppl.* **2**, pp. 1–212.

Mackey, J. H., Kopp, M., Tynan, E. C., and Yen, T. F. (1969). *In* "Electron Spin Resonance of Metal Complexes," pp. 33–57. Plenum Press, New York.

Mackey, J. H., Boss, J. W., and Kopp, M. (1970). *Phys. Chem. Glasses* **11**, 205.

Maruani, J. (1964). *In* "Electron Magnetic Resonance and Solid Dielectrics" (R. Servant and A. Charru, eds.), pp. 303–307. North–Holland Publ., Amsterdam.

Milberg, M. E., O'Keefe, J. G., Verhelst, R. A., and Hooper, H. O. (1972). *Phys. Chem. Glasses* **13**, 79–84.

Mozzi, R. L., and Warren, B. E. (1970). *J. Appl. Crystallogr.* **3**, 251.

Müller-Warmuth, W. (1965a). *Glastech. Ber.* **38**, 121–133.

Müller-Warmuth, W. (1965b). *Glastech. Ber.* **38**, 405.

Peterson, G. E., and Kurkjian, C. R. (1972). *Solid State Commun.* **11**, 1105–1107.

Peterson, G. E., Kurkjian, C. R., and Carnevale, A. (1974a). *Phys. Chem. Glasses* **15**, 52–58.

Peterson, G. E., Kurkjian, C. R., and Carnevale, A. (1974b). *Phys. Chem. Glasses* **15**, 59–64.

Phillips, W. A. (1972). *J. Low Temp. Phys.* **1**, 351.

Pollak, H., De Coster, M., and Amelinckx, S. (1962). *In* "Mossbauer Effect" (D. M. J. Compton and D. Schoen, eds.), p. 298. Wiley, New York.

Poole, C. P. (1967). "Electron Spin Resonance." Wiley (Interscience), New York.

Purcell, T., and Weeks, R. A. (1969). *Phys. Chem. Glasses* **10**, 198–208.

Reinecke, T. L., and Ngai, K. L. (1975). *Phys. Rev. B* **12**, 3476–3478.

Resing, H. A. (1972). *In* "Advances in Molecular Relaxation Processes," p. 199. Elsevier, Amsterdam.

Rhee, C., and Bray, P. J. (1971). *Phys. Chem. Glasses* **12**, 165–174.

Rubinstein, M., and Resing, H. A. (1976). *Phys. Rev. B* **13**, 959–968.

Rubinstein, M., and Taylor, P. C. (1974). *Phys. Rev. B* **9**, 4258–4276.

Rubinstein, M., Resing, H. A., Reinecke, T. L., and Ngai, K. L. (1975). *Phys. Rev. Lett.* **34**, 1444–1447.

Ruby, S. L. (1972). *J. Non-Cryst. Solids* **8–10**, 78.

Ruby, S. L., Gilbert, L. R., and Wood, C. (1971). *Phys. Lett.* **37A**, 453.

Sands, R. H. (1955). *Phys. Rev.* **99**, 1222–1226.

Schreurs, J. W. H. (1967). *J. Chem. Phys.* **47**, 818–830.

Sidorov, T. A., and Tyul'kin, B. A. (1967). *Dokl. Akad. Nauk SSSR* **175**, 872.

Silsbee, R. H. (1961). *J. Appl. Phys.* **32**, 1459.

Silver, A. H., and Bray, P. J. (1958). *J. Chem. Phys.* **29**, 984–990.

Snyder, L. C., and Basch, H. (1972). "Molecular Wave Functions and Properties." Wiley, New York.

Snyder, L. C., Peterson, G. E., and Kurkjian, C. R. (1976). *J. Chem. Phys.* **64**, 1569–1573.

Street, R. A., and Mott, N. F. (1975). *Phys. Rev. Lett.* **35**, 1293–1297.

Strong, S. L., and Kaplow, R. (1968). *Acta Cryst.* **B24**, 1032.

Svanson, S. E., and Johansson, R. (1970). *Acta Chem. Scand.* **24**, 755.

Symons, M. C. R. (1970). *J. Chem. Phys.* **53**, 468.

Szeftel, J., and Alloul, H. (1975). *Phys. Rev. Lett.* **34**, 657–660.

Taneja, S. P., Kimball, C. W., and Shaffer, J. C. (1973). *In* "Mössbauer Effect Methodology" (I. J. Gruverman, ed.), Vol. 8, p. 41. Plenum Press, New York.

Taylor, P. C., and Bray, P. J. (1968). Lineshape Program Manual, Brown Univ., Providence, Rhode Island, unpublished. This manual is available from the authors.

Taylor, P. C., and Bray, P. J. (1970). *J. Magn. Resonance* **2**, 305–331.

Taylor, P. C., and Bray, P. J. (1972a). *Bull. Am. Ceram. Soc.* **51**, 234–239.

Taylor, P. C., and Bray, P. J. (1972b). *J. Phys. Chem. Solids* **33**, 43–58.

Taylor, P. C., and Friebele, E. J. (1974). *J. Non-Cryst. Solids* **16**, 375–386.

Taylor, P. C., and Griscom, D. L. (1971). *J. Chem. Phys.* **55**, 3610–3611.

Taylor, P. C., Griscom, D. L., and Bray, P. J. (1971). *J. Chem. Phys.* **54**, 748–760.

Taylor, P. C., Bishop, S. G., and Mitchell, D. L. (1973). Rep. NRL Progress, March, pp. 16–32.

Taylor, P. C., Baugher, J. F., and Kriz, H. M. (1975). *Chem. Rev.* **75**, 205–240.

Taylor, P. C., Rubinstein, M., and Friebele, E. J. (1976a). *Proc. NATO Summer School Phys. Struct. Disordered Solids* (S. S. Mitra and B. Bendow, eds.), pp. 661–697. Plenum Press, New York.

Taylor, P. C., Strom, U., and Bishop, S. G. (1976b). Unpublished research.

Townes, C. H., and Dailey, B. P. (1949). *J. Chem. Phys.* **17**, 782.

Tucker, R. F. (1962). "Advances in Glass Technology," pp. 103–114. Plenum Press, New York.

Van Kranendonk, J. (1954). *Physica* **20**, 781–800.

Warren, B. E. (1972). *Sov. Phys.-Crystallogr.* **16**, 1106 [*Kristallografiya* **16**, 1264 (1971)].

Warren, B. E., Krutter, H., and Morningstar, O. (1936). *J. Am. Ceram. Soc.* **19**, 202.

Watanabe, I., Inagaki, Y., and Shimizu, T. (1975). *J. Non-Cryst. Solids* **17**, 109–120.

Weeks, R. A. (1956). *J. Appl. Phys.* **27**, 1376.

Weeks, R. A. (1967). *In* "Interaction of Radiation with Solids" (A. Bishay, ed.), pp. 55–93. Plenum Press, New York.

Weeks, R. A. (1974). *In* "Introduction to Glass Science" (L. D. Pye, H. J. Stevens, and W. C. LaCourse, eds.), pp. 137–171. Plenum Press, New York.

Weeks, R. A., and Nelson, C. M. (1960). *J. Appl. Phys.* **31**, 1555.

Weeks, R. A., and Sonder, E. (1963). *In* "Paramagnetic Resonance" (W. Low, ed.), Vol. 2, p. 869. Academic Press, New York.

Wong, J., and Angell, C. A. (1971). *Appl. Spectrosc. Rev.* **4**, 97–232.

Wong, J., and Angell, C. A. (1976). "Vitreous State Spectroscopy" Dekker, New York.

Zachariasen, W. H. (1932). *J. Am. Chem. Soc.* **54**, 3841.

Dielectric Characteristics of Glass

MINORU TOMOZAWA

Materials Engineering Department
Rensselaer Polytechnic Institute
Troy, New York

I. Introduction

Electrical properties of glass has been a research subject for over a century. The phenomenon of anomalous electrical absorption in insulators puzzled numerous prominent scientists in the 19th century such as Curie, Hopkins, and Kohlrausch. In recent years, primarily because of the rapid development of electrical engineering, renewed interests exist on the electrical properties of glasses.

This time, however, the scope of interest is of a much wider range, including the so-called semiconducting glasses such as chalcogenide glasses and glasses containing transition metal oxides as well as thin-film forms of glassy materials.

In this chapter, mainly the low-frequency ac and dc characteristics of ionic conducting oxide glasses will be reviewed, although there is a striking similarity in behaviors between ionic conducting glasses and electronic conducting glasses, and it is likely that the major portion of the discussion in this chapter is applicable to the electronic conducting glasses too. Special attention will be focused on the dielectric relaxations of oxide glasses and their structural interpretations. In the beginning, a short description of the general theories of dielectric relaxations will be presented. This will be followed by experimental observations on glasses and various proposed interpretations. Finally, the most plausible explanations will be presented.

There have been numerous review articles on the electrical properties of glasses; the most recent and thorough one is by Owen (1963). In the present review, the emphasis will be on the development made after Owen's article appeared. The earlier literatures on the subject include those by Stanworth (1950), Morey (1954), Stevels (1957), Sutton (1960), and Mazurin (1962, English translation 1965). General information on dielectrics are found in books by Debye (1929), Fröhlich (1949), von Hippel (1954), Anderson (1964), McCrum et al. (1967), and Zheludeve (1971).

In the discussion of the electrical properties of matter, various units and symbols are used, often confusing the readers. In this chapter MKS units will be employed, and for clarification a list of symbols is included. Because of numerous constants and variables to be defined, sometimes the same symbol is used for two different quantities. Of course this was done only when the distinction can be easily made from the context and no confusion arises.

Symbols

A	Area of specimen	E, $\mathbf{E}$	Electric field
A	A constant in Eq. (40)	E_0	Amplitude of electric field
A'	A constant in Eq. (85)	E_1, E_2	Electric fields in layers 1 and 2, respectively
A_1	Preexponential factor of the dc conductivity equation, Eq. (1)	$E_1{}^0$, $E_2{}^0$	Initial electric field in layer 1 and 2, respectively
B^0	Mobility when no ionic interaction exists	$E_{1,\mathrm{st}}$, $E_{2,\mathrm{st}}$	Steady state electric field in layer 1 and 2, respectively
b	Distance between sites 1 and 2	ΔE_σ	Activation energy of dc conductivity
C	Initial (or average) alkali ionic charge concentration	ΔE_D	Activation energy of diffusion
C'	Integration constant	ΔE_{fm}	Activation energy of dielectric loss maximum
C_0	Geometric capacitance defined by Eq. (20)	e	Base of natural logarithm
C_V	Capacitance defined by Eq. (26)	e	Ionic charge (positive)
D	Diffusion coefficient [Eq. (2)]	$\mathbf{F}$	Force
D, $\mathbf{D}$	Displacement [Eq. (5)]	f	Parameter in the Nernst–Einstein relation
D_0	Amplitude of displacement		
D^0	Diffusion coefficient when no ionic interaction exists	f	Frequency

f_m	Frequency corresponding to the dielectric loss maximum	P_{O^-}	Distribution of O^-
G	Shear modulus	$P_{O^-}^0$	Distribution of O^- in the absence of the external field
G	Amplitude of distribution variation defined by Eq. (147)	P_0	Probability of transition in the absence of field in conducting path model
G_0	DC conductance		
$G(\tau)$	Distribution function of relaxation time defined by Eq. (35)	P_1, P_3	Probability of transition from A to B and from B to A, respectively, in the field direction in conducting path model
g	Gap between guard electrode and guarded electrode		
$g(\tau)$	Distribution function of the relaxation time defined by Eq. (39)	P_2, P_4	Probability of transition from A to B, and from B to A, respectively, in direction opposed to field in conducting path model
H	Height of energy barrier		
I	Electric current		
i	Current density	ΔP	$P - C$
i_1	Amplitude of current density	p	Quantity defined by Eq. (127)
i_c	Current density due to conduction	Q	Quantity defined by Eq. (165)
$J, \mathbf{J}$	Flux	q	Fractional excess charge concentration defined by $(P/C) - 1$
j	Imaginary constant, $\sqrt{-1}$	q	Probability of defect migration
K	Reciprocal of Debye length	q^*	Factor dependent upon valence of charge carriers; defined by Eq. (62)
K_1	Quantity defined by Eq. (62)		
k	Boltzmann's constant		
L, L_1, L_2	Thickness of specimen	R	Quantity defined by Eq. (164)
l	Caribration factor for effective electrode radius	R	Gas constant
		r	Quantity defined by Eq. (128)
M^*	Complex "electric modulus"	r	effective radius of the electrode
M'	Real part of "electric modulus"	r_1, r_2, r_3	Radii of the electrode defined by Fig. 2
M''	Imaginary part of "electric modulus"	$r, \mathbf{r}$	Distance in radial direction
m	Frequency dependent variable defined by Eq. (66), or (66')	r	Distance between two ions
		r	Quantity defined by Eq. (128)
N	Number of charged particles in unit volume	r_c	Radius of cation
		r_O	Radius of oxygen ion
N_1, N_2	Number of charged particles in position 1 and 2, respectively	r_D	Radius of "doorway" in glass
		T	Temperature in K
N_A	Number of equivalent sites around non-bridging oxygen ion	t	Time
		U	Energy barrier to migration
n	Constant in Eq. (40)	u_1, u_2	Thickness fraction of layer 1 and 2, respectively
n	Defect concentration	V	Voltage
n_0	Total number of charge carrier	V_0	Amplitude of voltage
n_A, n_B	Number of charge carriers in A and B, respectively	V_1	Applied voltage
		$v, \mathbf{v}$	Velocity
P	Distribution (local concentration) of charge carriers	W	Energy required to form a defect
P_1	Amplitude of charge concentration variation	x	x-coordinate taken perpendicular to specimen area
P_{Na^+}	Distribution of Na^+	x	Number of bridging oxygen around a site
$P_{Na^+}^0$	Distribution of Na^+ in the absence of the external field	y	Number of bridging oxygen around a site
P_{Na^+}'	Perturbation of Na^+ distribution	Z_1, Z_2	Valence

α — Constant defined by Eq. (69)

$\alpha_1, \alpha_2, \ldots$ — Experimentally determined constants, defined by Eq. (36)

β — Finite displacement factor

β — Constant in Eq. (47)

$\beta_1, \beta_2, \ldots$ — Experimentally determined constants, defined by Eq. (36)

Γ — Jump frequency

Γ_0 — Most probable jump frequency

$\Gamma(n)$ — Gamma function of n

γ — Jump distance

δ — Phase difference between electric field and displacement

ε^* — Complex dielectric constant

ε' — Dielectric constant

ε'' — Dielectric loss factor

ε_0 — Dielectric constant of vacuum

ε_s — Dielectric constant at zero frequency

ε_∞ — Dielectric constant at high frequency

$\varepsilon_1, \varepsilon_2$ — Dielectric constant of layer 1 and 2, respectively, at high frequency

$\Delta\varepsilon$ — Relaxation strength (height of dielectric dispersion)

θ — Angle between direction of field and radius vector $\mathbf{r}$ in Eq. (153)

θ — Angle between path direction and jump direction in conducting path model

λ — Jump distance in conducting path model

ν — Vibrational frequency

ν_A — Vibrational frequency of alkali ion

ν_{Os} — Vibrational frequency of normal nonbridging oxygen ion

ν'_{Os} — Vibrational frequency of alkali-free nonbridging oxygen ion

ν''_{Os} — Vibrational frequency of nonbridging oxygen ion with doubly occupied sites

ν_{OB} — Vibrational frequency of normal bridging oxygen

ν'_{OB} — Vibrational frequency of bridging oxygen around alkali ion

ζ — Proportionality constant defined by Eq. (55)

π — 3.14159...

ρ — Dimensionless parameter defined by Eq. (56)

σ — DC conductivity

σ_1, σ_2 — DC conductivity of layer 1 and 2, respectively

σ^0 — Conductivity when no ionic interaction exists

σ_a — Apparent conductivity

τ — Relaxation time

τ_0 — Most probable relaxation time

$\phi(t)$ — Relaxation function

ϕ_0 — Constant defined by Eq. (30)

ϕ — Angle between path direction and direction of applied field in conducting path model

ψ — Potential around a central oxygen ion

ψ^0 — Potential around a central oxygen ion in the absence of external field

ψ' — Perturbation of potential

χ — Quantity defined by Eq. (157)

ω — Angular frequency

ω_{ep} — Angular frequency corresponding to a dielectric constant in the electrode polarization

ω_c — Angular frequency at which two different thickness specimens have the same dielectric constants in the electrode polarization

ω_{12}, ω_{21} — Probability of particle transition from 1 to 2 and from 2 to 1 respectively

ω_0 — Probability of particle transition in the absence of external field

II. Basic Information

A. DC Conduction and the Nernst–Einstein Relation

Conventional oxide glasses which contain a few percent or more alkali oxide have been established by the early works by Warburg (1884) and others to be electrolytic conductors, conducting species being exclusively the alkali ions. The radial distribution analysis by Warren (1937) and Warren and Biscoe (1938) indicates that an alkali ion such as sodium stays in a site surrounded by oxygen. It is normally assumed that the alkali ions are near the oxygen ions to maintain charge neutrality.

When a dc electric field is applied to the glass, the alkali ions with positive charge e will move towards the negative electrode, giving rise to a dc conduction. The temperature dependency of the dc conductivity σ follows the Arrhenius-type equation

$$\sigma = A_1 \exp(-\Delta E_\sigma / RT) \tag{1}$$

where the preexponential factor A_1 and an activation energy of conduction ΔE_σ are temperature-independent constants, R is the gas constant, and T the absolute temperature.

Since the electric conduction is caused by the ionic motion, there is an interrelation between the ionic diffusion coefficient D and the dc electric conductivity σ, i.e.,

$$\sigma = Ce^2 D/kT \tag{2}$$

where C is the alkali ionic charge concentration, k Boltzmann's constant, and e the charge of the ion under consideration (Mott and Gurney, 1940). In reality the experimental observations indicate a slight deviation from the prediction of this Nernst–Einstein equation and often the coefficient f is introduced as in

$$\sigma = Ce^2 D/fkT \tag{3}$$

The magnitude of f is normally found to range from 0.2 to about 0.7 (Doremus, 1962; Terai, 1969; Engel and Tomozawa, 1975). The origin of this discrepancy is the subject of controversy but will not be discussed further here. Instead, it will be assumed for convenience that the Nernst–Einstein relation is valid in glasses.

B. Dielectric Constants

When a constant dc voltage is applied on two parallel conducting plates which are separated by a narrow gap of vacuum, the charge will be built up on the plate instantly. However, when the vacuum layer is replaced by a dielectric such as a glass, an additional charge will be built up on the plate and the latter part will be time dependent, being controlled by the rate of polarization of glass. Thus, the total charge on the plate becomes the time-dependent quantity. The electric flux associated with the total charge on the surface of the plate is called the electric displacement **D** and is numerically equal to the total charge density on the plate.

When a low-frequency periodic field given by

$$\mathbf{E} = \mathbf{E}_0 e^{j\omega t} \tag{4}$$

where ω is the angular frequency, is applied across the glass, the charge on the plate will vary in the periodic manner, but its phase will lag slightly behind that of the applied electric field. Thus, the corresponding electric displacement $\mathbf{D}$ is given by

$$\mathbf{D} = \mathbf{D}_0 e^{j(\omega t - \delta)} \tag{5}$$

where δ represents the difference in phase. These two quantities $\mathbf{E}$ and $\mathbf{D}$ are related by

$$\mathbf{D} = \varepsilon^* \varepsilon_0 \mathbf{E} \tag{6}$$

where ε_0 is the dielectric constant of vacuum and is equal to 8.8554×10^{-12} F/m and ε^* is the complex dielectric constant and is defined by

$$\varepsilon^* = \varepsilon' - j\varepsilon'' \tag{7}$$

ε', ε'' are the dielectric constant and (dielectric) loss factor, respectively. In general $\mathbf{D}$ and $\mathbf{E}$ are vector quantities but are treated as scalers when only isotropic material such as glass is considered. Substituting Eqs. (4) and (5) in Eq. (6),

$$\varepsilon_0 \varepsilon^* = (D_0/E_0)e^{-j\delta} = (D_0/E_0)(\cos \delta - j \sin \delta) \tag{8}$$

therefore

$$\tan \delta = \varepsilon''/\varepsilon' \tag{9}$$

The quantity $\tan \delta$ is called the loss angle or dissipation factor and the $\sin \delta$ the power factor. When δ is small, $\tan \delta \doteqdot \sin \delta \doteqdot \delta$. Thus, the measurement of dielectric constant and loss factor is the measurement of the relation between the applied field and the polarization of the dielectric.

Since the electric current I is defined as the rate of time variation of the charge on the plate

$$I = \dot{D}A = j\omega\varepsilon_0\varepsilon^* EA = (j\omega\varepsilon_0\varepsilon' + \omega\varepsilon_0\varepsilon'')V(A/L) \tag{10}$$

where A and L are area and thickness of the specimen, respectively, and V is the voltage given by

$$V = EL = V_0 e^{j\omega t} \tag{11}$$

Since the current and the voltage is related by

$$I = YV \tag{12}$$

where Y is the admittance,

$$\text{Re } Y = \omega\varepsilon_0\varepsilon''(A/L)$$
$$\text{Im } Y = \omega\varepsilon_0\varepsilon'(A/L) \tag{13}$$

In an ideal dielectric material, (when polarization follows the field without delay) the admittance consists entirely of the imaginary part, namely the current precedes the voltage by a phase $\pi/2$. In real dielectrics, however, the phase difference is $\pi/2 - \delta$ because of the slight delay of polarization behind the field or the finite value of the dielectric loss, ε''.

C. Alternate Expressions

Some investigators prefer to use alternate expressions to describe the dielectric characteristics of materials. Macedo et al. (1972) used an "electric modulus" M^* defined by

$$M^* = 1/\varepsilon^* \tag{14}$$

analogous to the mechanical modulus, which is defined as the ratio of stress over strain. Combining with Eq. (7),

$$M^* = \frac{\varepsilon'}{\varepsilon'^2 + \varepsilon''^2} + j\frac{\varepsilon''}{\varepsilon'^2 + \varepsilon''^2} = M' + jM'' \tag{15}$$

Thus, M' and M'' behave in a similar manner to ε' and ε''. Because of the analogy between the mechanical modulus and the electric modulus, it is convenient to use the electric modulus when the direct comparison of the mechanical and electrical relaxation is desired.

Another effect of using the electric modulus is to suppress the effect of the electrode polarization, since the large value of ε' appears in the denominator. Therefore, the electric modulus is useful when the electrode polarization is regarded as nuisance. On the other hand, the detailed information on the electrode polarization will be lost by the use of the electric modulus.

It does not appear to this author that the advantage of one expression over the other is overwhelming. Therefore, the more conventional expression of the dielectric constant ε^* will be employed throughout the remainder of this chapter.

D. Absorption Current and Dielectric Constants

When a constant dc voltage V is applied at time $t = 0$ across a specimen, a time-dependent current will flow

$$I(t) = V\phi(t) \tag{16}$$

where $\phi(t)$ is called relaxation function (or decay function) and represents the manner in which the current changes with time (beginning time $t = 0$). In general, the voltage itself may be changing with time. In this case use is made of the Curie–Hopkins superposition principle, which states that "each variation of voltage produces the same change in current as if it alone were acting" (Manning and Bell, 1940). In equation form this is expressed as (Gross, 1941)

$$I(t) = \int_{t'=-\infty}^{t} dV(t')\phi(t - t') = \int_{-\infty}^{t} \frac{dV(t')}{dt} \phi(t - t') \, dt' \qquad (17)$$

It is customary to separate out two terms from the integral, one is the finite leakage current and the other is the instantaneous current due to instantaneous polarization. The leakage current is proportional to V and is given by

$$I = G_0 V \qquad (18)$$

where G_0 is the steady-state dc (or leakage) conductance. The instantaneous polarization current is given by

$$I = C_0 \frac{\partial V}{\partial t} \qquad (19)$$

where C_0 is the geometric capacitance given by

$$C_0 = (A/L)\varepsilon_0 \varepsilon_\infty \qquad (20)$$

where ε_∞ is the relative dielectric constant of the specimen at high frequency. Thus

$$I(t) = G_0 V + C_0 \frac{\partial V}{\partial t} + \int_{-\infty}^{t} \frac{\partial V(t')}{\partial t'} \phi(t - t') \, dt' \qquad (21)$$

In this and subsequent equations, therefore, $\phi(t)$ represents the relaxation function corresponding to the anomalous charging current. By changing the variable using

$$t - t' = \theta \qquad (22)$$

the current is given by

$$I(t) = G_0 V + C_0 \frac{\partial V}{\partial t} - \int_{0}^{\infty} \frac{dV(t - \theta)}{d\theta} \phi(\theta) \, d\theta \qquad (23)$$

When the applied voltage is sinusoidal

$$V(t) = V_0 e^{j\omega t} \qquad (24)$$

and the current is given by

$$I(t) = G_0 V + j\omega C_0 V + j\omega V \int_0^\infty e^{-j\omega\theta} \phi(\theta) \, d\theta \tag{25}$$

On the other hand the current through a dielectric is related to the dielectric constant of the material by Eq. (10),

$$I = j\omega C_V \varepsilon^* V$$

where C_V is the capacitance when the dielectric is replaced by vacuum, i.e.,

$$C_V = (A/L)\varepsilon_0 \tag{26}$$

Comparing Eqs. (25) and (10)

$$\varepsilon^* = \varepsilon_\infty + \frac{\sigma}{j\omega\varepsilon_0} + \frac{1}{C_V} \int_0^\infty e^{-j\omega t} \phi(t) \, dt \tag{27}$$

where σ is the dc conductivity. In this equation the integration variable has been changed to t. Equivalently,

$$\varepsilon' = \varepsilon_\infty + \frac{1}{C_V} \int_0^\infty \cos \omega t \, \phi(t) \, dt \tag{28}$$

and

$$\varepsilon'' = \frac{\sigma}{\omega\varepsilon_0} + \frac{1}{C_V} \int_0^\infty \sin \omega t \, \phi(t) \, dt \tag{29}$$

Equations (28) and (29), together with Eq. (16) are often called the fundamental equations of dielectric absorption. These equations establish the relation between the absorption current and the dielectric constants. Thus it is possible, at least in principle, to obtain ε' and ε'' from the absorption current measurement. In particular when the relaxation function is given by a simple exponential function of the form

$$\phi(t) = \phi_0 e^{-t/\tau} \tag{30}$$

where the preexponential factor ϕ_0 and the relaxation time τ are constants, the dielectric constant and the dielectric loss are given by the followings known as Debye equations.

$$\varepsilon^* = \varepsilon_\infty + \frac{\sigma}{j\omega\varepsilon_0} + \frac{\phi_0}{C_V} \int_0^\infty e^{-j\omega t} e^{-t/\tau} \, dt$$

$$= \varepsilon_\infty + \frac{\sigma}{j\omega\varepsilon_0} + \frac{\phi_0}{C_V} \frac{\tau}{1 + j\omega\tau}$$

$$= \varepsilon_\infty + \frac{\sigma}{j\omega\varepsilon_0} + (\varepsilon_s - \varepsilon_\infty) \frac{1}{1 + j\omega\tau} \tag{31}$$

where ε_s is the dielectric constant at $\omega = 0$, and $(\varepsilon_s - \varepsilon_\infty)$ is called the relaxation strength or the step height of the dielectric constant. Thus

$$\varepsilon' = \varepsilon_\infty + (\varepsilon_s - \varepsilon_\infty) \frac{1}{1 + \omega^2 \tau^2} \tag{32}$$

$$\varepsilon'' = \frac{\sigma}{\omega \varepsilon_0} + (\varepsilon_s - \varepsilon_\infty) \frac{\omega \tau}{1 + \omega^2 \tau^2} \tag{33}$$

In reality, however, the relaxation function does not show a simple exponential time dependency and the corresponding dielectric constants deviate from Debye's equations, usually showing a broader frequency distribution. This is illustrated in Fig. 1, where the Debye's distribution (solid line) is compared with the broader distribution (dotted line) which is experimentally observed. It is customary to explain this broad distribution in terms of the distribution of the relaxation time. Namely,

$$\varepsilon^* = \varepsilon_\infty + \frac{\sigma}{j\omega\varepsilon_0} + (\varepsilon_s - \varepsilon_\infty) \int_0^\infty \frac{G(\tau)\, d\tau}{1 + j\omega\tau} \tag{34}$$

where $G(\tau)$ is the distribution function of the relaxation time and satisfies the relation

$$\int_0^\infty G(\tau)\, d\tau = 1 \tag{35}$$

Numerous attempts have been made to represent the experimental behavior of the dielectric characteristics by appropriate distribution functions as well as by simple functions (Wagner, 1913; Yager, 1936; Fuoss and Kirkwood 1941; Cole and Cole, 1941; Davidson and Cole, 1951).

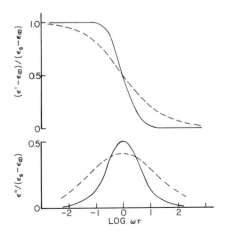

Fig. 1. Normalized dielectric constant ε' and dielectric loss ε''. Solid lines—Debye's theory; dotted line—a typical experimental observation.

Correspondingly, attempts were made to represent the relaxation function by appropriate simple functions, so that dielectric characteristics could be calculated using Eqs. (28) and (29). Some of these functions will be discussed below, together with the useful approximations to evaluate the dielectric characteristics.

1. WHITEHEAD AND BANÔS EQUATION[†]

It is assumed that the relaxation function can be expressed as the sum of the simple exponential functions, i.e.,

$$\phi(t) = \beta_1 e^{-\alpha_1 t} + \beta_2 e^{-\alpha_2 t} + \cdots \tag{36}$$

where $\alpha_1, \alpha_2, \ldots, \beta_1, \beta_2, \ldots$ are constants, then the dielectric constant and dielectric loss are given as

$$\varepsilon' \simeq \varepsilon_\infty + \frac{\alpha_1 \beta_1}{\omega^2 + \alpha_1{}^2} + \frac{\alpha_2 \beta_2}{\omega^2 + \alpha_2{}^2} + \cdots \tag{37}$$

$$\varepsilon'' \simeq \frac{\sigma}{\omega \varepsilon_0} + \frac{\omega \beta_1}{\omega^2 + \alpha_1{}^2} + \frac{\omega \beta_2}{\omega^2 + \alpha_2{}^2} + \cdots \tag{38}$$

Here the constants $\alpha_1, \alpha_2, \ldots, \beta_1, \beta_2, \ldots$ can be obtained by graphically analyzing the experimental log $\phi(t)$ versus t curve. Accuracy of the approximation will improve if numerous terms are considered. In an ultimate case the above approximation is equivalent to

$$\phi(t) = \int_0^\infty g(\tau) e^{-t/\tau} \, d\tau \tag{39}$$

where $g(\tau)$ is distribution function of the relaxation time.

2. HAMON'S APPROXIMATION[‡]

First it is assumed that the relaxation function $\phi(t)$ is given by

$$\phi(t) = At^{-n} \tag{40}$$

where A and n are constants. Then from Eqs. (28) and (29),

$$\varepsilon' = \varepsilon_\infty + (A/C_V)\omega^{n-1}\Gamma(1 - n)\cos[(1 - n)\pi/2] \qquad \text{for } 0 < n < 1 \tag{41}$$

$$\varepsilon'' = \frac{\sigma}{\omega \varepsilon_0} + \frac{A}{C_V} \frac{\omega^{n-1} \pi}{2 \sin(n\pi/2)\Gamma(n)} \qquad \text{for } 0 < n < 2 \tag{42}$$

where $\Gamma(n)$ is the gamma function of n. The loss factor ε'' can be expressed in terms of the relaxation function of the anomalous charging current $\phi(t_1)$

[†] Whitehead and Banôs (1932).
[‡] Hamon (1952).

at a particular time t_1 as

$$\varepsilon'' = \frac{\sigma}{\omega\varepsilon_0} + \frac{\phi(t_1)}{\omega C_V} \tag{43}$$

provided that ω and t_1 are related by

$$\omega t_1 = \left[\frac{\pi}{2\sin(n\pi/2)\Gamma(n)}\right]^{-1/n} \tag{44}$$

The expression on the right-hand side of Eq. (44) is almost independent of n in the range $0.3 < n < 1.2$ and, to an accuracy within $\pm 3\%$, can be taken as having the mean value of 0.63. Thus,

$$\varepsilon'' \simeq \frac{\sigma}{\omega\varepsilon_0} + \frac{\phi(0.63/\omega)}{\omega C_V} \tag{45}$$

or if $\sigma_a(t)$ is defined as the apparent conductivity including the dc conductivity and the conductivity corresponding to the anamalous charging current

$$\varepsilon'' \simeq \frac{\sigma_a(0.1/f)}{2\pi f \varepsilon_0} \tag{46}$$

where $2\pi f$ is substituted for ω. This approximate expression is valid even when n is not a constant but is changing with time in the range of $0.3 \sim 1.2$.

3. WILLIAMS AND WATTS' EQUATION[†]

Recently the following electric field decay function was found to describe best the dielectric behavior of variety of materials.

$$E(t) = E(0)e^{-(t/\tau)^\beta} \tag{47}$$

where E is the electric field and τ and β are constants. In terms of the relaxation function

$$\phi(t) \propto -(\partial/\partial t)\left[e^{-(t/\tau)^\beta}\right] \tag{48}$$

This expression, however, does not give a simple relation by which dielectric constant is obtained from the dc measurements.

III. Experimental Observations

There have been numerous experimental investigations on the low-frequency dielectric behavior of glasses. These investigations can be classified into two groups by their experimental methods. One method is to study

[†] Williams and Watts (1970); Moynihan et al. (1973).

charging–discharging current as a function of time and calculate the dielectric loss as a function of frequency using expressions discussed in Section II,D. The other is to measure the dielectric characteristics as a function of frequency using a capacitance bridge. Both measurements can be performed at various temperatures and as expected give the identical results.

The former, charging–discharging current measurement on dielectrics has been conducted quite sometime (Curie, 1888; Richardson, 1925; Joffe, 1928) but the conversion of the current to the dielectric loss was only recently made. Namikawa *et al.* (Namikawa and Asahara, 1966; Namikawa, 1974, 1975a,b) made an extensive use of this method to obtain the dielectric loss of various glasses. The latter capacitance bridge measurement on glass has been conducted in the past quarter century by many glass scientists (Volger *et al.*, 1953; Taylor, 1957; 1959; Heroux, 1958; Prod'homme, 1960; Charles, 1961, 1962, 1963, 1966; Isard, 1962; Snow and Deal, 1966; Hansen and Splann, 1966; Splann Mizzoni, 1973; Macedo *et al.*, 1972; Provenzano *et al.*, 1972, Guidee, 1972; Hakim and Uhlmann, 1973; Boesh and Moynihan, 1975; Kim and Tomozawa, 1976a).

In this section, some of the important techniques of measurement and the highlights of experimental observations by various investigators will be described.

A. Electrode

One of the important problems in the dc or low-frequency ac measurement is the choice of appropriate electrodes. Since the alkali ions are the charge carrying species in glasses under consideration, nonblocking electrodes have to be able to supply and accept the ions. This requirement becomes increasingly important at high temperature or low frequency, especially when the properties which were not influenced by the electrodes are desired.

Thus, an alkali–amalgam or sodium nitrate melt are appropriate electrodes. However, these liquid electrodes are difficult to handle, especially when an accurate determination of a contact area and the elimination of the surface conduction and fringe effect are needed. Thus, often the noble metal electrodes such as Ag or Au are used either in evaporated film form or in paste form. The use of these electrodes are acceptable in a limited temperature and frequency range. [The effects of electrodes on the dielectric measurements of glasses have been investigated by many, including Volger *et al.* (1953) Taylor (1957), Isard (1962), and Kim and Tomozawa (1976a).] Commonly, the three electrode arrangements shown in Fig. 2 are used to eliminate the surface conduction and the fringe effect and the effective area is calculated using the following equation for the effective radius r [after Amey and Hamburger (1949)].

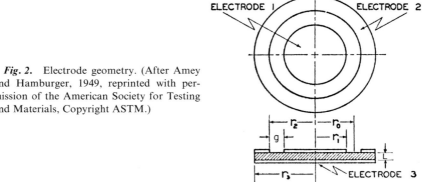

Fig. 2. Electrode geometry. (After Amey and Hamburger, 1949, reprinted with permission of the American Society for Testing and Materials, Copyright ASTM.)

$$r = r_1 + \tfrac{1}{2}g - l, \quad l/L = (2/\pi) \ln \cosh(\pi g/4L) = 1.466 \log \cosh(0.7858g/L)$$

where L is the specimen thickness, g is the gap width between guard electrode and guarded electrode $(r_1 - r_2)$, r_1 is the guarded electrode radius, and r_2 is the guard electrode interior radius.

B. Charging–Discharging Current

A typical circuit diagram for measurement by this method is shown in Fig. 3. When a constant voltage is applied on a glass specimen with evaporated metal electrodes at a low temperature (e.g., room temperature), the charging current flows, decreasing with time and reaching a constant steady-state value. At this stage if the voltage is suddenly removed and the electrodes are short-circuited, the discharge current flows in the reverse direction. These charging–discharging currents are reproducible. An example of this measure-

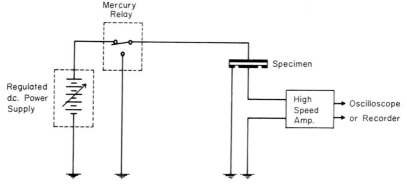

Fig. 3. Absorption current measuring circuit. (After Namikawa, 1974.)

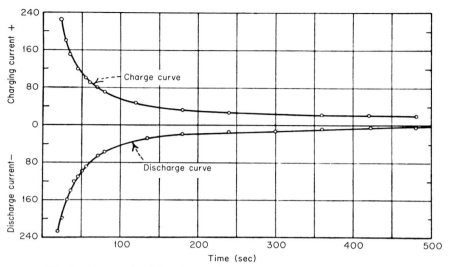

Fig. 4.   An example of charging and discharging currents. (After Guyer, 1933.)

ment is shown in Fig. 4. Often the charging current follows "Curie's law" (Manning and Bell, 1940) which states

1. *Proportionality to the voltage.* The ordinates of the charging-current curve as a function of the time are rigorously proportional to the voltage.

2. *Law of thickness.* For the same voltage, the ordinate of the charging-current curves are in inverse proportion to the thickness of the dielectric.

3. *Law of superposition.* Each variation of voltage between the two faces of the sample produces the same change in current as if it alone were acting.

In addition, the discharge current has been found to be identical to the time-dependent part of charging current (anomalous charging current), i.e., charging current minus a constant steady state (or leakage) current. Thus, it is easier to use the discharge current to obtain $\phi(t)$ and consequently the dielectric loss factor.

When a constant voltage is applied for an extended period of time or at a higher temperature, the steady-state charging current starts to decrease again. This can best be seen when plotted against log t scale, as shown in Fig. 5. Here the log of the normalized apparent conductivity σ_a is given against the log of the normalized charging time. This diagram indicates that there are two stages of the apparent conductivity variation; one labeled as A (so called anomalous charging current) takes place in a short time while the other labeled as B takes place after prolonged charging time.

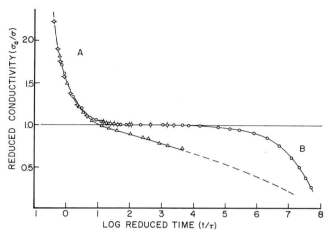

Fig. 5. Normalized charging current (normalized conductivity versus normalized charging time); Au evaporated electrode (○), Au paste electrode (△); different circular symbols refer to different measuring temperatures; soda–lime glass.

Similar behavior was also observed by Kinser and Hench (1969). Figure 5 also shows that there is a pronounced effect of electrode material on the process B, while very little is observed for process A. For this reason process B is called electrode polarization. When a perfectly nonpolarizing electrode is employed, this electrode polarization would not be observed, the apparent conductivity remaining constant indefinitely. On the other hand when a blocking electrode was used at high temperature only charging current corresponding to the electrode polarization will be observed.

C. Capacitance Bridge Method

In the frequency range of 10 Hz ~ 100 kHz the capacitance bridge is most frequently used to obtain the dielectric constant and dielectric loss (or conductivity). This method is based upon the balancing of the resistive and capacitive components of the specimen with the known variable, standard components. An example of the circuit diagram is shown in Fig. 6.

Numerous experimental observations on the dielectric properties of glasses obtained by capacitance bridge techniques also indicates that there are two types of dispersions at low frequencies. An example is shown in Fig. 7 which shows the dielectric constant of a soda–lime glass. The large dielectric constant at the lower frequency, which is labeled as B, is often called electrode polarization while the dispersion at slightly higher frequency (labeled as A) corresponds to the dielectric relaxation, often called migration loss. These two dispersions naturally correspond to what has been observed

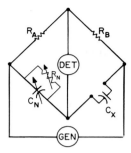

Fig. 6. ac capacitance bridge circuit in parallel circuits. DET = dectector, GEN = generator, C_x = unknown capacitor, C_N = standard capacitor, R_N = standard resistor, R_A, R_B = ratio arms.

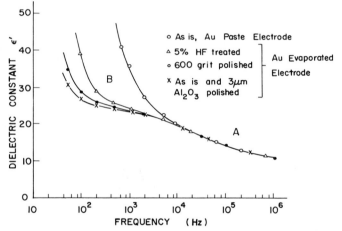

Fig. 7. Dielectric constant of soda–lime glass with various surface conditions; Au electrode at 247 ± 1°C. (After Kim and Tomozawa, 1976a.)

in charging–discharging current measurement, the lower frequency electrode polarization corresponding to the longer time charging current. The apparent conductivity corresponding to Fig. 7 is shown in Fig. 8. As was the case in the charging–discharging current measurement Figs. 7 and 8 indicate that the electrode polarization is influenced by the glass surface conditions as well as by the electrodes (metal-evaporated electrode or paste electrode). The effect of the kind of metals (Au or Ag) is small, as can be seen from the similarity of Figs. 7 and 9. The dielectric relaxation due to ion migration on the other hand appears insensitive to the glass surface and electrodes, at least in the high-frequency side of the dispersion.

To see the dielectric relaxation due to ion migration (dispersion labeled A in Figs. 5 and 7) clearly, it is customary to remove the dc conductivity contribution by substracting the frequency-independent or time-independent

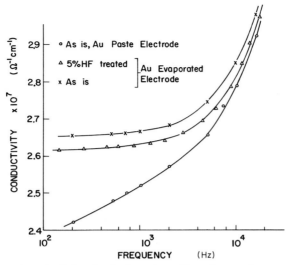

Fig. 8. Apparent conductivity of soda–lime glass with various surface conditions; Au electrode at 247°C. (After Kim and Tomozawa, 1976a.)

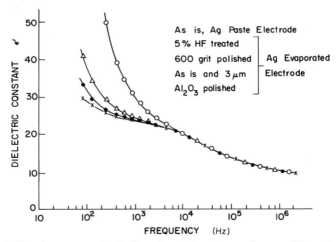

Fig. 9. Dielectric constant of soda–lime glass with various surface conditions; Ag electrodes at 247 ± 1°C. (After Kim and Tomozawa, 1976a.)

portion of the conductivity from the dielectric loss. This procedure is equivalent to the use of the discharging curve instead of the charging curve in the evaluation of $\phi(t)$ in Eq. (29). An example of such calculation using the capacitance bridge measurement is shown in Fig. 10 for a soda–lime glass

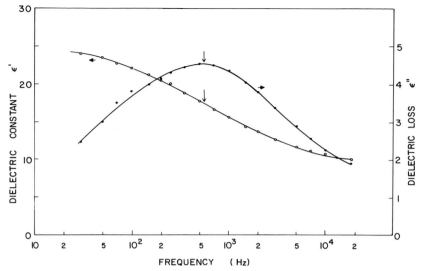

Fig. 10. An example of experimental observation of dielectric constant and dielectric loss in glass. Soda–lime glass, Au evaporated film electrode at 164°C.

(Kim, 1976), where the equivalence of the dielectric constant dispersion and the dielectric loss maximum is demonstrated. The arrows in the diagram indicate the frequency position of the inflection point of the dielectric constant and that of the dielectric loss maximum. The equivalence of two measurements—the charging–discharging and the capacitance bridge— were most clearly demonstrated by Namikawa (1975a) for this type of dielectric relaxation using a $Li_2O–CaO–SiO_2$ glass. His results are shown in Fig. 11.

D.   Main Features of the Dielectric Characteristics of Glasses

From the foregoing description, it should be clear that two major dispersions are observed for glasses in the low-frequency range. One is the electrode polarization and the other is the dielectric relaxation due to ionic migration. In addition to these two types of dispersions which appear to exist in all glasses, there is dielectric relaxation which is observed only in some selected glasses. One such dielectric relaxation is observed in phase-separated glasses with dispersed high-conductivity second phase (Charles, 1963; Namikawa, 1975b). This originates from the Maxwell–Wagner–Sillars type inhomogeneous structure (Maxwell, 1892; Wagner, 1914; Sillars, 1937) and is well understood. The other is the dielectric relaxation due to mixed alkali effect (Namikawa, 1974). Although this dielectric relaxation is probably

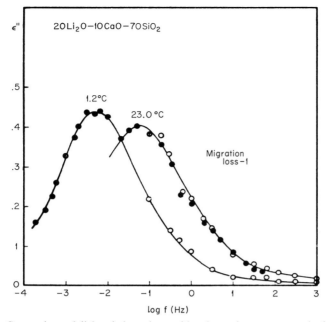

Fig. 11. Comparison of dielectric loss observed by absorption current method and capacitance bridge method. ○ = ac bridge method; ● = absorption current method. (After Namikawa, 1975a.)

similar in origin to the dielectric relaxation due to ion migration, it will be excluded from the discussion here and in the remainder of this chapter attention will be focused on two, apparently universal, dispersions, i.e., electrode polarization and the dielectric relaxation due to ion migration. Following are the main features of these dispersions, observed by numerous investigators. A correct theory or mechanism will have to be able to explain these features.

1. ELECTRODE POLARIZATION

The dielectric characteristics in the electrode polarization are dependent upon the thickness and the surface conditions of glasses as well as the electrode material, indicating that the phenomenon originates at the interface between glass and electrode. When a dc or ac low frequency is applied to a glass with a blocking electrode at high temperature, a nonuniform electric field is expected to develope due to the space charge. Proctor and Sutton (1959, 1960) and Sutton (1964a,b) investigated the electric field distribution in glass as a function of charging time and temperature using aluminum electrodes. As expected the high electric field developed near electrodes. An example of their results for an alkali lead silicate glass is shown in Fig. 12. A similar result was observed in fused silica (Cohen, 1957). The high field near the electrodes corresponds to the high-resistance layer produced by the depletion or accumulation of movable charges. Thus the resulting conduction current is strongly influenced by the amount of charge transport across this high-resistance layer. The suggested charged species moving across this high-resistance layer under the influence of the high electric field are alkali ions from the surface hydroxide layer (Kim and Tomozawa, 1976a), a proton from water (Hetherington *et al.*, 1965; Carlson, 1974b), nonbridging oxygen (Carlson *et al.*, 1972; Carlson, 1974a), and the oxidized electrode metal ions.

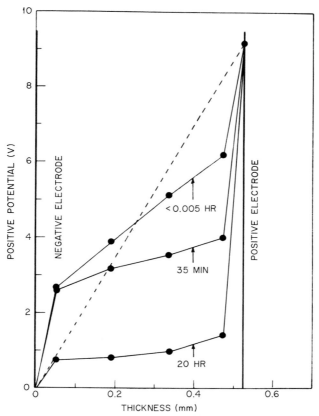

Fig. 12. Space–potential distribution in glass; applied potential +9.2 V, temperature 410°C. (After Proctor and Sutton, 1960.)

2. DIELECTRIC RELAXATION DUE TO ION MIGRATION

The most striking feature of the dielectric relaxation such as shown in Figs. 10 and 11 is the temperature dependence of the peak frequency f_m. Numerous observations indicate that the peak frequency has the same activation energy as the dc conductivity. Thus, it is most likely that the same motion of the same charged species is responsible both for dc conduction and the dielectric relaxation and the expression of the type

$$f_m = \text{const } \sigma$$

is expected to follow. Isard (1962) found that the peak frequency f_m is approximately given by

$$f_m = \sigma/2\pi\varepsilon_0\varepsilon_\infty \tag{49}$$

which corresponds to $\omega\tau = 1$, where ω is the angular frequency ($=2\pi f$) and the relaxation time $\tau = \varepsilon_0\varepsilon_\infty/\sigma$, which is the relaxation time of the space charge.

Alternatively, Nakajima (1972) and Namikawa et al. (Namikawa and Kumata, 1966; Namikawa 1974) found that the following relation agrees with the experimental results of numerous glasses

$$f_m = \sigma/2\pi\varepsilon_0 \, \Delta\varepsilon \tag{50}$$

where $\Delta\varepsilon$ is the relaxation strength or the step height of the dielectric dispersion. An example of their analysis is shown in Fig. 13. These two relations, Eqs. (49) and (50), give a nearly identical value for the peak frequency, since $\Delta\varepsilon \simeq \varepsilon_\infty$ for most glasses.

The other important features of the dielectric relaxation are that the loss peak is broad and asymmetric, the shape of which is temperature independent; the step height of dispersion increases with increasing concentration of the charge carrier and decreases with increasing temperature.

It is important to realize that the dielectric characteristics being discussed here are quite universal phenomena, not limited to ionic conducting glasses only. The similarity in dielectric behavior between ionic conducting glasses and electronic conducting glasses has been pointed out by Isard (1970). The similar dielectric behavior, i.e., the electrode polarization and the conductivity related dielectric relaxation, are also observed in a variety of materials other than glasses such as concentrated aqueous solutions (Moynihan et al., 1971; Ambrus et al., 1972), fused salts (Macedo and Weiler, 1969), oxide ceramics (Radzilowski et al., 1969), as well as single crystals of alkali halides (Kao et al., 1970). For example, Radzilowski et al. (1969) measured the dielectric properties and the diffusion coefficient of Na^+ of an aligned polycrystalline β-alumina and found that in the direction perpendicular to

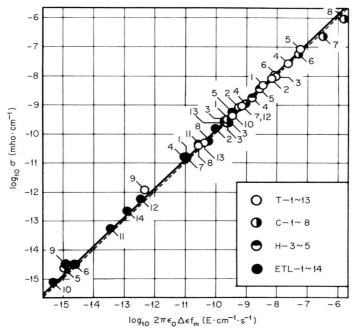

Fig. 13. Correlation between conductivity and dielectric relaxation. $\bigcirc$ = Taylor (1957, 1959); $\mathbb{O}$ = Charles (1962; 1963); $\ominus$ Heroux (1958); $\bullet$ = measured at Electrotechnical Laboratory, MITI, Japan. (After Nakajima, 1972.)

the c axis, the activation energy for the sodium diffusion coefficient was 3.81 kcal/mole while that for the peak frequency of the dielectric relaxation was 3.70 kcal/mole. Since the activation energy of diffusion coefficient is expected, from the Nernst–Einstein relation, to be greater than that of the dc conduction by RT which was 0.25 kcal/mole in this case, it can be seen that the activation energies for dc conduction and the dielectric relaxation are approximately the same for this material, similar to the phenomenon observed for glasses. The peak frequency of the dielectric relaxation was found to be related to dc conductivity by Eq. (49). Similar observations were made on single-crystal alkali halides including pure NaCl, KCl, and NaCl doped with impurities such as Ca^{2+} and Ba^{2+} (Kao *et al.*, 1970). One of the two observed dielectric relaxation times in these materials had the same activation energy as that for dc conduction as shown in Table I. The observed relaxation time was approximately the same as that calculated from the dc conductivity using Eq. (49).

TABLE I

Comparison of Activation Energies
for Single-Crystal Alkali Halides

Sample	ΔE_σ (eV)[a]	ΔE_{ac} (eV)[b]
NaCl	1.13	1.09
KCl	0.99	1.04
NaCl + Ca^{2+}	1.00	0.97
NaCl + Ba^{2+}	0.99	1.00

[a] ΔE_σ is the activation energy for dc conduction.
[b] ΔE_{ac} is the activation energy for dielectric relaxation; 1 eV $\simeq$ 23 kcal/mole.

The universal nature of the charging–discharging current has been recognized for a long time (Curie, 1888; Joffe, 1928). Correspondingly, the universal nature of the dielectric characteristics is also recognized (Sutton, 1964a; Jonscher, 1975).

It is also well known that there is a corresponding mechanical relaxation (Higgins *et al.*, 1972). Thus, the successful mechanism to explain the phenomenon has to be a universal one being applicable to a variety of materials as well as having a corresponding mechanism to explain the mechanical relaxation.

IV. Proposed Mechanisms

Many investigators have attempted to explain the observed dielectric behavior of glasses. Since there are apparently two different kinds of dispersions in the frequency range of interest, it is convenient to discuss them separately. In the following, various proposed models for the electrode polarization and the conductivity related dielectric relaxation will be discussed.

A. *Electrode Polarization*

1. Space Charge Mechanism

This is defined as the case when charge carriers migrate through a glass specimen in an electric field at a rate faster than they can be discharged or compensated at the glass–electrode interface. This phenomenon can lead to the apparent increase of the dielectric constant. Since in most oxide glasses alkali ions are primary charge carriers and the noble metal electrodes are often used in the electrical measurement, this charge accumulation (or deficiency) is expected to take place under appropriate frequency and temperature conditions. Therefore, it is reasonable to attempt to attribute the observed electrode polarization in glass to the space charge phenomenon. (Jaffe, 1952; MacDonald, 1953, 1970, 1971, 1974; Friauf, 1954; Proctor and Sutton, 1959; Kinser and Hench, 1969; Doi, 1972). There are numerous theoretical analyses of this space charge mechanism. Some of the theories were specifically developed for glasses in which only cations (alkali ions) are mobile. In the following, the theory by Beaumont and Jacobs (1967) will be described, with the slight modification appropriate for glasses containing alkali.

The flux J of alkali ions through the glass specimen, with concentration and potential gradients, in the positive direction is given by

$$J = -D\frac{\partial P}{\partial x} + \frac{DPe}{kT}E \tag{51}$$

where D is the diffusion coefficient of alkali ions, e is the charge of a cation, k is the Boltzmann's constant, T is the absolute temperature, and P and E are concentration of charge carriers and the electric field, respectively. The Nernst–Einstein equation which relates the electrical conductivity σ and the diffusion coefficient D is

$$\sigma = (Ce^2/kT)D$$

and is assumed to hold, where C is the uniform (or average) concentration of charge.

From the continuity condition, for a constant D,

$$\frac{\partial P}{\partial t} = D\frac{\partial^2 P}{\partial x^2} - \frac{De}{kT}\frac{\partial}{\partial x}(PE) \tag{52}$$

In this equation, the generation and recombination of carriers have been neglected. The electric field is related to the charge distribution by Poisson's equation

$$\frac{\partial E}{\partial x} = \frac{e(P - C)}{\varepsilon_0\varepsilon_\infty} \tag{53}$$

where ε_0 is the dielectric constant of free space and ε_∞ is the relative dielectric constant of the glass at high frequency. One boundary condition for the solution of the differential Eq. (52) is

$$\int_{-L/2}^{L/2} E(x, t)\, dx = -V_1 e^{j\omega t} \tag{54}$$

where $V_1 e^{j\omega t}$ is the sinusoidal applied voltage and L is the thickness of the glass specimen. Another boundary condition specifies the charge transport across the specimen–electrode surface. It is assumed that the current at the interface is proportional to the excess (or deficiency) concentration of charge over that existing for zero applied voltage. Namely,

$$J(-L/2, t) = -\xi \, \Delta P \qquad \text{and} \qquad J(L/2, t) = +\xi \, \Delta P \qquad (55)$$

where ξ is a proportionality constant, and ΔP is equal to $P - C$. A constant, ξ, is related to dimensionless parameter ρ by

$$\xi = \rho(D/L) \qquad (56)$$

For a completely blocking electrode $\rho = 0$ and for a completely nonblocking electrode $\rho = \infty$. Thus boundary conditions become

$$\text{at } x = -L/2 \qquad D\frac{\partial P}{\partial x} - \frac{DPe}{kT}E = +\rho\frac{D}{L}\Delta P$$

$$\text{at } x = L/2 \qquad D\frac{\partial P}{\partial x} - \frac{DPe}{kT}E = -\rho\frac{D}{L}\Delta P \qquad (57)$$

In order to linearize the differential equations, it is customary to neglect higher order harominics and to assume that P and E are given by

$$P = C + P_1 e^{j\omega t} \qquad (58)$$

$$E = E_1 e^{j\omega t} \qquad (59)$$

Substituting these expressions into (52), and retaining only the linear terms,

$$D\frac{\partial^2 P_1}{\partial x^2} = j\omega P_1 + \frac{DeC}{kT}\frac{\partial E_1}{\partial x} \qquad (60)$$

Combined with the Poisson's equation

$$\partial E_1/\partial x = P_1 e/\varepsilon_0 \varepsilon_\infty \qquad (53a)$$

Eq. (60) becomes

$$D\frac{\partial^2 P_1}{\partial x^2} = P_1(j\omega + K_1^2 D) \qquad (61)$$

$$K_1^2 = \frac{Ce^2}{kT\varepsilon_0\varepsilon_\infty} = q^* K^2 \qquad (62)$$

The quantity $1/K$ is called the Debye length and q^* is a constant determined by the valence of the charge carrier, being $\frac{1}{2}$ for alkali ion containing glasses (Moore, 1962).

Boundary conditions become

$$V_1 = -\int_{-L/2}^{L/2} E_1 \, dx \tag{63}$$

and

$$D\frac{\partial P_1}{\partial x} - \frac{DCe}{kT} E_1 = \rho \frac{D}{L} P_1 \qquad \text{at} \quad x = -L/2$$

$$D\frac{\partial P_1}{\partial x} - \frac{DCe}{kT} E_1 = -\rho \frac{D}{L} P_1 \qquad \text{at} \quad x = L/2 \tag{64}$$

Since P_1 must be antisymmetric, the solution to Eq. (62) is

$$P_1 = \alpha \sin hmx \tag{65}$$

with

$$m^2 = K_1{}^2 + j(\omega/D) \tag{66}$$

and α is a constant determined by the boundary conditions. Integrating the Poisson Eq. (53),

$$E_1 = (\alpha e \cosh mx/m\varepsilon_0\varepsilon_\infty) + C' \tag{67}$$

where the integration constant C' is determined by the boundary condition Eq. (64). Thus,

$$E_1 = \frac{\alpha e}{m\varepsilon_0\varepsilon_\infty} \cosh mx \frac{\alpha kT}{Ce} \cosh \frac{mL}{2} \left(\frac{Ce^2}{mkT\varepsilon_0\varepsilon_\infty} - \frac{\rho}{L} \tanh \frac{mL}{2} - m \right) \tag{68}$$

Substituting (68) in (63) yields

$$\alpha = -V_1 \left[\sinh \frac{mL}{2} \left(\frac{2e}{m^2\varepsilon_0\varepsilon_\infty} + \frac{j\omega LkT}{DmCe} \coth \frac{mL}{2} + \frac{kT\rho}{Ce} \right) \right]^{-1} \tag{69}$$

The current density $i(x, t)$ in the glass will be the sum of the displacement current and the current due to flux, thus

$$i(x, t) = \varepsilon_0\varepsilon_\infty \frac{\partial E}{\partial t} + e\left(-D\frac{\partial P}{\partial x} + D\frac{Pe^2}{kT} E \right) \tag{70}$$

The current density flowing into the specimen is obtained by taking the space average of $i(x, t)$ over the whole specimen giving

$$\langle i \rangle = 1/L \int_{-L/2}^{L/2} i(x, t) \, dx = i_1 e^{j\omega t} \tag{71}$$

where

$$i_1 = -j\omega\varepsilon_0\varepsilon_\infty \frac{V_1}{L} - \sigma \frac{V_1}{L} - \frac{eD}{L} [P_1(L/2) - P_1(-L/2)] \tag{72}$$

The negative sign indicates the direction of the current. Namely, when V_1 is positive in a positive x direction, the current flows in a negative x direction.

The corresponding complex dielectric constant is given using Eq. (10)

$$\varepsilon_0\varepsilon^* = \varepsilon_0\varepsilon' - j\varepsilon_0\varepsilon''$$

$$= \varepsilon_0\varepsilon_\infty - j\frac{\sigma}{\omega} - j\frac{eD}{\omega V_1}[P_1(L/2) - P_1(-L/2)] \tag{73}$$

Here ε_∞ is the dielectric constant due to other mechanism than to the electrode polarization. Thus,

$$\varepsilon_0\varepsilon' = \varepsilon_0\varepsilon_\infty + \mathrm{Im}\{eD/\omega V_1)[P_1(L/2) - P_1(-L/2)]\} \tag{74}$$

$$\varepsilon_0\varepsilon'' = \sigma/\omega + \mathrm{Re}\{eD/\omega V_1)[P_1(L/2) - P_1(-L/2)]\} \tag{75}$$

Using Eqs. (65) and (69)

$$(eD/\omega V_1)[P_1(L/2) - P_1(-L/2)]$$

$$= \left(-\frac{2eD}{\omega}\right)\bigg/\left(\frac{2e}{m^2\varepsilon_0\varepsilon_\infty} + \frac{j\omega LkT}{DmCe}\coth\frac{mL}{2} + \frac{kT\rho}{Ce}\right)$$

$$= \left(-2\varepsilon_0\varepsilon_\infty\frac{D^3K_1^{\,4}}{\omega^3}\frac{(2+\rho)}{L^2} + j2\varepsilon_0\varepsilon_\infty\frac{D^2K_1^{\,3}}{\omega^2 L}\right)\bigg/\left(1 + (2+\rho)^2\frac{D^2K_1^{\,2}}{\omega^2 L^2}\right) \tag{76}$$

Here since we are interested only in the low-frequency range, the approximations employed were

$$\omega/2K_1^{\,2}D \ll 1 \tag{77}$$

$$m \simeq K_1 + j(\omega/2K_1 D) \tag{78}$$

and

$$\coth(mL/2) \simeq 1 \tag{79}$$

Also, since $1/K_1$ is short

$$K_1 L \gg 1 \tag{80}$$

and in blocking and partially blocking electrodes

$$K_1 L \gg \rho \tag{81}$$

Therefore

$$\varepsilon_0\varepsilon' = \varepsilon_0\varepsilon_\infty + \frac{2\varepsilon_0\varepsilon_\infty(D^2K_1^{\,3}/\omega^2 L)}{1 + (2+\rho)^2(D^2K_1^{\,2}/\omega^2 L^2)} \tag{82}$$

$$\varepsilon_0\varepsilon'' = \frac{\sigma}{\omega} - \frac{2\varepsilon_0\varepsilon_\infty(D^3K_1^{\,4}/\omega^3)[(2+\rho)/L^2]}{1 + (2+\rho)^2(D^2K_1^{\,2}/\omega^2 L^2)} \tag{83}$$

2. OTHER MECHANISMS

In addition to the space charge mechanism, several other models were also suggested. Mitoff and Charles (1972), comparing the observed large dielectric constant at low frequency with the theory of the space charge, concluded that the electrochemical reaction at the electrode–specimen interface is responsible for the phenomenon.

Macedo *et al.* (1972), suggested that, in addition to the space charge, the highly resistive electrode oxide layer or a thin air gap between the sample and the electrode contributes to the surface capacitance.

B. *Dielectric Relaxation*

1. RANDOM POTENTIAL ENERGY MODEL

Stevels (1957) and Taylor (1959) proposed a model in which there are potential energy barriers of various heights. It was assumed that for dc conduction the highest energy barrier has to be overcome, while for ac conduction the migration across the limited distance overcoming the lower energy barrier is responsible. In this way migration of charge carriers, (e.g., alkali ions) can give rise to both dc and ac loss phenomena. Furthermore, this mechanism is considered to be consistent with the random network structure of oxide glasses. The schematic diagram of this model is shown in Fig. 14.

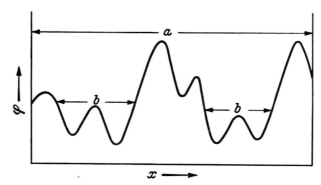

Fig. 14. A random potential energy model by Stevels (1957).

2. DEFECT MODEL

Charles (1961) postulated a structural model which can account for both polarization and dc conduction processes in glass. It is considered that an alkali metal ion in silica network stays in the vicinity of a nonbridging oxygen ion to which it is ionically associated. When an alkali ion obtains enough energy, it can migrate through the network. Charles postulated that there are a number of equivalent positions of stability for an alkali ion around each nonbridging oxygen atom. If, for example, a model of the type shown in Fig. 15 is considered, an alkali ion can migrate from a site around a given nonbridging oxygen ion to a site around an adjacent (or second) oxygen ion which now has two alkali ions around it. This is regarded as a "defect". One of the two alkali ions around the second nonbridging oxygen ion can move on to the site around the next (or third) nonbridging oxygen ion. When this alkali ion, which is moving from the second to the third nonbridging oxygen ion, is different from the alkali ion which arises from the first nonbridging ion, the orientation of the second nonbridging oxygen ion–alkali ion pair is different from the original direction. Thus in the process of alkali ion migration, both (orientation) polarization and conduction takes place.

Quantitatively, the theory by Charles starts with the orientation polarization theory by Frölich (1949) who postulated that a particle with charge e can take one of two equivalent sites 1 and 2 separated by a distance b and an energy of height H. When no electric field is applied, and if it is assumed that there are N charged particles in unit volume, the number of particles in positions 1 and 2 are equal, i.e.,

$$N_1 = N_2 = N/2 \qquad (84)$$

where N_1 and N_2 are the number in position 1 and 2 per unit volume, respectively. The probability of a particle transition from 1 to 2, ω_{12}, is the same as that for the transition from 2 to 1, ω_{21}, i.e.,

$$\omega_{12} = \omega_{21} = \omega_0 = vA' \exp(-H/kT) \qquad (85)$$

where v is the vibrational frequency of the charged particle and A' is a constant.

Under a field E in the direction of 1 to 2, the probabilities of transition will be modified, and since $ebE \ll kT$

$$\omega_{12} = \omega_0(1 + ebE/2kT), \qquad \omega_{21} = \omega_0(1 - ebE/2kT)$$

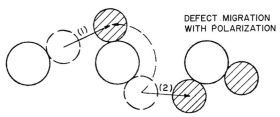

DEFECT MIGRATION
WITH POLARIZATION

Fig. 15.   Defect model by Charles (1961); large circle is oxygen, small circle is sodium.

The rate of change of the numbers of particles in positions 1 and 2 are

$$dN_1/dt = -N_1\omega_{12} + N_2\omega_{21} = -dN_2/dt \tag{86}$$

thus, the rate of change of the difference in numbers is expressed as

$$d(N_2 - N_1)/dt = -(N_2 - N_1)(\omega_{12} + \omega_{21}) + N(\omega_{12} - \omega_{21}) \tag{87}$$

and by using expressions for ω_{12} and ω_{21}

$$\frac{1}{2\omega_0} \frac{d(N_2 - N_1)}{dt} = -(N_2 - N_1) + \frac{NebE}{2kT} \tag{88}$$

This equation has a solution for a static field,

$$N_2 - N_1 = (ebEN/2kT)(1 - e^{-2\omega_0 t}) \tag{89}$$

The current density i is given by

$$i = \frac{d(N_2 - N_1)}{dt} be = \frac{Ne^2b^2E}{kT} \omega_0 e^{-2\omega_0 t} \tag{90}$$

The complex dielectric constant ε^* is given using Eq. (27),

$$\varepsilon_0\varepsilon^* = \varepsilon_0\varepsilon_\infty + \frac{e^2b^2N}{2kT} \frac{1}{1 + j(\omega/2\omega_0)} \tag{91}$$

This shows that the dielectric loss has a maximum at $\omega = 2\omega_0$, and the relaxation strength (step height of the dielectric constant) is given by

$$\varepsilon_0\varepsilon_s - \varepsilon_0\varepsilon_\infty = e^2b^2N/2kT \tag{92}$$

To apply this result to glass, b is identified as a statistical average of the projections on the field direction of all possible reorientations of an alkali ion around a nonbridging oxygen ion.

The defect concentration n (the number of nonbridging oxygen ions with two alkali ions per unit volume of glass) in glass can be calculated using the procedure for estimating the defect concentration in ionic crystals and is given by

$$\frac{n}{C} = \left(\frac{N_A'}{C}\right)^{1/2} \left[\frac{v_{Os}^2}{v_{Os}'v_{Os}''} \left(\frac{v_{OB}}{v_{OB}'}\right)^{x-y}\right]^{1/2} \exp\left(-\frac{W}{2kT}\right) \tag{93}$$

where C is the number of alkali metal ions in a unit volume of glass, which is equal to the number of nonbridging oxygen ions. N_A' is the number of equivalent sites around nonbridging oxygen ions. v_{Os}, v_{Os}', v_{Os}'' are vibrational frequencies of normal nonbridging ions, alkali-free nonbridging oxygen ions, and nonbridging oxygen ions with doubly occupied sites, respectively, v_{OB} and v'_{OB} are the vibrational frequencies of normal bridging oxygen and bridging oxygen around alkali ions. W is the energy required to form a defect. In the group of alkali jump, it is assumed that an alkali ion moves from a site which is surrounded by x bridging oxygen atoms to a site which is surrounded by y bridging oxygen atoms.

The probability q per unit time of the defect migration is given by

$$q = v_A \exp(-U/kT) \tag{94}$$

where v_A is the vibrational frequency of alkali ion and U is the energy barrier to migration. Thus, the probability that any alkali ion transfers from one site to another accompanied by a reorientation is

$$\omega_0 = \frac{n}{C} q = v_A \left[\frac{N_A'}{C} \frac{v_{Os}^2}{v'_{Os}v''_{Os}} \left(\frac{v_{OB}}{v'_{OB}} \right)^{x-y} \right]^{1/2} \exp\left[-\frac{W/2+U}{kT} \right] \tag{95}$$

Because of the factor $(v_{OB}/v'_{OB})^{(x-y)/2}$, there will be a distribution in the value of ω_0, the most probable value being obtained when $x = y$. Thus the most probable relaxation time τ_0 is given by

$$\tau_0 = \frac{1}{2\omega_0} = \frac{1}{2} \left\{ v_A \left[\frac{N_A'}{C} \frac{v_{Os}^2}{v'_{Os}v''_{Os}} \right]^{1/2} \right\}^{-1} \exp\left(\frac{W/2+U}{kT} \right) \tag{96}$$

The dielectric loss maximum is expected at $\omega = 1/\tau_0$.

The diffusion coefficient of alkali ions D is assumed to be given by

$$D = (\gamma^2/12)\Gamma \tag{97}$$

where γ is the jump distance and Γ is the jump frequency. In this expression, a factor of $\frac{1}{12}$ is used instead of usual $\frac{1}{6}$ since approximately one-half of all possible jump directions are blocked by nonbridging oxygen ions. The most probable jump frequency Γ_0 is related to the relaxation time τ_0 by

$$\Gamma_0 = 1/\tau_0 \tag{98}$$

In the original paper by Charles the distribution in Γ was considered.

Combining Eqs. (97) and (98) with the Nernst–Einstein relation, the conductivity σ of the glass is given by

$$\sigma = \frac{Ce^2}{kT} D = \frac{\gamma^2 Ce^2}{12kT} \frac{1}{\tau_0} \tag{99}$$

This equation relates dc conductivity and the frequency of dielectric loss maximum. The activation energies of the conductivity and the dielectric relaxation are therefore expected to be closely related.

3.   Conducting Path Model

Nakajima (1972) proposed what has subsequently been called the "conducting path model" which can correctly predict the frequency of the dielectric loss maximum. In this model he assumed that there are only two stable sites adjacent to each site and that the movement of charge carriers is restricted to a zig–zag path turning at random direction as is shown in Fig. 16a. Under an applied field, probabilities of a charge carrier making a transition will differ from site to site because of the different angles between each jump direction and the applied field. Therefore, the polarization current due to charge accumulation will be included in the conduction current.

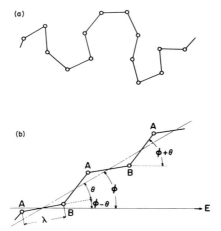

Fig. 16.   Conduction path model by Nakajima (1972). (a) Random zig-zag path; (b) regular zigzag path.

To simplify the calculation, a regular zig–zag path shown in Fig. 16b is considered, in which λ is the jump distance, ϕ is the angle between the path direction and the direction of an applied field E, θ is the angle between path direction and jump direction. If in this regular zig–zag model the angles ϕ and θ are varied to take all the possible values, the resultant current in the field direction could be equivalent to the current in the random zig–zag model.

In order to calculate the average current it is assumed that the heights of the potential energy barriers between sites are all equal (Fig. 17a). Therefore, in the absence of an applied field, the probability of transition of charge carriers, and the number of charge carriers are equal at every site.

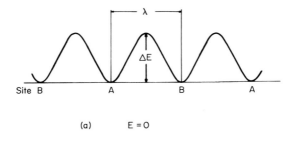

(a)   E = 0

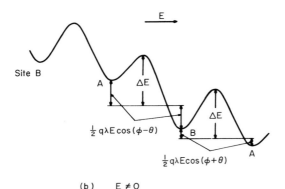

(b)   E ≠ 0

Fig. 17. Potential energy barrier in the regular zig–zag path. (After Nakajima, 1972.)

When a field E is applied the distribution of potential in the path varies as shown in Fig. 17b and the rates of change of the numbers of charge carriers in A and B are

$$\frac{dn_A}{dt} = -n_A P_1 - n_A P_4 + n_B P_2 + n_B P_3 \tag{100}$$

$$\frac{dn_B}{dt} = n_A P_1 + n_A P_4 - n_B P_2 - n_B P_3 \tag{101}$$

$$n_A + n_B = n_0 \tag{102}$$

where n_0 is the total number of charge carriers in the path, n_A and n_B are the numbers of charge carriers at any instance in A and B, respectively. P_1 and P_3 are the probabilities of transition from A to B and from B to A in the direction of the field, respectively, and P_2 and P_4 are those in the direction opposed to the field. These probabilities are given in the equations

$$P_1 = P(A \rightarrow B) = P_0 \exp\left[\frac{\lambda eE \cos(\phi - \theta)}{2kT}\right] \tag{103}$$

$$P_2 = P(A \leftarrow B) = P_0 \exp\left[-\frac{\lambda eE \cos(\phi - \theta)}{2kT}\right] \tag{104}$$

$$P_3 = P(B \rightarrow A) = P_0 \exp\left[\frac{\lambda eE \cos(\phi + \theta)}{2kT}\right] \tag{105}$$

$$P_4 = P(B \leftarrow A) = P_0 \exp\left[-\frac{\lambda eE \cos(\phi + \theta)}{2kT}\right] \tag{106}$$

where e is the charge of a carrier and P_0 is the probability of transition in the absence of the field.

The current in the direction of the field, contributed by a path direction making angle ϕ with the field direction, $i_{E,\phi}$, is given by

$$i_{E,\phi} = e\lambda \cos(\phi - \theta)(n_A P_1 - n_B P_2) + e\lambda \cos(\phi + \theta)(n_B P_3 - n_A P_4) \tag{107}$$

Substituting for n_A and n_B from Eqs. (100), (101), and (102) into Eq. (107) and averaging $i_{E,\phi}$ over the angle ϕ and θ, the average current i through a path is calculated as follows

$$i = \frac{n_0 e^2 \lambda^2 E P_0}{4kT} \left[1 + \exp(-4P_0 t)\right] \tag{108}$$

The first term in the right-hand side of Eq. (108) gives the steady-state conduction current and the second term the polarization current. Since there are numerous conduction paths in glass, the current density will be given by the above equation with n_0 being replaced by C, the number of charge carriers per unit volume. Therefore, σ and $\Delta\varepsilon$ are given by

$$\sigma = Ce^2\lambda^2 P_0/4kT \tag{109}$$

$$\varepsilon_0(\varepsilon_s - \varepsilon_\infty) = \varepsilon_0 \, \Delta\varepsilon = \int_0^\infty \frac{Ce^2\lambda^2 P_0}{4kT} e^{-4P_0 t} \, dt = \frac{Ce^2\lambda^2}{16kT} \tag{110}$$

Since $4P_0$ is equal to the reciprocal of the relaxation time, which corresponds to the dielectric loss maximum [cf. Eqs. (30), (33) and (108)], the following relation holds:

$$\sigma = 2\pi\varepsilon_0 \, \Delta\varepsilon f_m \tag{111}$$

which is identical with the experimentally observed relation, Eq. (50). This equation indicates that the activation energies of dc conductivity and the dielectric loss maximum are closely related.

4. Inhomogeneous Structure

Inhomogeneous structure can give rise to a dielectric relaxation. In many glass systems, glass-in-glass phase separation has been observed. Thus it appears reasonable to attempt to attribute the observed dielectric relaxation to the inhomogeneous structure, as was done by Isard (1962).

The theory of inhomogeneous dielectrics has been worked out for various microstructures. Maxwell (1892) considered a material made up of alternate layers perpendicular to the field with high and low conductivities. Wagner (1914) has discussed a material consisting of a dilute dispersion of highly conducting spheres in a matrix of low conductivity. Sillars (1937) extended the theory to the nonspherical dispersions.

Here the simplest inhomogeneous dielectric consisting of alternate laminae structure will be considered. Let us consider a two-layer dielectric with unit thickness in which the layers have the conductivities σ_1 and σ_2 and the dielectric constant at high-frequency ε_1 and ε_2 and the thickness fractions u_1 and u_2. An electric field E is applied perpendicularly to the planes of these layers. The initial distribution of the fields $E_1{}^0$ and $E_2{}^0$ in each layer satisfies the relationship

$$E_1{}^0/E_2{}^0 = \varepsilon_2/\varepsilon_1 \tag{112}$$

but this distribution varies with time; and under steady-state conditions, corresponding to the case when the conduction currents in the two layers are equal, the distribution satisfies the relation

$$\sigma_1 E_{1,\mathrm{st}} = \sigma_2 E_{2,\mathrm{st}} \tag{113}$$

where $E_{1,\mathrm{st}}$ and $E_{2,\mathrm{st}}$ are the steady-state electric field in layer 1 and 2, respectively. These relations show that a redistribution of the potential takes place in the two-layer structure when a constant voltage is applied. Under this condition, the current density is given by the sum of the conduction current density and the displacement current density. Thus

$$i = \sigma_1 E_1 + \varepsilon_1 \frac{dE_1}{dt} = \sigma_2 E_2 + \varepsilon_2 \frac{dE_2}{dt} \tag{114}$$

where the electric fields E_1 and E_2 are related to each other by

$$E_1 u_1 + E_2 u_2 = E \tag{115}$$

The solution of this equation for E_1 is given by

$$E_1 = \frac{\varepsilon_2 \sigma_1 - \varepsilon_1 \sigma_2}{(\varepsilon_1 u_2 + \varepsilon_2 u_1)(\varepsilon_1 u_2 + \varepsilon_2 u_1)} e^{-t/\tau} + \frac{E\sigma_2}{\sigma_1 u_2 + \sigma_2 u_1} \tag{116}$$

where τ is the relaxation time for the field distribution:

$$\tau = (\varepsilon_1 u_2 + \varepsilon_2 u_1)/(\sigma_1 u_2 + \sigma_2 u_1) \tag{117}$$

The expression for E_2 can be written in a similar form. The Eq. (116) gives the time dependency of the field in the layer 1. The steady state is reached at $t = \infty$

$$E_{1,\text{st}} = E\sigma_2/(\sigma_1 u_2 + \sigma_2 u_1) \tag{118}$$

The time dependency of the current is obtained from Eqs. (114) and (116).

$$i = \frac{(\varepsilon_2\sigma_1 - \varepsilon_1\sigma_2)^2 u_1 u_2 E}{(\varepsilon_1 u_2 + \varepsilon_2 u_1)^2(\sigma_1 u_2 + \sigma_2 u_1)} e^{-t/\tau} + \frac{\sigma_1\sigma_2 E}{\sigma_1 u_2 + \sigma_2 u_1} \tag{119}$$

The first term in Eq. (119) is known as the absorption (or polarization) current and was caused by the accumulation of transporting charge at the interface between the two layers and decays with time. The second term is the residual current which is related to the dc conductivity of the material. The discharge current in this material is given by

$$i = -\frac{(\varepsilon_2\sigma_1 - \varepsilon_1\sigma_2)^2 u_1 u_2 E}{(\varepsilon_1 u_2 + \varepsilon_2 u_1)^2(\sigma_1 u_2 + \sigma_2 u_1)} e^{-t/\tau} \tag{120}$$

Thus, the charge–discharge curve similar to Fig. 4 is obtained. Using the relation (27)

$$\varepsilon^* = \varepsilon_\infty + \frac{\sigma}{j\omega\varepsilon_0} + \frac{(\varepsilon_2\sigma_1 - \varepsilon_1\sigma_2)^2 u_1 u_2}{(\varepsilon_1 u_2 + u_2 u_1)^2(\varepsilon_1 u_2 + u_2 u_1)} \frac{\tau}{1 + j\omega\tau} \tag{121}$$

where σ is the dc conductivity of the specimen and is given by

$$\sigma = \sigma_1\sigma_2/(\sigma_1 u_2 + \sigma_2 u_1) \tag{122}$$

and ε_∞ is given by

$$\varepsilon_\infty = \frac{\varepsilon_1\varepsilon_2}{u_1\varepsilon_2 + u_2\varepsilon_1} \tag{123}$$

The relaxation strength (or step height of dielectric constant) $\Delta\varepsilon$ is given by

$$\Delta\varepsilon = \frac{(\varepsilon_2\sigma_1 - \varepsilon_1\sigma_2)^2 u_1 u_2}{(\varepsilon_1 u_2 + \varepsilon_2 u_1)(\sigma_1 u_2 + \sigma_2 u_1)^2} \tag{124}$$

To compare this model with the experimental observations, Isard assumed for simplicity that the layers of the material differ only in conductivity. Then

$$\frac{\tau}{(\varepsilon_0\varepsilon_\infty/\sigma)} = \frac{(p - 1)^2 r}{(p + r)^2} \tag{125}$$

and

$$\frac{\Delta\varepsilon}{\varepsilon_\infty} = \frac{(p - 1)^2 r}{(p + r)} \tag{126}$$

where p and r are the ratios of conductivities and the volume fractions respectively, i.e.,

$$p = \sigma_1/\sigma_2 \tag{127}$$

$$r = u_1/u_2 \tag{128}$$

To obtain agreement with the experimental observations, factors given by Eqs. (125) and (126) should be approximately equal to unity. By solving these simultaneous equations the values obtained are $r = u_1/u_2 = 2.62$ and $p = \sigma_1/\sigma_2 = 6.89$. Isard (1962) estimated that approximately $r = 1$ and $p = 10$ would give an agreement between theory and experiment.

5. SPACE CHARGE MECHANISM

The space charge mechanism was used by Joffe (1928) to explain the anomalous absorption current of various insulators. This mechanism is somewhat similar to the mechanism of inhomogeneous dielectrics in the sense that in both mechanisms the accumulation of the space charge is considered responsible for the absorption current. However, in this mechanism, no explicit structural assumption such as inhomogeneity or conducting paths is made. Instead, the space charge accumulates at or near the electrodes. The same mechanism has been proposed for the electrode polarization and the approximate solution at the low-frequency limit was given earlier. It was suggested (Doremus, 1973) that the same mechanism, under different boundary conditions, can explain the conductivity-related dielectric relaxation. What follows is analogous to the theory of mechanical relaxation based upon the same mechanism by Doremus (1970).

Basic equations used in the space charge mechanism are Eqs. (51), (52), and (53). In the frequency range where the conductivity related dielectric loss peak is observed, it is assumed that the excess ionic charge concentration is small relative to the average ionic charge concentration. Then, if the fractional excess charge q is defined as $q = (P/C) - 1$, the above condition means that $q \ll 1$. Another assumption used is that the association and dissociation of the cations are fixed, and anionic group is not a factor in the ionic transport in the glass. Under these conditions, Eqs. (51), (52) and (53) are simplified to

$$j = -CD\left(\frac{\partial q}{\partial x} + \frac{e}{kT}\frac{\partial V}{\partial x}\right) \tag{51a}$$

$$\frac{\partial q}{\partial t} = D\frac{\partial^2 q}{\partial x^2} - DK_1^2 q \tag{52a}$$

with

$$K_1^2 = Ce^2/kT\varepsilon_0\varepsilon_\infty$$

and

$$\partial^2 V/\partial x^2 = -Ceq/\varepsilon_0\varepsilon_\infty \tag{53a}$$

Equation (52a) is identical to the equation obtained by Volger *et al.* (1953).

Using this simplified diffusion conduction equation, under our appropriate boundary conditions, it is possible to obtain an expression for dielectric relaxation. The result (Doremus, 1973) shows a broad relaxation peak which is related to dc conductivity of the specimen. However, because of the very nature of the model, i.e., space charge accumulation at the specimen–electrode interface, the magnitude of the resulting relaxation peak becomes dependent upon the spcimen thickness. The experimental observation on the other hand, indicates that the dielectric relaxation under consideration is thickness independent.

6. OTHER MODELS

In addition to models discussed so far, there are several other models by which attempts were made to explain various features of the dielectric relaxation. One of the features which many investigators appear to regard as important is the extremely broad nature of the dielectric relaxation. The broad peak is normally associated with the distribution of the relaxation time. Thus, attempts were made to explain the broad distribution of the relaxation time in terms of the glass structure. However, since this distribution is temperature independent, the activation energy of these relaxation times has to be the same. To accomodate these experimental observations, Tomandl (1974) as well as Aitken and MacCrone (1975) proposed that there exist various lengths of chains through which charged species can migrate, the energy barrier of the jump being constant at all points of the chain. By assuming the distribution in the chain lengths, the distribution in the relaxation was obtained.

Boksay and Lengyel (1974), on the other hand, proposed a vacancy type mechanism which is somewhat similar to Nakajima's conducting path model. In Boksay and Lengyel's model, charge carriers migrate by vacancy mechanism and the pathways of vacancies have junctions and impasses. Thus polarization current, in addition to the dc conduction current, can flow in glass.

V. Discussion

Various models discussed in the previous section, at first glance, appear to be able to explain the basic features of the experimental observations. Also, some of the proposed models for the conductivity related dielectric loss peak appear to give similar predictions. However, in details, there are several important differences. In this section the models will be critically compared with the experimental results to see which model, if any, is the correct one. Again two dispersions, the electrode polarization and the conductivity related dielectric relaxation will be discussed separately.

A. Electrode Polarization

The fact that the Au and Ag electrodes gave the same results (cf. Figs. 7 and 9) eliminates the possibility of a glass–electrode reaction (Mitoff and Charles, 1972, 1973) since the two elements differ in their ability to react with glass. The reaction of the soda–lime glass with electrode materials under the influence of an ac electric field was investigated (Kim and Tomozawa, 1976b). At 400°C, Ag electrodes reacted with the glass while no reaction was observed for Au electrodes. Thus, if the glass–electrode reaction is causing the large dielectric constants observed, there should be a clear difference between data obtained with Ag electrodes and Au electrodes.

By the same reason, the contribution from the oxide layer (Macedo et al., 1972) is negligible, since the oxidation characteristics of Au and Ag are quite different.

The space charge mechanism can satisfactorily explain the observed phenomenon. It is known (Bach and Baucke, 1974) that the glass reacts with atmospheric water, producing an alkali–hydroxide layer and an interdiffusion layer. This alkali–hydroxide layer is loosely bonded to the glass surface and can be easily removed by wiping the glass surface with, for example, a moist tissue (Bach and Baucke, 1974). The metal evaporated film appears to stay on top of this loosely bonded alkali–hydroxide layer since the film can also be easily removed. On the other hand, the paste electrodes adhere strongly to the glass specimens, suggesting that the paste electrodes are directly bonded to the glass rather than to alkali–hydroxide layer. It is likely that the paste electrodes eliminate the loosely bonded alkali–hydroxide layer during the curing process (160°C, 1 hr. and 260°C, 1 hr). Thus, the paste electrode is expected to behave as a blocking electrode while the evaporated metal electrode will behave as a partially blocking electrode because of the surface alkali–hydroxide. With the evaporated metal electrodes the surface condition of the specimen will influence the charge transport across the specimen surface and consequently the dielectric characteristics. For example, the presence or absence of the surface hydroxide layers will influence the availability of sodium ions from the surface. However, a large amount of hydroxide does not necessarily indicate large surface flux because when the surface of glass reacts with atmosphere water an interdiffusion layer will be created (Doremus, 1976), and this interdiffusion layer has a higher electric resistivity than the bulk glass (Wikby, 1974). A schematic structure of the glass surface in this situation is shown in Fig. 18, where the flux of charge

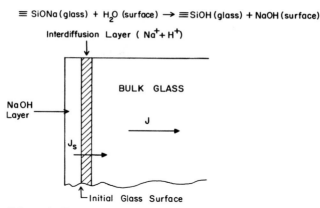

Fig. 18. Schematic diagram for glass surface structure. (After Tomozawa *et al.*, 1975.)

under a dc electric field is also indicated. Thus, the extensive reaction between glass surface and the atmosphere water will make the charge transport difficult, making the specimen behave as if it has the blocking electrode. These expectations were borne out as shown in Figs. 19 and 20. Fig. 19 compares the dielectric behavior of $Na_2O–SiO_2$ glass and $Na_2O–Al_2O_3–SiO_2$ glass and Fig. 20 shows the infrared spectrographs of the corresponding glasses in powdered form. $Na_2O–SiO_2$ glass, which is known to be more

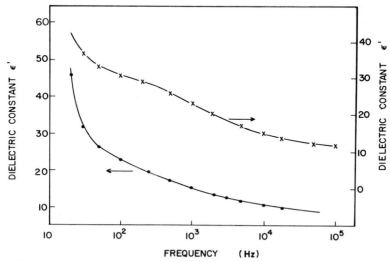

Fig. 19. Dielectric constant of $25Na_2O–75SiO_2$ (●) and $25Na_2O–15Al_2O_3–60SiO_2$ (×) glasses (vertical axis displaced for clarity); compositions in mole %; temperature $66 \pm 1°C$. (After Tomozawa *et al.*, 1975.)

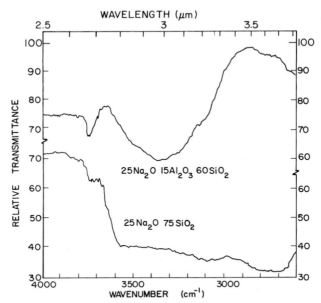

Fig. 20. ir transmittance of glasses corresponding to Fig. 19 (vertical axis displaced for clarity). (After Tomozawa *et al.*, 1975.)

hygroscopic, shows a stronger absorption at 3500 cm^{-1} indicating the presence of the larger amount of the surface water. This probably corresponds to the extensive glass surface–atmosphere water reaction and the high-resistance interdiffusion layer. Correspondingly, this glass shows more extensive electrode polarization; so much so that the conductivity related dispersion was completely smeared out (Fig. 19). The effect of the adsorbed water on the electrode polarization was pointed out also by Ono and Munakata (1974).

The effects of the specimen thickness and the frequency on the dielectric constant is shown in Figs. 21 and 22. As can be seen from Fig. 21, the thinner specimen gave the higher dielectric constant. At lower frequency the dielectric constant of the soda–lime glass changes approximately as $1/\omega^2$, as shown in Fig. 22.

The effect of temperature on the electrode polarization is shown in Fig. 23. Here the frequency corresponding to a certain constant value of the large dielectric constant (i.e., $\varepsilon' = 500$) is plotted against the reciprocal of temperature, together with the dc conductivity and the frequency of the conductivity related dielectric relaxation maximum. The slopes of these three curves are identical within experimental error, and yield an activation energy of 19.9 kcal/mole.

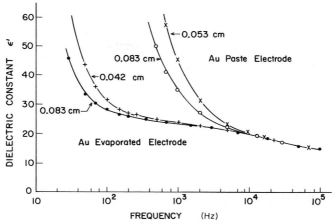

Fig. 21. Thickness dependence of dielectric constant for soda–lime glass at 247°C. Specimens were polished up to 600 grit. (Kim and Tomozawa, 1976a.)

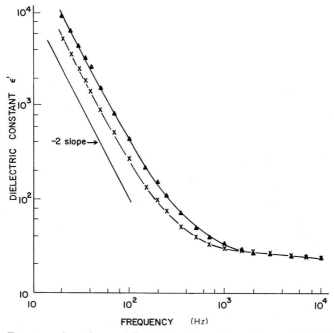

Fig. 22. Frequency dependency of dielectric constant for soda–lime glass at 312°C. Specimens were polished up to 600 grit (▲ = 0.039 cm thick, × = 0.090 cm thick). Au vacuum evaporated electrode. (After Kim and Tomozawa, 1976a.)

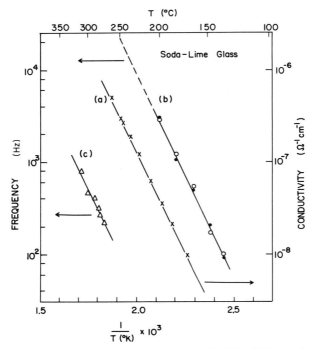

Fig. 23. Temperature dependence of electrode polarization ($\triangle$); dielectric relaxation ($\bullet$, $\bigcirc$); and dc conductivity ($\times$). Dielectric data were obtained with Au evaporated film electrode and dc conductivity with Na_2NO_3–$Cd(NO_3)_2$ eutectic melt electrode; $\triangle E = 19.9$ kcal/mole. Peak frequency of dielectric relaxation was obtained from both dielectric constant ($\bigcirc$) and dielectric loss ($\bullet$). (After Kim and Tomozawa, 1976a.)

It is possible to explain the observed experimental results by Eq. (82). For example, when the second term of the denominator of Eq. (82) is negligible compared with the first term, the dielectric constant is expected to be reciprocally proportional to thickness (Fig. 21) and change with $1/\omega^2$ (Fig. 22).

The angular frequency ω_{ep} corresponding to a constant dielectric constant in the electrode polarization region is given from Eqs. (82) and (62), using $\varepsilon' \gg \varepsilon_\infty$,

$$\omega_{ep} = \left[\frac{2\varepsilon_\infty K_1 L}{\varepsilon'} - (2 + \rho)^2 \right]^{1/2} \frac{\sigma}{\varepsilon_0 \varepsilon_\infty K_1 L} \qquad (129)$$

where ε' is the constant value chosen. Thus, it is clear that when ρ and K_1 are temperature independent, the activation energy calculated from the frequency–temperature relation of the large dielectric constant should be the same as that for the transport phenomenon (cf. Fig. 23). The variation of

K_1 and ρ with temperature is small compared with the exponential variation of σ with temperature.

In some cases, the thickness dependency of the dielectric constant was found to be reversed as the frequency is varied. For example, for 25 mole % Na_2O–SiO_2 glass with Au paste electrodes, the dielectric constant was larger for thinner specimen at higher frequency but was larger for thicker specimen at lower frequency. The measurement of the same glass with the Au-evaporated-film electrode did not show this reversal. These phenomena are shown in Fig. 24. The frequency corresponding to the crossover of the dielectric constants for two different thickness specimens with Au paste electrodes changed with the temperature, shifting to higher frequency at higher temperature. The temperature dependency of the crossover frequency is shown in Fig. 25 and yielded the activation energy of 13.9 kcal/mole, approximately the same activation energy as that of the dc conductivity of this glass. These features can also be explained by using Eq. (82).

For a partially blocking electrode with large surface charge flux ($\rho \gg 2$),

$$\varepsilon' = \frac{2\varepsilon_\infty[(\sigma/\varepsilon_0\varepsilon_\infty)^2/\omega^2 K_1 L]}{1 + [\rho^2(\sigma/\varepsilon_0\varepsilon_\infty)^2/\omega^2 K_1^2 L^2]} \tag{130}$$

Since $\rho \propto L$ (Eq. (56)), the thicker specimen should have a smaller dielectric constant at all frequencies. This is consistent with the experimental results obtained with Au-evaporated electrodes.

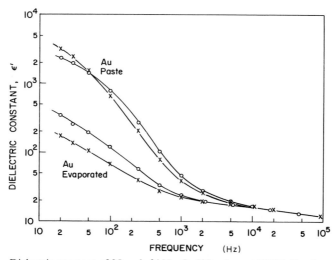

Fig. 24. Dielectric constant of 25 mole % Na_2O–SiO_2 glass at 109°C. Specimens were polished up to 600 grit ($\bigcirc$ = 0.059 cm thick; $\times$ = 0.128 cm thick). (Kim and Tomozawa, 1976a.)

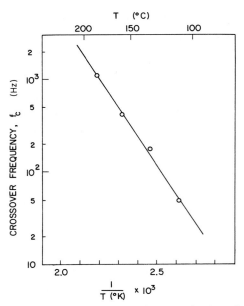

Fig. 25. Temperature dependence of crossover frequency for 25 mole % $Na_2O–SiO_2$ glass with Au paste electrode. (After Kim and Tomozawa, 1976a.)

For blocking electrodes ($\rho = 0$)

$$\varepsilon' = \frac{2\varepsilon_\infty(\sigma/\varepsilon_0\varepsilon_\infty)^2/\omega^2 K_1 L}{1 + 4(\sigma/\varepsilon_0\varepsilon_\infty)^2/\omega^2 K_1{}^2 L_2} \tag{131}$$

Thus when the second term of the denominator is negligible it is expected that the thicker specimen would give a smaller value of the dielectric constant, similar to the partial blocking electrode case. However, at extremely low frequency where the second term of the denominator becomes predominant, the thicker specimen is expected to have the larger dielectric constant. Namely, when the surface flux is not too large, the reversal of the thickness effect should take place depending upon the frequency, as shown in Fig. 24. With a blocking electrode ($\rho = 0$), the dielectric constants of specimens with different thickness L_1 and L_2 would crossover at an angular frequency ω_c given by

$$\omega_c = 2\pi f_c = \frac{2(\sigma/\varepsilon_0\varepsilon_\infty)}{K_1(L_1 L_2)^{1/2}} \tag{132}$$

The temperature dependency of the crossover frequency should yield the same activation energy as for alkali ion transport if K_1 is temperature

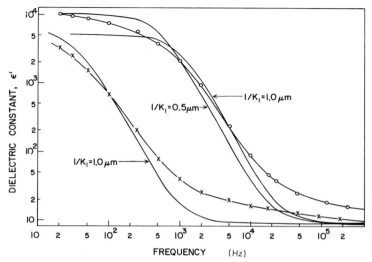

Fig. 26. Comparison of measured dielectric constants of 25 mole % $Na_2O–SiO_2$ glass (0.128 cm thick) with theory. (—) theory; (○) measured at 185°C; (×) measured at 109°C. (After Kim and Tomozawa, 1976a.)

independent. This analysis for 25 mole % $Na_2O–SiO_2$ glass with the Au paste electrode (Fig. 25), gave an activation energy of 13.9 kcal/mole which is slightly lower than the value obtained from the dc conductivity (14.9 kcal/mole). This small discrepancy may be due to the temperature-dependent K_1 in the expression.

So far, qualitative agreement between the theory of the space charge and experiment have been described. Quantitatively, there is some discrepancy between the theory and the experiment. To have a quantitative agreement, Debye length[†] has to be adjusted. The comparison of the experimental results with the theory is given in Fig. 26 for 25 mole % $Na_2O–SiO_2$ glass, where the theoretical curves were obtained by adding $\varepsilon_\infty = 9$ to the excess dielectric constant obtained from Eq. (131). The comparison shows that $1/K_1$ of ~ 1 μm is necessary for this glass at 109°C. Also the magnitude of $1/K_1$ appears to be slightly temperature dependent. A similar analysis of the experimental results for soda–lime glass gives a $1/K_1 \sim 250$ Å at 247°C. Use of Eq. (62), on the other hand, gives $1/K_1$ smaller than 1 Å. This discrepancy may suggest that the number of charge carriers in glass is much smaller than the nominal alkali concentration. In addition, the large value

---

[†] In the original paper (Kim and Tomozawa 1976a), $1/K_1$ was called the Debye length, while the correct definition of the Debye length $1/K$ for alkali ion containing glass is $(1/\sqrt{2})(1/K_1)$ (Moore, 1962).

of $1/K_1$ may result because of the high-resistance (interdiffusion) surface layer of glass produced by glass–water reaction (Wikby, 1974).

Thus it appears that the essential feature of the electrode polarization is explained by the space charge mechanism. However, by the very nature of the space charge mechanism, the dielectric characteristics is strongly influenced by the specimen surface conditions which influence the charge transport across the specimen surface. Thus, it is difficult to explain the electrode polarization quantitatively without the detailed knowledge on the structure and the properties of glass surface. On the other hand, the measurement of the electrode polarization can be a useful tool to study the glass surface (Tomozawa et al., 1975).

B.   Dielectric Relaxation

Stevels' model (Section IV,B,1) with various heights of potential energy, although it is attractive in view of the random network structure of glasses, leads to different activation energies for dc and ac conductions, while experimentally the same activation energy has been observed.

Both Charles' defect model (Section IV,B,2) and Nakajima's conducting path model (Section IV,B,3) give an expression in which the peak frequency of the dielectric relaxation is related to the dc conductivity. In particular Nakajima's expression

$$f_m = \sigma/2\pi\varepsilon_0 \, \Delta\varepsilon \tag{111}$$

was shown by Namikawa et al. (Namikawa, 1969; 1974; 1975a,b; Namikawa and Asahara, 1966; Namikawa and Kumata, 1968; Yamamoto et al., 1974) and Tsuchiya and Moriya (1973; 1974a,b; 1975), to predict correctly at least in the order of magnitude, the peak frequency. Barton (1966, 1967) also showed equivalently that the experimentally observed relaxation strength $\Delta\varepsilon$ is approximately proportional to $\sigma\tau/\varepsilon_0$, where τ is the relaxation time corresponding to the peak frequency, namely $2\pi f_m \tau = 1$. However, these observations do not necessarily support Eq. (50) or (111) (Nakajima's model) over Eq. (49) or (125) (Isard's model), since, as was pointed out earlier, in general $\Delta\varepsilon \simeq \varepsilon_\infty$. Also a fair amount of error can exist in the experimental evaluation of the peak frequency f_m since the evaluation of the dielectric loss involves the subtraction of the dc conduction, which constitutes a substantial fraction of the total loss.

The better, perhaps more reliable, test of the models can be made by evaluating the temperature dependency of various measured quantities. Especially, the activation energies of conductivity and the peak frequency of the dielectric relaxation can be evaluated comparatively accurately since normally three or more data points are used to obtain an activation energy.

Since $\Delta\varepsilon \propto 1/T$, both Charles' model and Nakajima's model lead to the relation

$$f_m \propto \sigma T \tag{133}$$

Therefore, if these models were correct a slight difference should exist between the activation energy for dc conduction and that for the peak frequency. The difference in activation energy should be $\simeq RT$ which is ~ 1 kcal/mole at $T \simeq 500$ K. Namikawa (1957b) believes that this difference is within experimental error. By now, however, there are enough experimental data to enables us to see whether or not a slight difference in activation energies exists. In Fig. 27, the data on the activation energy compiled by Namikawa (1975b) are shown. Although there is a certain amount of scatter in the data the average (solid line) appears to indicate that there is no difference in activation energy. A careful experiment in our laboratory on this particular point gave an exact agreement, shown as a circle in the diagram (Kim, 1976). To see the effect of RT on the activation energy, the activation energies were compared for the diffusion coefficient D and the dc conductivity σ which were related by the analogous expression,

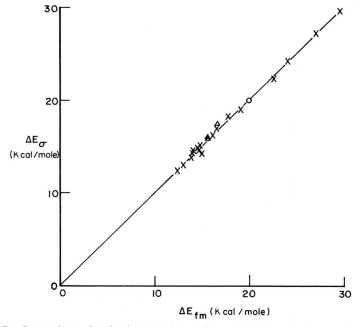

Fig. 27. Comparison of activation energies of dc conductivity and dielectric loss peak. $\bigcirc$ = Kim (1976), $\times$ = Charles (1963, 1966), $\triangle$ = Heroux (1958). Compiled by Namikawa (1975b.)

i.e., the Nernst–Einstein relation, which can be written as

$$D \propto \sigma T \tag{134}$$

The difference of RT in activation energy is expected for D and σ. The comparison of the activation energies for these quantities is shown in Fig. 28. In this case most data fall on the dotted line which indicates a difference of 1 kcal/mole. These analyses indicate that the difference on the order of RT in activation energies can be resolved and, therefore, there is no such difference between the activation energy of dc conductivity and that of the dielectric relaxation peak frequency. Thus, the expression obtained by Charles' and Namikawa's models appears not applicable to the experimental observation. In fact all models based upon the orientation polarization type mechanism, in which the only restoring energy from the polarization is the thermal energy, will give rise to the similar expression and not be adequate.

Isard's inhomogenous model (Section IV,B,4), on the other hand, predicts the relation of the type

$$f_m = \sigma/2\pi\varepsilon_0\varepsilon_\infty \tag{125a}$$

which is identical with the experimentally observed Eq. (49). Since ε_∞ is temperature independent, the activation energies of the peak frequency and

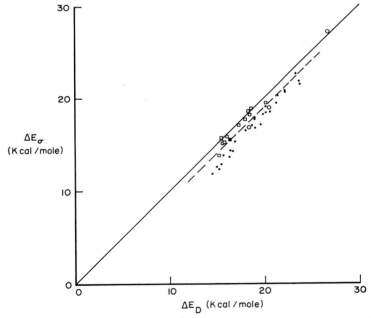

Fig. 28. Comparison of activation energies of dc conductivity and diffusion coefficient. ● = Terai (1969), □ = Heckman *et al.* (1967), ○ = Friauf (1957).

the conductivity are expected to be exactly identical, consistent with experimental observations. However, in this model there are several assumptions, some of which if realized are almost fortuitous. For example, to obtain the agreement with the experimental results Isard (1962) assumed two-phase structures with 50/50 volume fractions and one order of magnitude difference in conductivities. It is true that some glasses are phase separated and can satisfy these requirements. However the dielectric relaxation under consideration is, in general, insensitive to the phase separation and is observed in a glass in which phase separation is clearly absent. In Fig. 29 x-ray small-angle scattering of a commercial soda–lime glass (Na_2O 15 wt %) is compared with that of fused silica in absolute scale. The absolute intensity scale was calibrated by attenuation method (Shaffer and Hendricks, 1974). There is a slight uncertainty in the absolute intensity measurement and the scattered intensity of fused silica has been reported in the range of 5 ~ 15 e.u. (Weinberg, 1963; Bates et al., 1974). The scattered intensity of fused silica is usually attributed to density fluctuation (Weinberg, 1963; Zarzycki, 1974). The scattered intensity of soda–lime glass is observed to be smaller than that for fused silica; and it is unlikely that this glass has a phase separated structure. Assuming that this has a two-phase structure with different Na_2O contents, in order to have an order of magnitude difference in conductivity,

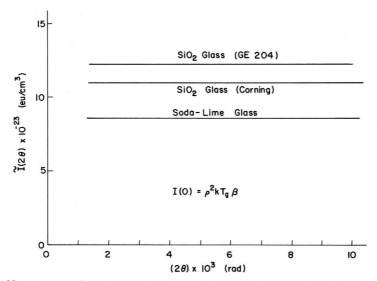

Fig. 29.   X-ray small angle scattering of SiO_2 glasses and a soda–lime glass in absolute intensity scale. The scattered intensity due to density fluctuation is given by $I(0) = \rho^2 k T_g \beta$, where $I(0)$ is the scattered intensity at zero angle, ρ is the electron density, k Boltzmann's constant, T_g the glass transition temperature, and β isothermal compressibility.

it is necessary to have a difference in Na_2O content of ~ 10 wt % (Shand, 1958). By assuming 50/50 volume fractions of 10% Na_2O and 20% Na_2O phases (with corresponding variations in SiO_2 content) the electron density difference was calculated to compare with the scattered intensity. It turned out that with this structure it is impossible to have a phase greater than 10 Å in diameter, even when the contribution of the density fluctuations to the scattered intensity was neglected. Thus, it is unlikely that this glass has an inhomogeneous structure proposed by Isard (1962).

The space charge mechanism proposed by Joffe (1928) and by Doremus (1973) is a universal mechanism, which is not confined to glass. However, since it attributes the dielectric relaxation to the phenomenon at a specimen surface, it leads to a thickness dependent relaxation strength (or peak height), while experimental results show a thickness independent peak height. Thus, this model is also inadequate.

Other proposed models such as truncated chain structure, can explain some features (e.g., a broad peak), but they fail to discuss the important experimental observation, namely the peak frequency—dc conductivity relation. If the high conducting chains are truncated charge transfer between chain ends would be difficult compared with charge motion in a chain, leading to a higher activation energy for dc conduction than for ac conduction. Isard (1970), to overcome this difficulty, proposed a fluctuating potential energy barrier. Still, it appears unlikely in this model that the activation energy is exactly the same both in dc and ac conductions. Thus, all the proposed models appear unsatisfactory.

As was pointed out earlier, the observed dielectric relaxation appears universal, the same phenomenon being observed in polycrystalline ceramics as well as in single crystals. Therefore, the correct mechanism has to reflect this universal nature. If the model requires a particular structure which is unique to glasses, that model would probably not be a correct one. Mechanisms such as the conducting path model, the defect model, and the inhomogeneous model are particular to glasses. It is hard to imagine these mechanisms in a material such as single-crystal NaCl. The only mechanism with the required universal characteristics is the space charge mechanism, since it can exist in any material as long as a movable charge carrier exists. However, as was indicated earlier, the space charge mechanism has the drawback of predicting the thickness-dependent dielectric loss. The correct model has to predict the thickness-independent loss peak while maintaining the versatility of the space charge mechanism. The thickness dependence of the space charge mechanism originates from the fact that the space charge is created at the specimen surface. The thickness-independent dielectric loss indicates that the phenomenon we are observing is the bulk phenomenon rather than the surface related phenomenon.

Any charged species has Coulombic interaction with other charged species. Thus, when an electric field is applied to a material containing charges the Coulombic interaction will be affected, producing a situation similar to the space charge but in much finer scale. It might be possible to call this process the "local space charge." This mechanism is essentially the same as the phenomenon in electrolyte solutions and can give rise to a dielectric relaxation related to the conductivity of the specimen. The details of the theory on electrolyte solutions have been worked out by Debye and Falkenhagen (Falkenhagen, 1934; Harned and Owen, 1958). It had been suggested earlier by Owen (1963) that this Debye–Falkenhagen theory may be applicable to the dielectric relaxation of glass. Isard (1962) on the other hand, dismissed this model since the concentration of alkali ions in glass is far greater than the charge concentration in aqueous solution for which the electrolyte theory is normally found applicable. In the following the Debye–Falkenhagen theory will be reproduced with a proper modification applicable to glasses.

1. DEBYE–FALKENHAGEN THEORY

a. Ionic Atmosphere. According to the theory of ionic solution by Debye and Hückel (Bockris and Reddy, 1970), the distribution of ions is affected by forces acting on them. Since Coulomb forces act between all pairs of ions, the presence of an ion at a given point in the solution will affect the space distribution of the other ions in its immediate vicinity. For example, each negative ion will be surrounded by an "atmosphere" which contains on the average more positive ions than negative ions. In the absence of the external field, the concentration in the "atmosphere" will have a spherical symmetry around each ion. [In the present example of glass containing alkali (e.g., sodium) ions, the distribution of alkali ions around an oxygen (O^-) site may not be perfectly spherically symmetric, but, on an average of numerous oxygen (O^-) sites, alkali distribution can be considered spherically symmetric.] This is schematically shown in Fig. 30, using Na^+ as an example.

An oxygen ion is chosen as a reference point. Then the potential around the oxygen ion is expected to be spherically symmetric. The potential ψ^0 in a small volume element at a distance r from the central oxygen ion, in the absence of the external field, is

$$\nabla^2 \psi^0 = -(e/\varepsilon_0\varepsilon_\infty)(P^0_{Na^+} - P^0_{O^-}) \tag{53b}$$

where $P^0_{Na^+}$ and $P^0_{O^-}$ represent the distribution of Na^+ and O^- at the point around the oxygen ion, respectively, and e is the charge of the sodium ion $(+ve)$. This leads to the potential ψ^0 and the Na^+ distribution $P^0_{Na^+}$ around

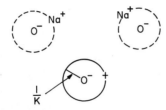

Fig. 30.   Schematic diagram of ionic "atmospheric" in Na containing glass.

the oxygen ion,

$$\psi^0 = -ee^{-Kr}/\varepsilon_0\varepsilon_\infty r \tag{135}$$

$$P_{Na^+}^0 \simeq C[1 - (\psi^0 e/kT)] = C[1 + (e^2/\varepsilon_0\varepsilon_\infty kT)(e^{-Kr}/r)] \tag{136}$$

where C is the average concentration of Na^+ (or O^-), r is the distance from the oxygen ion, k Boltzmann's constant, T the absolute temperature, and K is the reciprocal of average radius of ionic atmosphere often called the Debye length, and is defined by

$$K^2 = 2Ce^2/\varepsilon_0\varepsilon_\infty kT = 2K_1^2 \tag{62}$$

 b.   Differential Equations.   When an external field is applied the local potential ψ as well as Na^+ distribution P_{Na^+} is disturbed. (The distribution of O^- remains unchanged since O^- is immobile.) Thus

$$\psi = \psi^0 + \psi' \tag{137}$$

$$P_{Na^+} = P_{Na^+}^0 + P_{Na^+}' \tag{138}$$

where primed terms represent the effects of the external field. The potentials and the distributions are related according to Poisson's equation in the following manner,

$$\nabla^2\psi' = -(e/\varepsilon_0\varepsilon_\infty)P_{Na^+}' \tag{53c}$$

Under these conditions, the flux $\mathbf{J}$ of Na^+ ions is given by

$$\mathbf{J} = P_{Na^+}\mathbf{v} = P_{Na^+}\mathbf{F}B^0 \tag{139}$$

where $\mathbf{v}$ is the velocity and B^0 is the mobility of Na^+ when no ionic interaction exists or the mobility at infinite dilution. Force $\mathbf{F}$ on an Na^+ ion is given by

$$\mathbf{F} = \mathbf{E}e - e\,\nabla\psi_{Na^+}'(0) - e\,\nabla\psi - kT\,\nabla\ln P_{Na^+} \tag{140}$$

where $\mathbf{E}$ is the applied field. The second term on the right-hand side of Eq. (140) is the force on the Na^+ ion due to perturbation of its own atmosphere and is small compared with other terms. The third term is the force acting on the Na^+ ion due to the potential of O^- ion (as origin) and its atmosphere. The last term is the force due to the concentration gradient. Substituting Eqs. (139) and (140) in the continuity equation,

$$-\partial P_{Na^+}/\partial t = \mathbf{V} \cdot \mathbf{J} \tag{141}$$

the following equation results,

$$\partial P_{Na^+}/\partial t = -B^0 Ee \cdot \mathbf{V}P_{Na^+} + eB^0 C \, \nabla^2\psi + D^0 \, \nabla^2 P_{Na^+} \tag{142}$$

where D^0 is the diffusion coefficient of Na^+ ions at infinite dilution and is equal to $B^0 kT$. The equalities and approximations used were

$$\text{term containing } \nabla\psi'_{Na^+}(0) \simeq 0, \qquad \mathbf{V} \cdot \mathbf{E} = 0$$

$$eB^0 P_{Na^+} \, \nabla^2\psi \simeq eB^0 C \, \nabla^2\psi, \qquad eB^0 \, \mathbf{V}P_{Na^+} \cdot \mathbf{V}\psi \simeq 0$$

Further modification is made with the use of Eqs. (137) and (138). Since

$$P_{Na} \gg P'_{Na^+}$$

and at equilibrium without an external applied field

$$\partial P_{Na^+}/\partial t = 0 = eB^0 C \, \nabla^2\psi^0 + D^0 \, \nabla^2 P^0_{Na^+} \tag{143}$$

Eq. (142) becomes

$$\partial P_{Na^+}/\partial t = -B^0 Ee \cdot \mathbf{V}P^0_{Na^+} + eB^0 C \, \nabla^2\psi' + D^0 \, \nabla^2 P'_{Na^+} \tag{144}$$

When the electric field $\mathbf{E}$ is directed along the x axis

$$B^0 Ee \cdot \mathbf{V}P^0_{Na^+} = B^0 Ee(\partial P^0_{Na^+}/\partial x) \tag{145}$$

Using Eqs. (53c) and (145), Eq. (144) becomes

$$\frac{\partial P_{Na^+}}{\partial t} = -B^0 Ee \frac{\partial P^0_{Na^+}}{\partial x} - \frac{e^2 B^0 C}{\varepsilon_0 \varepsilon_\infty} P'_{Na^+} + D^0 \, \nabla^2 P'_{Na^+} \tag{146}$$

When a small ac field $E = E_0 e^{j\omega t}$ is applied, it is assumed that the variation of the Na^+ distribution is also periodic. Thus,

$$P_{Na^+} = P^0_{Na^+} + G e^{j\omega t} \tag{147}$$

where

$$G e^{j\omega t} = -(\varepsilon_0 \varepsilon_\infty/e) \, \nabla^2\psi' \tag{148}$$

Substituting these relations in Eq. (146)

$$\nabla^2 G - m^2 G = (E_0 e/kT)(\partial P^0_{Na^+}/\partial x) \tag{149}$$

where

$$m^2 = (j\omega/D^0) + (e^2 C/\varepsilon_0 \varepsilon_\infty kT) \qquad (66a)$$

or

$$m^2 = K_1{}^2(1 + j\omega\tau) \qquad (150)$$

where K_1 was defined by Eq. (62), and

$$\tau = 1/D^0 K_1{}^2 = \varepsilon_0\varepsilon_\infty/\sigma^0 \qquad (151)$$

σ^0 is the conductivity corresponding to D^0. Substituting Eqs. (136) and (148) in Eq. (149)

$$\nabla^4\psi' - m^2\nabla^2\psi' = -(E_0 e^2/kT\varepsilon_0\varepsilon_\infty)K_1{}^2(\partial(e^{-Kr}/r)/\partial x)e^{j\omega t} \qquad (152)$$

This has the solution for small values of **r**

$$\psi' = \frac{E_0 e^2}{6kT\varepsilon_0\varepsilon_\infty} \frac{K^2}{(K + m)} \mathbf{r}e^{j\omega t}\cos\theta \qquad (153)$$

where θ is the angle between the direction of the field and the radius vector **r**. In the case of a solution with only two kinds of ions, such as oxygen (O^-) and sodium (Na^+) the identical expression as Eq. (153) is applicable to the perturbation of the potential around a sodium ion, also.

From this equation, the perturbation of the field at the position of Na^+ for $\theta = 0$ may be obtained as

$$\Delta E = \Delta E_0 e^{j\omega t} = -\nabla\psi'(r \simeq 0) = -(E_0 e^2/2kT\varepsilon_0\varepsilon_\infty)[K^2/(K + m)]e^{j\omega t} \quad (154)$$

c. Dielectric Characteristics. Velocity of Na^+ ion $ve^{j\omega t}$ is given by

$$ve^{j\omega t} = (E_0 eB^0 + \Delta E_0 eB^0)e^{j\omega t} \qquad (155)$$

where the effect of electrophoresis is neglected since O^- is immobile. Using Eq. (154)

$$ve^{j\omega t} = E_0[eB^0 - (e^3 B^0 K/3kT\varepsilon_0\varepsilon_\infty)\chi]e^{j\omega t} \qquad (156)$$

where

$$\chi = \frac{1}{2}\frac{K}{K + m} = \frac{1/2}{1 + (1/\sqrt{2})(1 + j\omega\tau)^{1/2}} \qquad (157)$$

The current density due to conduction is given by

$$i_c = Cve^{j\omega t} = Ce^2 E_0 e^{j\omega t}[B^0 - (e^2 B^0 K/3kT\varepsilon_0\varepsilon_\infty)\chi] \qquad (158)$$

The total current density i is equal to the sum of the conduction current and the displacement current, i.e.,

$$i = j\omega\varepsilon_0\varepsilon_\infty E + Ce^2 E[B^0 - (e^2 B^0 K/3kT\varepsilon_0\varepsilon_\infty)\chi] \qquad (159)$$

Using Eq. (13),

$$\varepsilon_0\varepsilon' = \varepsilon_0\varepsilon_\infty - \frac{Ce^4B^0K}{3kT\varepsilon_0\varepsilon_\infty\omega}\,\text{Im}\,\chi \tag{160}$$

$$\varepsilon_0\varepsilon'' = Ce^2\left[\frac{B^0}{\omega} - \frac{e^2B^0K}{3kT\varepsilon_0\varepsilon_\infty\omega}\,\text{Re}\,\chi\right] \tag{161}$$

Thus,

$$\varepsilon_0\varepsilon' = \varepsilon_0\varepsilon_\infty + \frac{Ce^4B^0K}{3kT\varepsilon_0\varepsilon_\infty}\frac{[Q - \omega\tau(\sqrt{2} - R)]/\sqrt{2}}{\omega(1 + \omega^2\tau^2)} \tag{162}$$

$$\varepsilon_0\varepsilon'' = \frac{\sigma}{\omega} + \frac{Ce^4B^0K}{3kT\varepsilon_0\varepsilon_\infty}\frac{1}{\omega}\left[\frac{\sqrt{2} - 1}{\sqrt{2}} - \frac{(\sqrt{2} - R + \omega\tau Q)/\sqrt{2}}{1 + \omega^2\tau^2}\right] \tag{163}$$

where

$$R = [1 + (1 + \omega^2\tau^2)^{1/2}]^{1/2}/\sqrt{2} \tag{164}$$

$$Q = [-1 + (1 + \omega^2\tau^2)^{1/2}]^{1/2}/\sqrt{2} \tag{165}$$

and dc conductivity σ is given by

$$\sigma = \sigma^0\left(1 - \frac{e^2K}{3kT\varepsilon_0\varepsilon_\infty}\frac{\sqrt{2} - 1}{\sqrt{2}}\right) \tag{166}$$

In normalized form

$$\frac{\varepsilon_0\varepsilon' - \varepsilon_0\varepsilon_\infty}{\varepsilon_0\varepsilon_s - \varepsilon_0\varepsilon_\infty} = \frac{2\sqrt{2}}{3\sqrt{2} - 4}\frac{Q - \omega\tau(\sqrt{2} - R)}{\omega\tau(1 + \omega^2\tau^2)} \tag{167}$$

and

$$\frac{\varepsilon_0\varepsilon''}{\varepsilon_0\varepsilon_s - \varepsilon_0\varepsilon_\infty} = \frac{(4 - 2\sqrt{2})/(3\sqrt{2} - 4)}{\omega\tau}\left[1 - \frac{\sqrt{2} - R + \omega\tau Q}{(1 + \omega^2\tau^2)(\sqrt{2} - 1)}\right] \tag{168}$$

where the relaxation strength is given by

$$\varepsilon_0\varepsilon_s - \varepsilon_0\varepsilon_\infty = \varepsilon_0\,\Delta\varepsilon = (e^2K/3kT)(3\sqrt{2} - 4)/4 \tag{169}$$

Equations (167) and (168) are plotted in Figs. 31 and 32. The peak is observed at approximately $\omega\tau \simeq 2.1$. Thus, the frequency corresponding to the loss maximum f_m is given by

$$f_m \simeq \frac{2.1}{2\pi\tau} = \frac{2.1\sigma^0}{2\pi\varepsilon_0\varepsilon_\infty} \propto \frac{\sigma}{\varepsilon_0\varepsilon_\infty} \tag{170}$$

The final results of the calculations in Eqs. (167) and (168) show that the broad dielectric loss peak is obtained because of the asymmetric poten-

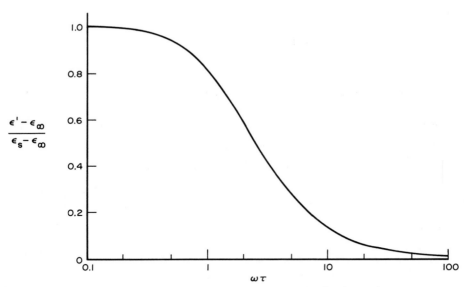

Fig. 31. Dielectric constant according to Debye and Falkanhagen theory.

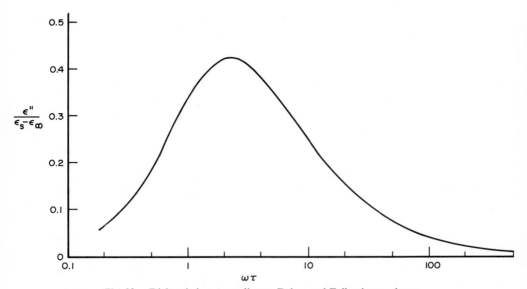

Fig. 32. Dielectric loss according to Debye and Falkenhagen theory.

tial around a charge in glass. And the peak frequency is related to the dc conductivity by Eq. (170).

The theory of Debye and Falkenhagen is based upon the Debye–Hückel theory of electrolyte which can best explain the phenomena in dilute electrolyte but fails to explain quantitatively the features of concentrated aqueous electrolyte. It is argued (Isard, 1962), therefore, that glasses which contain several mole % of Na_2O can not possibly follow the Debye–Falkenhagen theory. However, there is an important difference between the aqueous solution for which the electrolyte theory has been tested and glasses under consideration. That is the magnitude of the dielectric constant ε_∞. Water has an unusually high dielectric constant $\varepsilon_\infty \simeq 80$ while most glasses have $\varepsilon_\infty \simeq 10$. Since the Coulombic interaction energy between ions (valence Z_1 and Z_2) separated by a distance r is proportional to $Z_1Z_2e^2/r\varepsilon_\infty$, it is expected that the nonbridging oxygen ion (O^-) and sodium ion (Na^-) interact more strongly than ions in water. Because of this difference the limit of the applicability of the theory may be different. It is possible, for example, that only a small fraction of the total Na^+ concentration is free, while remaining ions form pairs, triplets, etc. (Bockris and Reddy, 1970). Then the glass is similar to the weak electrolytes and the concentration limitation of the theory may not be so stringent in glasses as in aqueous solutions. In fact, there is an opinion that glass behaves like weak electrolytes (Myuller, 1960). Furthermore, a remarkably successful explanation of the activation energy of dc conductivity of glasses by Anderson and Stuart (1954) postulates that the activation energy consists of two terms of the form

$$\Delta E_\sigma = [\beta Z_1 Z_2 e^2/\varepsilon_\infty(r_c + r_O)] + 4\pi G r_D(r_c - r_D)^2 \qquad (171)$$

where β is the finite displacement factor, r_c and r_O are the radii of cation and oxygen ion respectively, r_D is the radius of "doorway" in glass, and G is shear modulus. The first term on the right-hand side of this equation represents the energy required to remove a cation away from a site adjacent to oxygen anion and the second term represents the elastic energy needed to move the cation in the glass matrix. The magnitude of the first term accounted for the major portion (80%) of the total activation energy for Na_2O–SiO_2 glasses (Anderson and Stuart, 1954). This suggests again that only a small fraction of Na^+ is free in the absence of the electric field. Under these circumstances the concentration of the sodium ions C in the equation should be regarded as the concentration of the free ions. This will give a much greater Debye length.

Furthermore, the Debye–Falkenhagen theory has its counterpart in the mechanical relaxation (Hall, 1952). Thus this mechanism appears promising, although to claim the quantitative agreement between theory and the experiment further refinement in the theory and the various experimental tests are needed. For example, to test the possibility of glass being a weak electrolyte, it would be valuable to study the conductivity–composition relationship at extremely low alkali content or the high field conductivity (Harned and Owen, 1958).

VI. Summary

Direct current as well as low-frequency ac measurements show that there are two dielectric dispersions (and associated dielectric relaxations) in glasses. One at the lower frequency is often called electrode polarization and is related to the phenomenon at glass–electrode interface. This was explained by the space charge mechanism. The other dielectric relaxation has the same activation energy as that of dc conductivity of glasses. Although numerous mechanisms have been proposed for this conductivity related dielectric relaxation, none appears satisfactory. Since the similar dielectric relaxation has been observed in numerous other materials including alkali–halide single crystals, the correct mechanism has to be a universal one. It was proposed that this dielectric relaxation was caused by the local Coulombic interaction between the mobile charge (e.g., Na^+) and the oppositely charged immobile charge (e.g., $-O^-$), similarly to the Debye–Falkenhagen effect in the electrolytes.

ACKNOWLEDGMENTS

Comments on the manuscript by Professor R. H. Doremus and Dr. C. Kim are greatly appreciated. This work was supported by the National Science Foundation under Grant No. DMR 73-02536.

References

Aitken, J., and MacCrone, R. K. (1975). *In* "Mass Transport Phenomena in Ceramics" (A. R. Cooper and A. H. Haner, eds.). Plenum Press, New York.
Ambrus, J. H., Moynihan, C. T., and Macedo, P. B. (1972). *J. Phys. Chem.* **76**, 3287.
Amey, W. G., and Hamburger, F. (1949). *Proc. ASTM* **49**, 1079.
Anderson, J. C. (1964). "Dielectrics". Van Nostrand-Reinhold, Princeton, New Jersey.
Anderson, O. L., and Stuart, D. A. (1954). *J. Am. Ceram. Soc.* **37**, 573.
Bach, H., and Baucke, F. G. K. (1974). *Phys. Chem. Glasses* **15**, 123.
Barton, J. L. (1966). *Verres Ref.* **20**, 328.
Barton, J. L. (1967). *C. R. Acad. Sci. Paris* **264**, 1139.
Bates, J. B., Hendricks, R. W., and Shaffer, L. B. (1974). *J. Chem. Phys.* **61**, 4163.
Beaumont, J. H., and Jacobs, P. W. M. (1967). *J. Phys. Chem. Solids* **28**, 657.
Bockris, J. O'M., and Reddy, A. K. N. (1970). "Modern Electrochemistry," Vol. 1. Plenum Press, New York.
Boesch, L. P., and Moynihan, C. T. (1975). *J. Non-Cryst. Solids* **17**, 44.
Boksay, Z., and Lengyel, B. (1974). *J. Non-Cryst. Solids* **14**, 79.

Carlson, D. E. (1974a). *J. Am. Ceram. Soc.* **57**, 291.
Carlson, D. E. (1974b). *J. Am. Ceram. Soc.* **57**, 461.
Carlson, D. E., Hang, K. W., and Stockdale, G. F. (1972). *J. Am. Ceram. Soc.* **55**, 337.
Charles, R. J. (1961). *J. Appl. Phys.* **32**, 1115.
Charles, R. J. (1962). *J. Am. Ceram. Soc.* **45**, 105.
Charles, R. J. (1963). *J. Am. Ceram. Soc.* **46**, 235.
Charles, R. J. (1966). *J. Am. Ceram. Soc.* **49**, 55.
Cohen, J. (1957). *J. Appl. Phys.* **28**, 795.
Cole, K. S., and Cole, R. H. (1941). *J. Chem. Phys.* **9**, 341.
Curie, M. J. (1888). *Ann. Chim. Phys.* **17**, 385.
Davidson, D. W., and Cole, R. H. (1951). *J. Chem. Phvs.* **19**, 1484.
Debye, P. (1929). "Polar Molecules." Dover, New York.
Doi, A. (1972). *J. Non-Cryst. Solids* **11**, 235.
Doremus, R. H. (1962). *In* "Modern Aspects of the Vitreous State" (J. D. MacKenzie, ed.). Butterworths, London.
Doremus, R. H. (1970). *J. Appl. Phys.* **41**, 3366.
Doremus, R. H. (1973). "Glass Science." Wiley, New York.
Doremus, R. H. (1976). *J. Non-Cryst. Solids* **19**, 137.
Engel, J. R., and Tomozawa, M. (1975). *J. Am. Ceram. Soc.* **58**, 183.
Falkenhagen, H. (1934). "Electrolytes." Oxford Univ. Press (Clarendon), London and New York.
Friauf, R. J. (1954). *J. Chem. Phys.* **22**, 1329.
Friauf, R. J. (1957). *Phys. Rev.* **105**, 843.
Fröhlich, H. (1949). "Theory of Dielectrics; Dielectric Constant and Dielectric Loss," Oxford Univ. Press (Clarendon), London and New York.
Fuoss, R. M., and Kirkwood, J. G. (1941). *J. Am. Chem. Soc.* **63**, 385.
Gross, B. (1941). *Phys. Rev.* **59**, 748.
Guidee, P. (1972). *Verres Refract.* **26**, 103, 138.
Guyer, E. M. (1933). *J. Am. Ceram. Soc.* **16**, 607.
Hakim, R. M., and Uhlmann, D. R. (1973). *Phys. Chem. Glasses* **14**, 81.
Hall, L. H. (1952). *J. Aroust. Soc. Am.* **24**, 704.
Hamon, B. V. (1952). *Proc. I.E.E. London* **99**, pt. IV, 151.
Hansen, K. W., and Splann, M. T. (1966). *J. Electrochem. Soc.* **113**, 895.
Harned, H. S., and Owen, B. B. (1958). "The Physical Chemistry of Electrolytic Solutions," Van Nostrand-Reinhold, Princeton. New Jersey.
Heckman, R. W., Ringlieu, J. A., and Williams, E. L. (1967). *Phys. Chem.* **8**, 145.
Heroux, L. (1958). *J. Appl. Phys.* **29**, 1639.
Hetherington, C. T., Jack, K. H., and Ramsay, M. W. (1965). *Phys. Chem. Glasses* **6**, 6.
Higgins, T. J., Macedo, P. B., and Volterra, V. (1972). *J. Am. Ceram. Soc.* **55**, 488.
Isard, J. O. (1962). *Proc. Inst. Elec. Eng. Suppl. 22*, **109**, Part B, 440.
Isard, J. O. (1970). *J. Non-Cryst. Solids* **4**, 357.
Jaffe, G. (1952). *Phys. Rev.* **85**, 354.
Joffe, A. F. (1928). "Physics of Crystals." McGraw-Hill, New York.
Jonscher, A. K. (1975). *Nature (London)* **253**, 717.
Kao, K. C., Whitham, W., and Calderwood, J. H. (1970). *J. Phys. Chem. Solids* **31**, 1019.
Kim, C. (1976). Ph.D. Thesis, Rensselaer Polytechnic Inst., Troy, New York.
Kim, C., and Tomozawa, M. (1976a). *J. Am. Ceram. Soc.* **59**, 127.
Kim, C., and Tomozawa, M. (1976b). *J. Am. Ceram. Soc.* **59**, 321.
Kinser, D. L., and Hench, L. L. (1969). *J. Am. Ceram. Soc.* **52**, 638.
MacDonald, J. Ross. (1953). *Phys. Rev.* **92**, 4.
MacDonald, J. Ross. (1970). *Trans. Faraday Soc.* **66**, 943.

MacDonald, J. Ross. (1971). *J. Chem. Phys.* **54**, 2026.

MacDonald, J. Ross. (1971). *J. Appl. Phys.* **45**, 73.

Macedo, P. B., and Weiler, R. A. (1969). *In* "Molten Salts: Characterization and Analysis" (G. Mamantov, ed.), pp. 377–408. Dekker, New York.

Macedo, P. B., Moynihan, C. T., and Bose, R. (1972). *Phys. Chem. Glasses* **13**, 171.

Manning, M. F., and Bell, M. E. (1940). *Rev. Mod. Phys.* *12*, 215.

Maxwell, J. C. (1892). "Treatise on Electricity and Magnetism." Oxford Univ. Press (Clarendon), London and New York.

Mazurin, O. V. (1962, English translation 1965). Electrical Properties of Glass, *In* "Structure of Glass," Vol. 4, p. 15. Consultants Bureau, New York.

McCrum, N. G., Read, B. E., and Williams, C. T. (1967). "Anelastic and Dielectric Effects in Polymeric Solids." Wiley, New York.

Mitoff, S. P., and Charles, R. J. (1972). *J. Appl. Phys.* **43**, 927.

Mitoff, S. P., and Charles, R. J. (1973). *J. Appl. Phys.* **44**, 3786.

Moore, W. J. (1962). "Physical Chemistry." Prentice-Hall, Englewood Cliffs, New Jersey.

Morey, G. W. (1954). "The Properties of Glass." Van Nostrand-Reinhold, Princeton, New Jersey.

Mott, N. F., and Gurney, R. W. (1940). "Electronic Processes in Ionic Crystals." Oxford Univ. Press, London and New York.

Moynihan, C. T., Bressel, R. D., and Angell, C. A. (1971). *J. Chem. Phys.* **55**, 4414.

Moynihan, C. T., Boesch, L. P., and Laberge, N. L. (1973). *Phys. Chem. Glasses* **14**, 122.

Myuller, R. L. (1960). *In* "Structure of Glass" (Porai-Koshits, Vol. 2, ed.), pp. 215–219. Consultants Bureau, New York.

Nakajima, T. (1972). *In* "1971 Annual Report, Conference on Electrical Insulation and Dielectric Phenomena." Nat. Acad. of Sci.

Namikawa, H. (1969). *Yogyo-Kyokai-Shi* **77**, 18.

Namikawa, H. (1974). *J. Non-Cryst. Solids* **14**, 88.

Namikawa, H. (1975a). *Yogyo-Kyokai-Shi* **83**, 500.

Namikawa, H. (1975b). *J. Non-Cryst. Solids* **18**, 173.

Namikawa, H., and Asahara, Y. (1966). *J. Ceram. Assoc. J.* **74**, 205.

Namikawa, M., and Kumata, K. (1968). *J. Ceram. Assoc. J.* **76**, 10.

Owen, A. E. (1963). *Progr. Ceram. Sci.* **3**, 77.

Ono, S., and Munakata, M. (1974). *Yogyo-Kyokai-Shi* **82**, 511.

Proctor, T. M., and Sutton, P. M. (1959). *J. Chem. Phys.* **30**, 212.

Proctor, T. M., and Sutton, P. M. (1960). *J. Am. Ceram. Soc.* **43**, 173.

Prod'homme, L. (1960). *Verres Ref.* **14**, 69, 124.

Provenzano, V., Boesch, L. P., Volterra, V., Moynihan, C. T., and Macedo, P. B. (1972). *J. Am. Ceram. Soc.* **55**, 492.

Radzilowski, R. H., Yao, Y. F., and Kummer, J. T. (1969). *J. Appl. Phys.* **40**, 4716.

Richardson, S. W. (1925). *Proc. Roy Soc.* **A107**, 101.

Shaffer, L. S., and Hendricks, R. W. (1974). *J. Appl. Crystallogr.* **7**, 159.

Shand, E. B. (1958). "Glass Engineering Handbook." McGraw-Hill, New York.

Sillars, R. W. (1937). *J. Inst. Elect. Eng.* (*London*) **80**, 378.

Snow, E. H., and Deal, B. E. (1966). *J. Electrochem. Soc.* **113**, 263.

Splann Mizzoni, M. (1973). *J. Electrochem. Soc.* **120**, 1592.

Stanworth, J. E. (1950). "Physical Properties of Glass." Oxford Univ. Press (Clarendon), London and New York.

Stevels, J. M. (1957). *In* "Handbuch der Physik" (S. Flügge, ed.), Vol. 20, p. 350. Springer-Verlag, Berlin.

Sutton, P. M. (1960). *Progr. Dielec.* **2**, 113.

Sutton, P. M. (1964a). *J. Am. Ceram. Soc.* **47**, 188.

Sutton, P. M. (1964b). *J. Am. Ceram. Soc.* **47**, 219.
Taylor, H. E. (1957). *J.* Soc. Glass Technol. **41**, 350T.
Taylor, H. E. (1959). *J. Soc. Glass Technol.* **43**, 124T.
Terai, R. (1969). *Phys. Chem. Glasses* **10**, 146.
Tomandl, G. (1974). *J. Non-Cryst. Solids* **14, 101**.
Tomozawa, M., Kim, C., and Doremus, R. H. (1975). *J. Non-Cryst. Solids* **19**, 115.
Tsuchiya, T., and Moriya, T. (1973). *Yogyo-Kyokai-Shi* **81**, 303.
Tsuchiya, T., and Moriya, T. (1974a). *Yogyo-Kyokai-Shi* **82**, 147.
Tsuchiya, T., and Moriya, T. (1974b). *Yogyo-Kyokai-Shi* **82**, 518.
Tsuchiya, T., and Moriya, T. (1975). *Yogyo-Kyokai-Shi* **83**, 419.
Volger, J., Stevels, J. M., and van Amerongen, C. (1953). *Philips Res. Rep.* **8**, 452.
von Hippel, A. R. (1954). "Dielectric Materials and Applications." M.I.T. Press, Cambridge, Massachusetts.
Wagner, K. W. (1913). *Ann. Phys.* **40**, 817.
Wagner, K. W. (1914). *Arch. Electrotech.* **2**, 372.
Warburg, E. (1884). *Ann. Physik* **21**, 622.
Warren, B. E. (1937). *J. Appl. Phys.* **8, 645**.
Warren, B. E., and Biscoe, J. (1938). *J. Am. Ceram. Soc.* **21**, 259.
Weinberg, D. L. (1963). *Rev. Sci. Instrum.* **34**, 691.
Whitehead, J. B., and Banôs, A. Jr. (1932). *Trans. AIEE* **51**, 392.
Williams, G., and Watts, D. C. (1970). *Trans. Faraday. Soc.* **66**, 88.
Wikby, A. (1974). *Electrochim. Acta* **19**, 329.
Yager, W. A. (1936). *Physics* **7**, 434.
Yamamoto, K., Kumata, K., and Namikawa, H. (1974). *Yogyo-Kyokai-Shi* **82**, 538.
Zarzycki, J. (1974). *Proc.* Xth ICG, Kyoto, Japan pp. 12–28.
Zheludeve. I. S. (1971). "Physics of Crystalline Dielectrics," Vol. 1, 2. Plenum Press, New York.

Index

A

Alkali silicate glasses
  absorption currents, 297
  dielectric relaxation, 299, 302, 305, 323ff
  elastic properties, 213
  electrical conductivity, 305
  ESR of radiation defects, 237
  light scattering, 183, 191, 209
  optical absorption, 15–19, 67, 324
  photochromism, 92
  relaxation times, 192, 194
  x-ray scattering, 333
Aluminosilicate glasses
  light scattering, 198, 200
  optical absorption, 16
  photochromism, 95, 96
Amber glass, 47
Annealing
  birefringence and, 132, 135
  light scattering and, 190

B

Birefringence, 123–154
  measurement, 125, 133
  relaxation, 135
  theory, 140
Borate glasses
  coordination number, 262
  elastic properties, 213
  ESR of radiation defects, 239
  light scattering, 182, 206
  NMR, 29, 73, 256, 268, 269
  optical absorption, 29, 70

Boron silicate glasses
  birefringence, 131, 151
  microporous, 151
  optical absorption, 21, 30
  photochromism, 94
Brillouin scattering, 165, 167, 179, 202, 205

C

Chalcogenide glasses
  ESR
    optically induced, 252
    radiation defects and, 247
  Mössbauer spectra of Te, 274
  NMR
    of As, 264, 271
    of Ti, 266
  optical absorption, 79–82
Chromium ions
  ESR, 42
  optical absorption, 26, 36, 41
Cobalt ions, optical absorption, 48
Color, 35
Copper ions
  ESR, 249
  in photochromic glass, 98, 110
  optical absorption, 25, 36, 52
Crystallization and birefringence, 144

D

Dielectric relaxation
  absorption current and, 289, 296
  conducting path model, 315
  Debye theory, 291